The MAPLE BOOK

The MAPLE BOOK

FRANK GARVAN

CHAPMAN & HALL/CRC

A CRC Press Company

Boca Raton London New York Washington, D.C.

Library of Congress Cataloging-in-Publication Data

Garvan, Frank (Frank G.), 1955-
The Maple book / by Frank Garvan.
p. cm.
Rev. ed. of: Maple V primer. c1997.
Includes bibliographical references and index.
ISBN 1-58488-232-8 (alk. paper)
1. Maple (Computer file) 2. Algebra—Data processing. I. Garvan, Frank (Frank G.), 1955- Maple V primer. II. Title.

QA155.7.E4 G36 2001
510′.285′53042—dc21 2001043596

Direct all inquiries to CRC Press LLC, 2000 N.W. Corporate Blvd., Boca Raton, Florida 33431.

Visit the CRC Press Web site at www.crcpress.com

International Standard Book Number 1-58488-232-8
Library of Congress Card Number 2001043596
Printed in the United States of America 4 5 6 7 8 9 0
Printed on acid-free paper

To my parents,
Kevin and Clare

PREFACE

MAPLE® is a very powerful interactive computer algebra system. It is used by students, educators, mathematicians, statisticians, scientists, and engineers for doing numerical and symbolic computation. MAPLE has many strengths: (1) it can do exact integer computation, (2) it can do numerical computation to any (well, almost) number of specified digits, (3) it can do symbolic computation, (4) it comes with many built-in functions and packages for doing a wide variety of mathematical tasks, (5) it has facilities for doing two- and three-dimensional plotting and animation, (6) it has a worksheet-based interface, (7) it has facilities for making technical documents, and (8) MAPLE is a simple programming language, which means that users can easily write their own functions and packages.

The present book is a greatly expanded version of an earlier book, *The MAPLE V Primer*, by the author. A lot has happened to MAPLE since. This book covers MAPLE 7, the latest version of MAPLE. The book is quite comprehensive. It should serve both as an introduction to MAPLE and as a reference. If you are learning MAPLE for the first time, it is advised that you work slowly through the book until at least Chapter 7. Keep the book open with you at the computer as you try the commands. All the examples of MAPLE commands used in the book, as well as supplementary files are available for downloading from `www.crcpress.com`. The files are also available from the author's site
`http://www.math.ufl.edu/~frank/maple-book/mbook.html`
See Section 12.3.

MAPLE is both an interactive computer algebra system and a programming language. An important goal of this book is to show you how to write simple MAPLE programs (or procedures). Chapter 7 is a tutorial for learning the MAPLE programming language. There are programming exercises for the reader to tackle. Their solution is given at the end of the chapter. Once the reader has learned how to program, he or she will appreciate the real power of MAPLE. Hopefully readers will learn to write their own programs and packages to suit their needs.

As you progress further into the book you will learn how to use MAPLE

® Maple is a registered trademark of
Waterloo Maple Inc.,
57 Erb Street West,
Waterloo, Ontario,
Canada N2L 5J2,
Phone: 1-800-267-6583, (519) 747-2373,
Fax: (519) 747-5284,
E-mail: info@maplesoft.com,
Web site: http://www.maplesoft.com.

for more advanced mathematics: differential equations, linear algebra, vector calculus, complex analysis, special functions, statistics, finite fields, group theory, combinatorics, and number theory. MAPLE has many packages that are not automatically loaded when a MAPLE session is begun. To load a *package*, one needs to use the `with(`*package*`)` function. One of the big changes in MAPLE 6 was the new *LinearAlgebra* package. All of MAPLE's packages are covered in the book to some degree. Some are covered in great depth, such as the *LinearAlgebra* package in Chapter 9 and the *stats* package in Chapter 16.

Additional MAPLE packages and worksheets are available free at The Maple Application Center page on the Web at `http://www.mapleapps.com`. See Appendix A for more information.

MAPLE has fabulous built-in help facilities. Help can be found either through the interactive Help menu or by using the `?` command. For instance, a very short introduction to MAPLE can be found by typing `?intro`.

MAPLE is available on Windows, Macintosh, UNIX, and Linux systems. The author would like to thank Waterloo Maple, Inc., for permission to include pictures of the MAPLE icons and buttons, and some portions of the text from the on-line help system. Special thanks go to Cynthia Wilson Garvan and Weir Hou, who helped a lot with Chapter 16, the chapter on statistics. The author thanks Bob Stern at CRC Press, for his encouragement and patience.

Frank Garvan (`frank@math.ufl.edu`)
Department of Mathematics
University of Florida
Gainesville, Florida

Contents

1. GETTING STARTED

1.1 Starting a MAPLE session

On most systems a MAPLE session is started by double clicking on the MAPLE icon. In the UNIX X Windows version, MAPLE is started by entering the command `xmaple`. In the command-line (tty) version, the Maple logo appears followed by the > prompt.

In most versions a window with menus will appear. See Figure 1.1 below. At the top are the menus File, Edit, View, Insert, Format, Spreadsheet, Options, Window and Help. Beneath are two rows of buttons. The first row of buttons is called the *tool bar* and contains 24 buttons:

Create a new worksheet.

Open an existing worksheet.

Open a specified URL.

Save the active worksheet.

Print the active worksheet.

Cut the selection to the clipboard.

Copy the selection to the clipboard.

Paste the clipboard contents into the current worksheet.

Undo the last operation.

Undo the last "undo."

Insert MAPLE commands.

Insert text.

Insert a new MAPLE input region after the cursor.

Remove any section enclosing the selection.

Enclose the selection in a subsection.

Go back in the hyperlink history.

Go forward in the hyperlink history.

Interrupt the current computation.

Set magnification to 100%.

Set magnification to 150%.

Set magnification to 200%.

Display nonprinting characters.

Resize the active window to fill the available space.

Restart.

The next row is called the *context bar* and contains five buttons:

Toggle the expression display between mathematical and MAPLE notation.

Toggle the expression display between inert text and executable MAPLE command.

Auto-correct the expression for syntax.

Execute the current expression.

Execute the worksheet.

The > prompt will be in the *worksheet* window. Don't worry about the buttons too much at this stage.

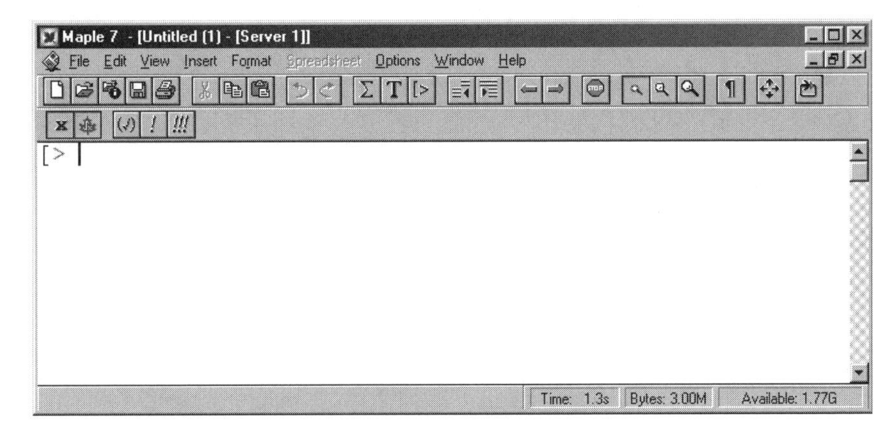

Figure 1.1 MAPLE worksheet window.

1.2 Different versions of MAPLE

The current version of MAPLE is MAPLE 7. The previous version was MAPLE 6. Before that, there was MAPLE V Release 5, MAPLE V Release 4, and way back in 1994, we had MAPLE V Release 3. This book covers MAPLE 7. The change from MAPLE 6 to MAPLE 7 was not a big one so most of the book applies to MAPLE 6. Occasionally we will point out differences between the earlier versions.

1.3 Basic syntax

In MAPLE the prompt is the symbol >. MAPLE commands are entered to the right of the prompt. Each command ends with either ":" or ";". If the colon is used, the command is executed but the output is not printed. When the semicolon is used, the output is printed. Try typing `105/25:` followed by a `Return` (or `Enter`).

```
> 105/25:
```

Observe that the output was not printed. Now type `105/25;`

```
> 105/25;
```

$$\frac{21}{5}$$

Below in Figure 1.2 is a rendering of how this looks in the worksheet window.

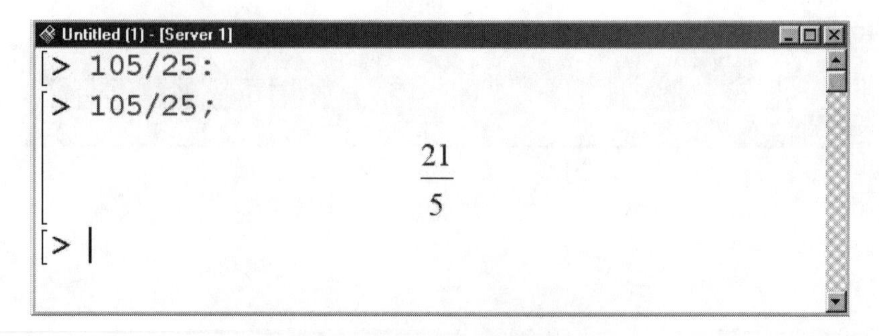

Figure 1.2 MAPLE commands with output.

Try these

```
> 105/25-1/5;
```

$$4$$

```
> %+1/5;
```

$$\frac{21}{5}$$

```
> %%;
```

$$4$$

Observe that MAPLE uses exact arithmetic. The percent sign `%` refers to the previous result. The double percent `%%` refers to the result before the previous result. It is possible to refer back 3 lines using `%%%`, but one cannot refer back any further. The percent sign `%` is called the ditto operator.

Warning: In MAPLE V Release 4 (and earlier versions), the ditto operator was the double quote character `"`. The two double quotes `""` were used to refer to the result before the previous result, and to refer back 3 lines one used triple double quotes `"""`.

One of the most common mistakes is to omit the semicolon or colon.

```
> 105/25
Warning, incomplete statement or missing semicolon
> 105/25;
syntax error, unexpected number
```

Don't panic! MAPLE has interpreted this to mean 105/25 105/25, hence the syntax error. MAPLE also gave a warning about the missing semicolon! If you forget the semicolon, simply type it on the next line.

```
> 105/25
> ;
```

$$21/5$$

See Section 1.3 for a method for editing mistakes.

Results can be assigned to variables using the colon-equals "`:=`".

```
> f:=%;
```

$$f := 21/5$$

```
> G:= -1/5;
```

$$G := -1/5$$

```
> f+g;
```

$$21/5 + g$$

```
> #Observe that Maple is case sensitive.
> f+G;
```

$$4$$

Note that comment lines begin with `#`. In the first line of our session we used the ditto operator `%`. Remember, if you are using MAPLE V Release 4 (or an earlier version), use `"` as the ditto operator.

1.4 Editing mistakes

MAPLE has built-in editing facilities. On most platforms, lines of input can be edited using the arrow keys and the mouse. Cutting and pasting is also possible with the mouse. In the Windows version, you can select input by highlighting with the mouse, then you can copy, cut, and paste by using `Control C`, `x`, and `v` as usual. In the command-line (or tty) version, MAPLE has two built-in editors: *emacs* and *vi*. Use the help command `?editing` for more information.

```
> 105/25
> 105/25;
syntax error, unexpected number
```

Just click the mouse after `105/25`, enter a semicolon, and press enter.

```
> 105/25;
```

$$21/5$$

The *vi* editor can be invoked using the `Esc` key.

1.5 Help

Ever since MAPLE V (Release 4) came out, MAPLE has had a fabulous interactive help facility. Click on Help and a menu should appear:

Introduction	
Help on Context	`Ctrl+F1`
New User's Tour	
What's New	
Using Help	
Glossary	
Topic Search ...	
Full Text Search ...	
History ...	
Save to Database ...	
Remove Topic ...	
Balloon Help ...	
Register Maple 7 ...	
About Maple 7 ...	

Select Full Text Search. A little window should appear. In the Word(s) box, type `floating point arithmetic` then click on `Search`. A search is then made of the interactive help manual. A list of topics should appear in the Matching Topics box. See Figure 1.3.

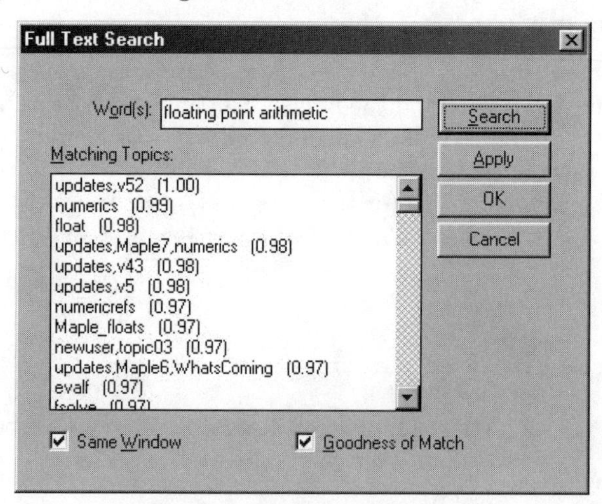

Figure 1.3 Full Text Search window.

Select `evalf` with the mouse, then click on `Apply`. A help window should appear with information on the `evalf` command. Click on `OK`.

Now go back to the Help menu and select Introduction. A new window should appear offering you a list of topics to explore.

If you know the name of a command, then you can select Topic Search in the Help menu.

To return to our original worksheet window, click on Window and select Untitled(1)-Server(1).

Help can also be accessed directly from the worksheet. Try

```
> ?evalf
```

The `evalf` help window should appear. In the command-line version, this information will appear below the cursor.

Now try selecting Balloon Help in the Help menu. Next move the cursor onto a button and a little bubble should appear, giving a brief description. Keep this option until you are familiar with the buttons and menus.

The command `?index` provides a list of categories: expression, function, misc, module, etc. For instance, `?index[function]` gives a list of MAPLE's standard library functions. For more information on navigating through the worksheet environment, see `?worksheet`.

1.6 A sample session and context menus

Open a new worksheet by pressing [new worksheet button]. Enter the following into the worksheet:

```
> Int(x/sqrt(1+x^4),x);
```

and hit return after you type ";". You should have something like this:

```
> Int(x/sqrt(1+x^4),x);
```

$$\int \frac{x}{\sqrt{1+x^4}}\,dx$$

The `Int` function is for calculating integrals. More information can be found in Section 5.7. Now click on the integral (above) with the right mouse button. A menu should appear:

Copy
Differentiate ▶
Integrate ▶
Evaluate
Complex Maps ▶
Integer Functions ▶
Simplications ▶
Conversions ▶
Plots ▶

This menu is called a *context menu*. When you click on MAPLE output, such a menu will appear. It won't always be the same menu. The menu depends on the type of object you click, hence the name context menu. Now select

Differentiate and click on [button]. Magically MAPLE has taken the derivative with respect to x:

```
> Int(x/sqrt(1+x^4),x);
```

$$\int \frac{x}{\sqrt{1+x^4}}\,dx$$

```
> R0 := diff(Int(x/sqrt(1+x^4),x),x);
```

$$R0 := \frac{x}{\sqrt{1+x^4}}$$

Naturally, MAPLE found that

$$\frac{d}{dx}\int \frac{x}{\sqrt{1+x^4}}\,dx = \frac{x}{\sqrt{1+x^4}}.$$

Now, click on the integral again and this time select Evaluate in the contex menu. This time MAPLE evaluates the integral:

```
> Int(x/sqrt(1+x^4),x);
```

$$\int \frac{x}{\sqrt{1+x^4}}\,dx$$

```
> R1 := value(Int(x/sqrt(1+x^4),x));
```

$$R1 := \frac{1}{2}\,arcsinh(x^2)$$

```
> R0 := diff(Int(x/sqrt(1+x^4),x),x);
```

$$R0 := \frac{x}{\sqrt{1+x^4}}$$

MAPLE found that

$$\int \frac{x}{\sqrt{1+x^4}}\,dx = \frac{1}{2}\sinh^{-1} x^2.$$

Click on the output with name `R0`, and a different context menu will appear:

Copy
Differentiate ▶
Integrate ▶
Factor
Simplify
Expand
Approximate ▶
Solve
Numerical Solve
Rationalize
Combine ▶
Collect ▶
Complex Maps ▶
Integer Functions ▶
Constructions ▶
Simplifications ▶
Conversions ▶
Plots ▶

Select `Plots` and press 2-D Plot. MAPLE produces a graph of the function $y = \frac{x}{\sqrt{1+x^4}}$. See Figure 1.4.

```
> smartplot(R0);
```

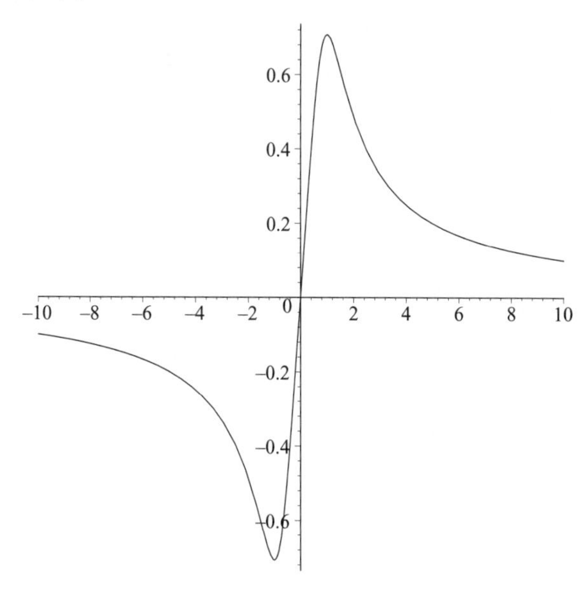

Figure 1.4 A smartplot.

We will learn a lot more about plotting in Chapter 6.

Before going on we should save our work. Click on File and a menu appears:

	New	Ctrl+o
	Open ...	
	Save	Ctrl+s
	Save As ...	
	Export As	▶
	Close	Ctrl+F4
	Save Settings	
✓	AutoSave Settings	
	Print ...	Ctrl+P
	Print Preview ...	
	Print Setup ...	
	Exit	Alt+F4

Click on Save. A **Save As** window appears. In the File name box type `ch1a.mws`. Then click on OK. The worksheet has been saved as the file *ch1a.mws*. Here *mws* is a file type which stands for MAPLE worksheet.

1.7 Palettes

So far we have seen how to enter MAPLE commands by typing after the MAPLE prompt >, and by using a context menu. Another method is to use palettes. Open a new worksheet by pressing [New]. Now click on View and a menu appears:

✓	Toolbar	
✓	Context Bar	
✓	Status Bar	
	Palettes	▶
	Zoom Factor	▶
	Bookmarks	▶
	Back	
	Forward	
	Hide Content	▶
	Show Invisible Characters	
✓	Show Section Ranges	Shift+F9
✓	Show Group Ranges	F9
	Show Object type	
	Expand All Sections	
	Collapse All Sections	

Select Palettes, slide to the right, and another menu appears:

Symbol Palette
Expression Palette
Matrix Palette
Vector Palette
Show All Palettes
Hide All Palettes

In MAPLE 7 there are four palettes: the **Symbol** palette, the **Expression** palette, the **Matrix** palette, and the **Vector** palette. In Chapter 9 we will use the **Matrix** and **Vector** palettes. For the time being let's select `Expression Palette`. A window should appear. See Figure 1.5.

Figure 1.5 The **Expression** palette.

Let's start with something simple. In the **Expression** palette press `a-b`.

```
> (%? - %?);
```

MAPLE has produced a template for an expression of the form $(a - b)$. Notice %?. Now type `105/25`.

```
> (105/25 - %?);
```

Notice that 105/25 has been entered where %? was. Now hit the `Tab` key.

```
> (105/25 - %?);
```

MAPLE is now waiting for us to type the second number. We type `1/5`.

```
> (105/25 - 1/5);
```

We hit `Return` (or `Enter`):

```
> (105/25 - 1/5);
```

$$4$$

Do you see how the **Expression** palette works? Many other types of expressions can be entered in this way. You should be able to figure out the possible expressions by looking at the buttons in the palette. Try each button and experiment.

To open the **Symbol** palette, click on View, select Palettes, slide right, and select `Symbol Palette`. See Figure 1.6.

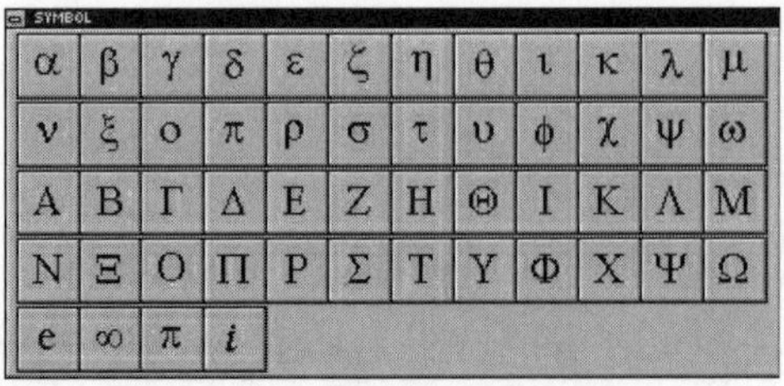

α	β	γ	δ	ε	ζ	η	θ	ι	κ	λ	μ
ν	ξ	ο	π	ρ	σ	τ	υ	φ	χ	ψ	ω
Α	Β	Γ	Δ	Ε	Ζ	Η	Θ	Ι	Κ	Λ	Μ
Ν	Ξ	Ο	Π	Ρ	Σ	Τ	Υ	Φ	Χ	Ψ	Ω
e	∞	π	i								

Figure 1.6 The **Symbol** palette.

The **Symbol** palette is used for entering Greek letters and some mathematical constants such as e and π. Try out some of the buttons.

1.8 Spreadsheets

Click on Insert. A menu should appear:

Text	`Ctrl+T`
Standard Math	`Ctrl+R`
Maple Input	`Ctrl+M`
Standard Math Input	`Ctrl+G`
Execution Group	▶
Plot	▶
Spreadsheet	
Paragraph	▶
Section	
Subsection	
Hyperlink...	
Object...	
Page Break	`Ctrl+Enter`

Select Spreadsheet. A spreadsheet appears in the worksheet:

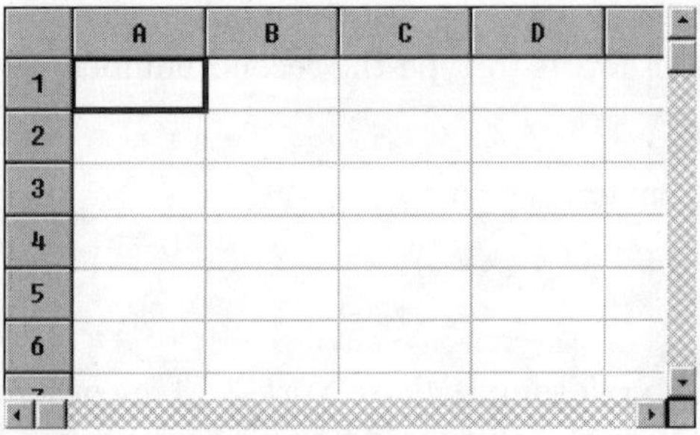

Figure 1.7 A MAPLE spreadsheet.

Notice that the upper left-most cell (A1) is highlighted. There are four new buttons in the context bar:

Fill a range of cells.

Evaluate all stale cells.

Accept the input and evaluate it.

Restore input to the previous value.

Type `n` and press enter. The symbol n should appear in cell A1. In cell A2 type `1` and press enter. Now click on cell A2 and select the first column of cells up to cell A9 by holding the mouse button down. Now click on . A **Fill** window should appear. Enter `1` for Step Size and press `OK`. The numbers 2, 3, ... , 8 should appear in cells A3, A4, ... , A9. Type `x^n-1` in cell B1. We now have $x^n - 1$ in cell B1. This is good, but we want to change it. Click on cell B1. Notice that `x^n-1` is in the edit field (the box to the right of the new buttons). Backspace over it and type `x^(~A1) - 1`. We still get $x^n - 1$ in cell B1. What is going on? Here `~A1` refers to value in cell A1 which is n, so that the value of cell B1 is linked to that of A1. We want to put $x^n - 1$ with $n = 1, 2, \ldots 8$ in the second column. Click on Spreadsheet. A menu should appear:

Evaluate Selection
Evaluate Spreadsheet
Row ▶
Column ▶
Fill ▶
Import Data ▶
Export Data ▶
Properties...
✓ Show Border
Resize To Grid

Select Fill, slide right, and select Down. Did you get the polynomials $x-1$, x^2-1, ... , $x^8 - 1$ in column B? You will probably want to resize the spreadsheet. Click in the bottom right corner, hold the mouse button down, and stretch the spreadsheet so you can see all the entries. Now we want to factor the polynomials in column B. Enter `factored polynomial` in cell C1. In cell C2 enter `factor(~B2)`. Select the column of cells C2, C3, ... , C9. From the Sreadsheet menu select Fill and then Down. Did you get the desired effect? You should now have a table:

n	$x^n - 1$	*factored polynomial*
1	$x - 1$	$x - 1$
2	$x^2 - 1$	$(x-1)(x+1)$
3	$x^3 - 1$	$(x-1)(x^2+x+1)$
4	$x^4 - 1$	$(x-1)(x+1)(x^2+1)$
5	$x^5 - 1$	$(x-1)(x^4+x^3+x^2+x+1)$
6	$x^6 - 1$	$(x-1)(x+1)(x^2+x+1)(x^2-x+1)$
7	$x^7 - 1$	$(x-1)(x^6+x^5+x^4+x^3+x^2+x+1)$
8	$x^8 - 1$	$(x-1)(x+1)(x^2+1)(x^4+1)$

For more information on MAPLE spreadsheets see `?worksheet,spreadsheet`. For programmers there is a spreadsheet package called *Spread*. See Section 17.7.19.

1.9 Quitting MAPLE

If you are done with your MAPLE session, click on [save icon]. The **Save As** window should appear. In the File name box type `ch1.mws` and click on OK. Your worksheet has now been saved. To quit MAPLE, go to the File menu and select Exit. Later you can reopen your worksheet by clicking on [open icon].

In the command-line version, the easiest way to quit a Maple session is to use `quit`.

```
> quit
```

2. MAPLE AS A CALCULATOR

2.1 Exact arithmetic and basic functions

As we noted in Section 1.3, MAPLE does exact arithmetic. Also, MAPLE does integer arithmetic to infinite precision. Try the following examples:

```
> 2/3 + 3/5;
```

$$\frac{19}{15}$$

```
> 7 - 11/15;
```

$$\frac{94}{15}$$

```
> 12^20;
```

$$3833759992447475122176$$

The basic arithmetic operations in MAPLE are

$+$	addition
$-$	subtraction
$*$	multiplication
^or $**$	exponentiation
$/$	division

MAPLE also has the basic mathematical functions (and much more) that are available on a scientific calculator.

`abs(x)`	absolute value $\lvert x\rvert$
`sqrt(x)`	square root $\sqrt{x}$
`n!`	factorial
`sin(x)`	sine
`cos(x)`	cosine
`tan(x)`	tangent
`sec(x)`	secant
`csc(x)`	cosecant
`cot(x)`	cotangent
`log(x)`	natural logarithm
also `ln(x)`	
`exp(x)`	exponential function e^x
`sinh(x)`	hyperbolic sine
`cosh(x)`	hyperbolic cosine
`tanh(x)`	hyperbolic tan

MAPLE has many other built-in mathematical functions. For instance, it has the inverse trig functions (`arcsin`, `arccos`, etc.), the Bessel functions (`BesselI`), the Riemann zeta function (`Zeta`), the gamma function (`GAMMA`), and the complete and incomplete elliptic integrals (`EllipticE`). For a complete listing, see `?index[functions]` or Section 15.1.

2.2 Floating-point arithmetic

MAPLE can do floating-point calculation to any required precision. This is done using `evalf`.

```
> tan(Pi/5);
```

$$\sqrt{5 - 2\sqrt{5}}$$

```
> evalf(%);
```

$$0.7265425273$$

Notice that `evalf` found $\tan(\pi/5)$ to 10 decimal places, which is the default. Also, note that in MAPLE, π is represented by `Pi`. There are two ways to change the default and increase the number of decimal places.

```
> E := exp(1);
```

$$E := \mathbf{e}$$

```
> evalf(E,20);
```

$$2.7182818284590452354$$

```
> Digits := 30;
```

$$30$$

```
> evalf(E);
```

$$2.71828182845904523536028747135$$

Here `E` is the mathematical constant e, which is represented in MAPLE by `exp(1)`. We found e to 20 digits using `evalf(E,20)`. The other method is to use the global variable `Digits` (whose default value is 10). After assigning `Digits := 30`, we found e correct to 30 digits simply by calling `evalf(E)`.

We can also find an approximation using a context menu (see Section 1.6). Right-click on **e** which is the output of `E := exp(1)`. A context menu appears:

Copy	
Approximate	▶
Complex Maps	▶
Integer Functions	▶
Conversions	▶
Plots	▶

Select `Approximate` and press [20].

```
> E := exp(1);
```

$$E := \mathbf{e}$$

```
> R0 := evalf(E,20);
```

$$R0 := 2.7182818284590452354$$

Under this context menu the number of digits can be 5, 10, 20, 50, or 100.

We reset the default and calculate $\sin(\pi/6)$.

```
> Digits := 10:
> evalf(sin(Pi/6));
```

$$0.5000000000$$

```
> convert(%,rational);
```

$$\frac{1}{2}$$

Notice that after we found the decimal approximation, we were able to convert it into an exact rational using `convert(%,rational)`. The `convert` function is used to convert expressions from one type to another. More on the `convert` function is to be found in section 4.6. The interested reader can find more using `?convert`. Below is a table of MAPLE's built-in mathematical constants.

`Catalan`	Catalan's constant (about .9159655942)
`gamma`	Euler's constant (about 0.5772156649)
`I`	complex number i ($i^2 = -1$)
`Pi`	π (about 3.141592654)

3. HIGH SCHOOL ALGEBRA

3.1 Polynomials and rational functions

3.1.1 Factoring a polynomial

MAPLE can do high school algebra. It can manipulate polynomials and rational functions of one or more variables quite easily.

```
> p := x^2+5*x+6;
```

$$p := x^2 + 5x + 6$$

```
> factor(p);
```

$$(x+3)(x+2)$$

```
> b := 1 - q^7 - q^8 - q^9 + q^15 + q^16 + q^17 - q^24;
```

$$b := 1 - q^7 - q^8 - q^9 + q^{15} + q^{16} + q^{17} - q^{24}$$

```
> factor(b);
```

$$-(q+1)(q^2+1)(q^2+q+1)(q^6+q^3+1)(q^4+1)$$
$$(q^6+q^5+q^4+q^3+q^2+q+1)(q-1)^3$$

To factor a polynomial or rational function, we use `factor`. We let $p = x^2+5x+6$ and found the factorization using `factor(p)`. This could have easily been done by hand. Factoring $b = 1 - q^7 - q^8 - q^9 + q^{15} + q^{16} + q^{17} - q^{24}$ is not so easy, but child's play for MAPLE.

We can also use a context menu to factor a polynomial.

```
> p;
```

$$x^2 + 5x + 6$$

Use the right mouse button to click on the polynomial. A context menu should appear. Select `Factor`.

```
> R0 := factor(x^2+5*x+6);
```

$$R0 := (x+3)(x+2)$$

3.1.2 Expanding an expression

To expand a polynomial use `expand`. The command `combine` is also useful for expanding certain expressions.

```
> p := (x+2)*(x+3);
```

$$(x+2)(x+3)$$

```
> expand(%);
```

$$x^2 + 5x + 6$$

```
> (1-q^8)*(1-q^7)*(1-q^6);
```

$$(1-q^8)(1-q^7)(1-q^6)$$

```
> expand(%);
```

$$1 - q^6 - q^7 + q^{13} - q^8 + q^{14} + q^{15} - q^{21}$$

```
> y := sqrt(x+2)*sqrt(x+3);
```

$$\sqrt{x+2}\sqrt{x+3}$$

```
> expand(y);
```

$$\sqrt{x+2}\sqrt{x+3}$$

```
> combine(y);
```

$$\sqrt{x+2}\sqrt{x+3}$$

```
> combine(y,radical);
```

$$\sqrt{x+2}\sqrt{x+3}$$

```
> combine(y,radical,symbolic);
```

$$\sqrt{x^2 + 5x + 6}$$

Notice we were not able to expand the expression $(x+2)^{1/2}(x+3)^{1/2}$ with `expand` and had to use `combine`, using two additional arguments, `radical` and `symbolic`.

3.1.3 Collecting like terms

In the last section y had the value $\sqrt{x+2}\,\sqrt{x+3}$.

```
> y;
```

$$\sqrt{x+2}\sqrt{x+3}$$

To start over we use the `restart` function.

```
> restart:
> y;
```

$$y$$

Now y is just y. See Section 3.1.10 for another way to restore y to its variable status. One can also restart by pressing the restart button in the tool bar.

The `collect` function is useful when looking at a polynomial in more than one variable.

```
> (x+y+1)*(x-y+1)*(x-y-1);
```

$$(x+y+1)(x-y+1)(x-y-1)$$

```
> p := expand(%);
```

$$p := x^3 - x^2y + x^2 - 2\,xy - x - y^2x + y^3 + y^2 - y - 1$$

```
> collect(p,x);
```

$$x^3 + (1-y)\,x^2 + \left(-1 - y^2 - 2\,y\right)x - y - 1 + y^3 + y^2$$

We let $p = (x+y+1)(x-y+1)(x-y-1) = x^3 - x^2y + x^2 - 2\,xy - x - y^2x + y^3 + y^2 - y - 1$. We used `collect(p,x)` to write p as a polynomial in x with coefficients that were polynomials in the remaining variable y. Similarly, try `collect(p,y)` to get p as a polynomial in y.

3.1.4 Simplifying an expression

The first thing you should try when presented with a complicated expression is `simplify`.

```
> 3*4^(1/2)+5;
```

$$3\sqrt{4}+5$$

```
> simplify(%);
```

$$11$$

```
> x^2;
```

$$x^2$$

```
> %^(1/2);
```

$$\sqrt{x^2}$$

```
> simplify(%);
```

$$\mathtt{csgn}(x)\,x$$

Notice we were able to simplify $3\sqrt{4}+5$ to 11. Of course, the value of $(x^2)^{1/2}$ depends on the sign of x. Here `csgn` is a function that returns 1 if x is positive and -1 otherwise. It is also defined for complex numbers. See `?csgn` for more information. If we know that $x > 0$, we can use `assume` to do further simplification ($x\sim$ replaces x).

```
> y:=((x-2)^2)^(1/2);
```

$$y := \sqrt{(x-2)^2}$$

```
> assume(x>2);
> simplify(y);
```

$$x\tilde{} - 2$$

To show the assumptions placed on a variable, we use the about function.

```
> about(x);
Originally x, renamed x~:
    is assumed to be:  RealRange(Open(2),infinity)
```

The output RealRange(Open(2),infinity) means the interval $(2, \infty)$. This translates into the assumption that $x > 2$.

To remove the assumption on x, we could use the restart function, but then we would lose the value of y. Instead we do the following.

```
> x :='x';
```

$$x := x$$

This restores x to its original status. See Section 3.1.9.

MAPLE 7 has a nifty new command called assuming. This allows us to do simplifications with temporary assumptions.

```
> y;
```

$$\sqrt{(x-2)^2}$$

```
> simplify(y) assuming x>2;
```

$$x - 2$$

The last command simplified y under the assumption that $x > 2$. Notice that the output is in terms of x and not $x\tilde{}$.

3.1.5 Simplifying radicals

To simplify expressions using radicals, we can use simplify and radsimp.

```
> y := x^3 + 3*x^2 + 3*x + 1;
```

$$y := x^3 + 3x^2 + 3x + 1$$

```
> simplify(y^(1/3));
```

$$((1+x)^3)^{1/3}$$

```
> radsimp(y^(1/3));
```

$$1 + x$$

```
> assume(x>-1);
> simplify(y^(1/3));
```

$$1 + x\tilde{}$$

```
> assume(x<-1);
```

```
> simplify(y^(1/3));
```

$$-\frac{1}{2}(x^{\sim}+1)(1+I\,3^{1/2})$$

```
> x := 'x':
```

Notice that `simplify` recognized y as a cube but failed to simplify $y^{1/3}$. The command `radsimp`, on the other hand, was able to simplify $y^{1/3}$ to $1+x$. If assumptions are given for x, then `simplify` is able to simplify the radical further. However, it should be noted that the value of the cube root depends on these assumptions, so care needs to be taken.

A cute MAPLE command is `rationalize`.

```
> 1/(1+sqrt(2));
```

$$\frac{1}{\sqrt{2}+1}$$

```
> rationalize(%);
```

$$\sqrt{2}-1$$

```
> (1-2^(2/3))/(1+2^(1/3));
```

$$\frac{1-2^{2/3}}{1+2^{1/3}}$$

```
> rationalize(%);
```

$$-2^{1/3}+1$$

```
> y:= z/(1 + sqrt(x));
```

$$y := \frac{z}{1+\sqrt{x}}$$

```
> rationalize(y);
```

$$\frac{z\,(-1+\sqrt{x})}{-1+x}$$

Notice that `rationalize` does a great job rationalizing a denominator not only for expressions involving square roots but for more complicated radicals as well. It can also handle symbolic expressions.

3.1.6 Working in the real domain

Sometimes MAPLE will return an expected complex number. We saw an instance of this in the last section. We reexamine the example.

```
> restart:
> y := (1+x)^3:
> simplify(y^(1/3)) assuming x<-1;
```

$$-\frac{1}{2}(x^{\sim}+1)(1+I\,3^{1/2})$$

Here I is MAPLE's notation for the complex number $i = \sqrt{-1}$. The unsuspecting precalculus or calculus student may not be expecting this complex cube root of y and would prefer to work in the real domain. Fortunately, there is a new package in MAPLE 7 for working in the real domain. Funnily enough the package is called *RealDomain*. To load this package we must use the `with` function.

```
> with(RealDomain):
Warning,
these protected names have been redefined and unprotected:
Im, Re, ^, arccos, arccosh, arccot, arccoth, arccsc, arccsch,
arcsec, arcsech, arcsin, arcsinh, arctan, arctanh, cos, cosh, cot,
coth, csc, csch, eval, exp, expand, limit, ln, log, sec, sech,
signum, simplify, sin, sinh, solve, sqrt, surd, tan, tanh
```

We redo the calculation of $y^{1/3}$:

```
> y := (1+x)^3:
> simplify(y^(1/3)) assuming x<-1;
```

$$1 + x$$

```
> simplify(y^(1/3));
```

$$1 + x$$

This time $y^{1/3}$ simplified to $1+x$. This is the only real cube root of y, assuming x is real.

Let's redo some calculations from Section 3.1.4, but this time in the real domain.

```
> with(RealDomain):
> x^2;
```

$$x^2$$

```
> %^(1/2);
```

$$\sqrt{x^2}$$

```
> simplify(%);
```

$$|x|$$

```
> y:=x^3+3*x^2+3*x+1;
```

$$x^3 + 3\,x^2 + 3\,x + 1$$

```
> simplify(y^(1/3));
```

$$signum(x^3 + 3x^2 + 3x + 1)^{2/3}\,((x+1)^3)^{1/3}$$

This time in the real domain we found $\sqrt{x^2} = |x|$, which is more palatable than $\text{csgn}(x)x$. Here $y^{1/3}$ should have simplified to $x + 1$, so I guess MAPLE still is not perfect.

```
> restart:
> sqrt(-1);
```

$$I$$

```
> with(RealDomain):
> sqrt(-1);
```

$$\mathit{undefined}$$

After restarting, MAPLE recognizes $\sqrt{-1}$ as the complex number i. When *RealDomain* is loaded, MAPLE considers $\sqrt{-1}$ as being undefined.

3.1.7 Simplifying rational functions

To simplify a rational function (i.e., a function that can be written as a quotient of two polynomials) we use the command `normal`. This has the effect of canceling any common factors between numerator and denominator. First we restore x and y's variable status.

```
> y:='y':  z:='z':
> a:= (x-y-z)*(x+y+z);
```

$$a := (x-y-z)(x+y+z);$$

```
> b :=(x^2-2*x*y-2*x*z+y^2+2*y*z+z^2)*(x^2-x*y+x*z-y*z);
```

$$b := (x^2-2xy-2xz+y^2+2yz+z^2)(x^2-xy+xz-yz)$$

```
> c:=a/b;
```

$$c := \frac{(x-y-z)(x+y+z)}{(x^2-2xy-2xz+y^2+2yz+z^2)(x^2-xy+xz-yz)}$$

```
> normal(c);
```

$$-\frac{(x+y+z)}{(x^2-yx+xz-yz)(-x+y+z)}$$

```
> simplify(c);
```

$$-\frac{(x+y+z)}{(x^2-yx+xz-yz)(-x+y+z)}$$

```
> factor(c);
```

$$\frac{(x+y+z)}{(x-y)(x+z)(x-y-z)}$$

Observe that `normal` and `simplify` had the same effect on the rational function c. We use `normal` for rational functions if we can do without the more expensive `simplify`. Also, we could have used `factor` to simplify c and get it into a nice form. It should be noted that `normal` is able to do this simplification without factoring, which is more expensive in terms of memory.

Some useful functions for manipulating rational functions are: numer, denom, rem, and quo. We let c be as above.

```
> numer(c);
```

$$-(-x+y+z)(x+y+z)$$

```
> denom(c);
```

$$(x^2-2xy-2xz+y^2+2yz+z^2)(x^2-xy+xz-yz)$$

```
> factor(%);
```

$$(-x+y+z)^2(x-y)(x+z)$$

The functions numer and denom select the numerator and denominator, respectively, of a rational function. After factoring the denominator of c, we see that there was simplification because of the common factor $(-x+y+z)$.

Many operations on rational functions can also be performed through a context menu.

```
> c;
```

$$\frac{(x-y-z)(x+y+z)}{(x^2-2xy-2xz+y^2+2yz+z^2)(x^2-xy+xz-yz)}$$

Click the right mouse button on our rational function above. A context menu should appear. Now try clicking on Factor, Simplify, Expand, Normal, Numerator, and Denominator.

The functions quo and rem give the quotient and remainder upon polynomial division.

```
> a:= 2*x^3+x^2+12;
```

$$a := 2x^3+x^2+12$$

```
> b := x^2 - 4;
```

$$b := x^2-4$$

```
> q := quo(a,b,x);
```

$$q := 2x+1$$

```
> r := rem(a,b,x);
```

$$r := 16+8x$$

```
> expand( a - (b*q + r) );
```

$$0$$

The command quo(a,b,x) gives the quotient q when a is divided by b as polynomials in x. The command rem(a,b,x) gives the remainder r so that

$$a = bq + r,$$

and the degree of r (as a polynomial in x) is less than the degree of b.

3.1.8 Degree and coefficients of a polynomial

In Section 3.1.3 the `collect` command was introduced to view polynomials. Two other useful functions are `coeff` and `degree`. Let p be as before.

```
> p:= y*(x+y+1)*(x-y+1)*(x-y-1):
> q := expand(%);
```

$$yx^3 - x^2y^2 + x^2y - 2\,y^2x - xy - y^3x + y^4 + y^3 - y^2 - y$$

```
> coeff(q,x,2);
```

$$-y^2 + y$$

```
> coeff(p,x,2);
```

$$y\,(y+1) + y\,(-y+1) + y\,(-y-1)$$

```
> expand(%);
```

$$-y^2 + y$$

```
> degree(q,x);
```

$$3$$

The command `coeff(q,x,2)` found the coefficient of x^2 in the polynomial q. The command `degree(q,x)` gave the degree of q as a polynomial in x. Observe also that when `coeff` was applied to the unexpanded form p, MAPLE still returned the correct value for the coefficient but in an unexpanded form.

Warning: In MAPLE V Release 4 (and earlier versions), `coeff` will either return an "incorrect" result or an error message, if it is applied to an unexpanded polynomial like p. So be careful when using `coeff` in these earlier versions of MAPLE.

Another useful and related function is `ldegree`.

```
> q := q - 2*x/y;
```

$$q := yx^3 - x^2y^2 + x^2y - 2\,y^2x - xy - y^3x + y^4 + y^3 - y^2 - y - 2\,\frac{x}{y}$$

```
> ldegree(q,x);
```

$$0$$

```
> ldegree(q,y);
```

$$-1$$

```
> c1:=1;
```

$$1$$

```
> c2:=0;
```

$$0$$

```
> degree(c1,x);
```

$$0$$

```
> degree(c2,x);
```

$$-\infty$$

The assignment `q := q - 2*x/y` subtracted $2x/y$ from q and assigned the result to q. `ldegree(q,x)` returns the degree of the lowest power of x in the polynomial q, which in our session was 0. Because of the term $2x/y$, `ldegree(q,y)` returned -1 as the lowest degree in the variable y. Also, observe that MAPLE returns 0 for the degree of a nonzero constant but returns $-\infty$ for the degree of the zero polynomial.

Warning: In MAPLE V Release 4 (and earlier versions), `degree` will return 0 for the zero polynomial.

3.1.9 Substituting into an expression

We can substitute into an expression using the command `subs`.

```
> p := (x+y+z)*(x-y+z)*(x-y-z);
```

$$p := (x + y + z)(x - y + z)(x - y - z)$$

```
> subs(x=1,p);
```

$$(1 + y + z)(1 - y + z)(1 - y - z)$$

To substitute $x = 1$ into p, we used the command `subs(x=1,p)`. Try substituting $x = 1$ and $y = 2$ into p using the command `subs(x=1,y=2,p)`.

3.1.10 Restoring variable status

In the last section we saw how `subs` is used to do substitution. There is another way to do this. We let p be as Section 3.1.8.

```
> p;
```

$$(x + y + z)(x - y + z)(x - y - z)$$

```
> x:=1:  y:=2:
> p;
```

$$(3 + z)(-1 + z)(-1 - z)$$

We are able to do the substitution by assigning $x := 1$ and $y := 2$. However, now p has changed. There is a way to restore x and y's variable status.

```
> x := 'x':    y := 'y':
> p;
```

$$(x + y + z)(x - y + z)(x - y - z)$$

The assignments `x := 'x'` and `y := 'y'` restored x and y to their variable status. It is neat that p was also restored to its original status.

3.2 Equations

3.2.1 Left- and right-hand sides

To assign a value to a variable, we use `:=`. The symbol `=` has a different meaning and is reserved for equations.

```
> eqn := x^2 - x = 1;
```

$$eqn := x^2 - x = 1$$

```
> R := solve(eqn,x);
```

$$R := \frac{1}{2}\sqrt{5} + \frac{1}{2},\quad \frac{1}{2} - \frac{1}{2}5^{1/2}$$

```
> simplify(R[1]*R[2]);
```

$$-\frac{1}{4}(\sqrt{5}+1)(\sqrt{5}-1)$$

```
> expand(%);
```

$$-1$$

We assigned to equation $x^2 - x = 1$ the name *eqn*. We solved the equation for x by typing `solve(eqn,x)`. We named the list of solutions R. The two solutions were $R[1]$ and $R[2]$. In this way we can manipulate the solutions. Observe that we computed the product of the roots to be -1 as expected.

The left and right sides of an equation can be manipulated using `lhs` and `rhs`.

```
> eqn;
```

$$x^2 - x = 1$$

```
> lhs(eqn);
```

$$x^2 - x$$

```
> subs(x=R[1],lhs(eqn));
```

$$\left(\frac{1}{2} + \frac{1}{2}\sqrt{5}\right)^2 - \frac{1}{2}\sqrt{5} - \frac{1}{2}$$

```
> expand(%);
```

$$1$$

The command `lhs(eqn)` gave us the left side of the equation. Then we were able to substitute $x = R[1]$ (the first root) into the left side of the equation, which simplified to 1 (as expected) using `expand`.

3.2.2 Finding exact solutions

MAPLE has the capability of solving systems of equations.

```
> restart:
> eqn1 := x^3+a*x=14;
```

$$eqn1 := x^3 + ax = 14$$

```
> eqn2 := a^2-x=7;
```

$$eqn2 := a^2 - x = 7$$

```
> solve({eqn1,eqn2},{x,a});
```

$$\{a = 3, x = 2\},$$
$$\{a = \mathit{RootOf}(\_Z^5 + 3\_Z^4 - 12\_Z^3 - 35\_Z^2 + 42\_Z + 119, \mathit{label} = \_L1),$$
$$x = \left(\mathit{RootOf}(\_Z^5 + 3\_Z^4 - 12\_Z^3 - 35\_Z^2 + 42\_Z + 119, \mathit{label} = \_L1)\right)^2 - 7\}$$

The syntax for solving systems of equations is `solve(S,X)` where S is a set of equations and X is the required set of variables. Observe that MAPLE was able to find the solution $x = 2$, $a = 3$. It also found that $a = z$, $x = z^2 - 7$ are solutions where z is any root of the following polynomial equation:

$$Z^5 + 3Z^4 - 12Z^3 - 35Z^2 + 42Z + 119 = 0.$$

The argument `label = _L1` gives the root a label. This is a way of distinguishing roots when using the `RootOf` function. As in the previous section, we can manipulate solutions. We select the first set of solutions and substitute them into the first equation.

```
> %[1];
```

$$\{a = 3, x = 2\}$$

```
> subs(%,eqn1);
```

$$14 = 14$$

3.2.3 Finding approximate solutions

In the last section we came upon the following quintic:

$$Z^5 + 3Z^4 - 12Z^3 - 35Z^2 + 42Z + 119 = 0.$$

Although naturally enough MAPLE is unable to find an exact explicit solution, it is able to find approximate solutions using `fsolve`.

```
> polyeqn := Z^5+3*Z^4-12*Z^3-35*Z^2+42*Z+119=0:
> a1 := fsolve(polyeqn,Z);
```

$$a1 := -3.136896207$$

```
> x1:= a1^2 -7;
```

$$x1 := 2.840117813$$

```
> subs({x=x1,a=a1},{eqn1,eqn2});
```

$$\{14.00000003 = 14,\ 7.0000000000 = 7\}$$

We used the command `fsolve(polyeqn,Z)` to find the approximate solution $Z \approx -3.136896207$. This implied that $a = -3.136896207$ and $x = a^2 - 7 = 2.840117813$ are approximate solutions to our system of equations in the previous section. We were able to check this using `subs`.

3.2.4 Assigning solutions

Once an equation or system of equations has been solved, we can use `assign` to assign a particular solution to the variable(s). We use the example given in Section 3.2.2.

```
> solve({x^3+a*x=14,a^2-x=7},{a,x}):
> %[1];
```

$$\{a = 3, x = 2\}$$

```
> assign(%);
> a; x;
```

$$3$$

$$2$$

To restore a and x to variable status we could use the method of Section 3.1.9 or use the `unassign` function.

```
> unassign('a','x');
> a,x;
```

$$a, \quad x$$

3.3 Fun with integers

3.3.1 Complete integer factorization

The command `ifactor` gives the prime factorization of an integer.

```
> 2^(2^5)+1;
```

$$4294967297$$

```
> ifactor(%);
```

$$(641)\,(6700417)$$

```
> ifactor(500326623506762117757 9);
```

$$(3)^2\,(13)\,(31)^3\,(67)\,(139)\,(320057)\,(481577)$$

3.3.2 Quotient and remainder

The integer analogs of quo and rem, the functions for finding the quotient and the remainder in polynomial division, are the functions iquo and irem. They work in the same way.

```
> a := 23;   b := 5;
```

$$a := 23$$
$$b := 5$$

```
> q := iquo(a,b);  r := irem(a,b);
```

$$q := 4$$
$$r := 3$$

```
> b*q+r;
```

$$23$$

We observe that if $q =$ iquo(a,b) and $r =$ irem(a,b), then

$$a = bq + r,$$

where $0 \leq r < b$ if a and b are positive.

Two related functions are floor and frac. The function floor(x) gives the greatest integer less than or equal to x and frac(x) gives the fractional part of x. Try

```
> x := 22/7;
> floor(x);
> frac(x);
> floor(-x);
> frac(-x);
```

3.3.3 Gcd and lcm

The greatest common divisor and the lowest common multiple of a set of numbers can be found using gcd and lcm.

```
> gcd(28743,552805);
```

$$11$$

```
> ifactor(28743);   ifactor(552805);
```

$$(3)(11)(13)(67)$$
$$(5)(11)(19)(23)^2$$

```
> lcm(21,35,99);
```

$$3465$$

We find that the gcd of 28743 and 552805 is 11. This can also be seen from the prime factorizations. The lcm of 21, 35, and 99 is 3465.

3.3.4 Primes

The ith prime can be computed with `ithprime`. The function `isprime` tests whether a given integer is prime or composite.

```
> ithprime(100);
```

$$541$$

```
> isprime(2^101-1);
```

$$\textit{false}$$

```
> 7*3^10 + 10;
```

$$413353$$

```
> isprime(%);
```

$$\textit{true}$$

We found that the 100th prime is 541, that $2^{101} - 1$ is composite, and that $7 \cdot 3^{10} + 10 = 413353$ is prime. Try making a table of the first 200 primes:

```
> matrix(20,10,[seq(ithprime(k),k=1..200)]);
```

For a positive integer n, `nextprime(`n`)` gives the smallest prime larger than n, and `prevprime(`n`)` gives the largest prime smaller than n.

```
> nextprime(1000);
```

$$1009$$

```
> prevprime(1000);
```

$$997$$

The next prime past 1000 is 1009 and the previous prime is 997.

3.3.5 Integer solutions

In Sections 3.2.1 and 3.2.2 we saw how to solve equations in MAPLE using `solve`. The integer analog of `solve` is `isolve`. We use this function if we are only interested in integer solutions. We use the example from Section 3.2.2. Remember to restore variable status to x and a first.

```
> x:='x':  a:='a':
> eqn1:= x^3+a*x=14:  eqn2 := a^2-x=7:
> isolve({eqn1,eqn2},{x,a});
```

$$\{a = 3, x = 2\}$$

This time we found the unique integer solution $a = 3$, $x = 2$ to the given system of equations.

3.3.6 Reduction mod p

MAPLE can do computations with integers modulo m.

```
> modp(117,13);
```

$$0$$

```
> modp(129,13);
```

$$12$$

```
> ifactor(129-12);
```

$$(3)^2\,(13)$$

```
> 117 mod 13;
```

$$0$$

```
> 129 mod 13;
```

$$12$$

```
> 1/17 mod 257;
```

$$121$$

```
> modp(121*17,257);
```

$$1$$

The functions for reduction modulo m are `modp` and `mod`. Given an integer a and a positive integer m, `modp(a,m)` reduces a modulo m. The syntax using `mod` is `a mod m`. In our MAPLE session, `modp(129,13)` returned 12, which means

$$129 \equiv 12 \pmod{13},$$

and this is indeed the case in as much as 13 divides the difference $129 - 12$. The call `129 mod 13` also reduced 129 modulo 13. When a and m are relatively prime, i.e., 1 is their greatest common divisor, `modp(1/a,m)` or `1/a mod m` returns the multiplicative inverse of a modulo m. We see that 121 is the inverse of 17 modulo 257, and indeed

$$(121)(17) \equiv 1 \pmod{257}.$$

3.4 Unit conversion

MAPLE 7 has new facilities for converting from one system of units to another. There are both command line and menu-driven facilities. In the tool bar click on Edit and then on Unit Converter. A **Unit Converter** window should open.

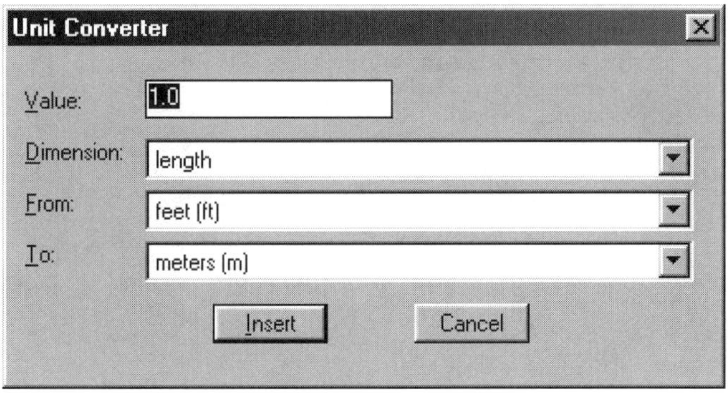

Figure 3.1 Menu-driven unit converter.

The window is already set up to do a simple example. Notice that 1.0 (Fig. 3.1) is in the Value box, Dimension is set to length, and we are ready to do a conversion from feet to meters. Click on `Insert`.

```
> convert( 1.0, 'units', 'ft', 'm' );
```

$$.3048000000$$

This means that

$$1.0\,\text{ft} = 0.3048\,\text{m}.$$

Let's try another conversion. Click on ▼ in the Dimension box and select `temperature`. Notice that the units have changed in the From and To boxes. Let's convert 100 degrees Fahrenheit to degrees Celsius. In the Value box type `100.0`, select `degrees Celsius (degC)` in the To box, and press `Insert`.

```
> convert( 100.0, 'temperature', 'degF', 'degC' );
```

$$37.7777778$$

This means that

$$100.0^{o}F \approx 37.778^{o}C.$$

We have seen two types of dimensions: length and temperature. There are many other dimensions available, including acceleration, angle, area, electric capacitance, force, magnetic flux, mass, power, pressure, speed, time, torque, volume, and work. A list of all dimensions can be obtained by loading the *Units* package and calling the `GetDimensions` function. Try

```
> with(Units):
> GetDimensions();
```

MAPLE 7 knows many systems of units, including SI, FPS, MKS, and CGS. See `?Units[System]` for more information.

The `convert` function can also be used to make conversion tables. We make a conversion table for meters, yards, kilometers, and miles:

```
> convert([m,yd,km,mile],conversion_table,output=grid,
   filter=evalf[6]);
```

		To:	m	yd	km	mi
Unit Name	Symbol					
meters	m		1.	1.09361	0.001	0.000621371
yards	yd		0.9144	1.	0.0009144	0.000568182
kilometers	km		1000.	1093.61	1.	.621371
miles	mi		1609.34	1760.	1.60934	1.

Here `evalf[6]` means to use `evalf` with 6 digits. From the table we see that to convert from miles to kilometers just multiply by 1.60934. For other examples see `?conversion,conversion_table`.

We can do MAPLE calculations using units. As an example we sum 12.1 feet and 4 meters.

```
> 12.1*Unit(ft)+4*Unit(m);
```

$$12.1\,[ft] + 4\,[m]$$

We can simplify this by loading the *Standard* function in the *Units* package:

```
> with(Units[Standard]):
> 12.1*Unit(ft)+4*Unit(m);
```

$$7.688080000\,[m]$$

This means that the sum of 12.1 feet and 4 meters is 7.68808 meters. Other MAPLE functions recognize these units.

```
> max(12.1*Unit(ft),4*Unit(m));
```

$$4\,[m]$$

This means that 4 meters is bigger than 12.1 feet. A different style of representing units can be used by loading the `Natural` function in the *Units* package. Try

```
> with(Units[Natural]):
> max(12.1*ft,4*m);
```

3.5 Trigonometry

3.5.1 Degrees and radians

To convert between degrees and radians we use `convert`.

```
> convert(72*degrees,radians);
```

$$2/5\,\pi$$

```
> convert(2/5*Pi,degrees);
```

$$72\ \textit{degrees}$$

To convert d degrees to radians we use `convert(d*degrees,radians)`. To convert r (in radians) to degrees we use `convert(r,degrees)`. We see that 72 degrees is $2\pi/5$ radians. Remember, we use `Pi` for π in MAPLE. Alternatively, we could convert degrees to radians by multiplying by $\pi/180$.

```
> 72*Pi/180;
```

$$2/5\,\pi$$

In MAPLE 7 we can use the `convert` function with the `units` option as we did in Section 3.4. Try

```
> convert(72,units,degrees,radians);
> convert(2*Pi/5,units,radians,degrees);
```

3.5.2 Trigonometric functions

In MAPLE, the trigonometric functions are `sin`, `cos`, `tan`, `sec`, `csc`, and `cot`. The arguments for all the trigonometric functions are in radians.

```
> sin(0);
```

$$0$$

```
> cos(0);
```

$$1$$

```
> tan(0);
```

$$0$$

```
> sin(Pi/2);
```

$$1$$

```
> cos(Pi/2);
```

$$0$$

```
> tan(Pi/2);
Error, (in tan) singularity encountered
> cot(Pi/2);
```

$$0$$

Remember, $\tan(\pi/2)$ is not defined.

The inverse trigonometric functions are `arcsin` `arccos`, `arctan`, `arcsec`, `arccsc`, and `arccot`.

```
> arcsin(1/2);
```

$$1/6\,\pi$$

```
> arcsec(-2);
```

$$2/3\,\pi$$

```
> arctan(1);
```

$$1/4\,\pi$$

```
> arcsin(sin(Pi/12));
```

$$1/12\,\pi$$

```
> arcsin(sin(Pi/12+Pi));
```

$$-1/12\,\pi$$

We found that

$$\begin{aligned} \sin^{-1}(1/2) &= \pi/6, & \sec^{-1}(-2) &= 2\pi/3 \\ \tan^{-1}(1/2) &= \pi/4, & \sin^{-1}(\sin(\pi/12)) &= \pi/12 \\ \sin^{-1}(\sin(13\pi/12)) &= -\pi/12 \end{aligned}$$

3.5.3 Simplifying trigonometric functions

Ever have trouble remembering the addition formulas for the trigonometric functions? Try the following:

```
> expand(sin(a+b));
```

$$\sin(a)\cos(b)+\cos(a)\sin(b)$$

```
> expand(cos(a+b));
```

$$\cos(a)\cos(b)-\sin(a)\sin(b)$$

```
> expand(tan(a+b));
```

$$\frac{\tan(a)+\tan(b)}{1-\tan(a)\tan(b)}$$

Now it all comes back to us:

$$\begin{aligned} \sin(a+b) &= \sin(a)\cos(b)+\cos(a)\sin(b) \\ \cos(a+b) &= \cos(a)\cos(b)-\sin(a)\sin(b) \\ \tan(a+b) &= \frac{\tan(a)+\tan(b)}{1-\tan(a)\tan(b)} \end{aligned}$$

To simplify a trigonometric expression, use `simplify`.

```
> y:=(1+sin(x)+cos(x))/(1+sin(x)-cos(x));
```

$$\frac{1+\sin(x)+\cos(x)}{1+\sin(x)-\cos(x)}$$

```
> simplify(y);
```

$$-\frac{\sin(x)}{\cos(x)-1}$$

We found that

$$\frac{1+\sin(x)+\cos(x)}{1+\sin(x)-\cos(x)}=\frac{\sin(x)}{1-\cos(x)}.$$

Can you show this result by hand?

Now try the following:

```
> expand(sin(5*x));
```

$$16\ \sin(x)\,(\cos(x))^4-12\ \sin(x)\,(\cos(x))^2+\sin(x)$$

```
> factor(%);
```

$$\sin(x)\left(4\ (\cos(x))^2+2\ \cos(x)-1\right)\left(4\ (\cos(x))^2-2\ \cos(x)-1\right)$$

This means that

$$\sin 5x=\sin x\,(4\cos^2 x+2\cos x-1)(4\cos^2 x-2\cos x-1).$$

By letting $x=\frac{2\pi}{5}=72^{\text{o}}$, derive a nice value for $\cos 72^{\text{o}}$.

4. DATA TYPES

4.1 Sequences

In MAPLE, sequences take the form

$$expr1,\ expr2,\ expr3,\ \ldots,\ exprn.$$

```
> x := 1,2,3;
```

$$x := 1, 2, 3$$

```
> y := 4,5,6;
```

$$y := 4, 5, 6$$

```
> x,y;
```

$$1, 2, 3, 4, 5, 6$$

We observe that in MAPLE, `x,y` concatenates the two sequences x and y. There are two important functions used to construct sequences: `seq` and the repetition operator `$`.

```
> f:='f':    seq(f(i), i=1..6);
```

$$f(1), f(2), f(3), f(4), f(5), f(6)$$

```
> seq(i^2, i=1..5);
```

$$1, 4, 9, 16, 25$$

```
> x:= 'x':
> x$4;
```

$$x, x, x, x$$

In general, `seq(f(i), i=1..n)` produces the sequence

$$f(1),\ f(2),\ \ldots,\ f(n)$$

and `x$n` produces a sequence of length n

$$x,\ x,\ \ldots,\ x$$

The `op` function can be used to create sequences.

```
> b:='b':   c:='c':
> L := a+b+2*c+3*d;
```

$$L := a + b + 2c + 3d$$

```
> op(%);
```

$$a,\ b,\ 2c,\ 3d$$

`op(expr)` produces a sequence whose elements are the operands in `expr`.

```
> nops(L);
```

$$4$$

```
> op(3,L);
```

$$2c$$

`nops(expr)` gives the length of the sequence `op(expr)` and `op(j,expr)` gives the jth term in the sequence `op(expr)`.

If s is a sequence, then the jth term of the sequence is $s[j]$.

```
> s := 1, 8, 27, 64, 125;
```

$$s := 1,\ 8,\ 27,\ 64,\ 125$$

```
> s[3];
```

$$27$$

4.2 Sets

We have already seen the set data type in Section 3.2.2 when solving systems of equations. In MAPLE, a *set* takes the form

$$\{expr1,\ expr2,\ expr3,\ \ldots,\ exprn\}.$$

In other words, a set has the form $\{S\}$ where S is a sequence. A set is a set in the mathematical sense — order is not important.

```
> y := 'y':   {x,y,z,y};
```

$$\{x,\ y,\ z\}$$

Observe that $\{x, y, z, y\} = \{x, y, z\}$. MAPLE can perform the usual set operations: union, intersection, and difference.

```
> a := {1,2,3,4}; b := {2,4,6,8};
```

$$a := \{1,\ 2,\ 3,\ 4\}$$
$$b := \{2,\ 4,\ 6,\ 8\}$$

```
> a union b;
```

$$\{1,\ 2,\ 3,\ 4,\ 6,\ 8\}$$

```
> a intersect b;
```

$$\{2,\ 4\}$$

```
> a minus b;
```

$$\{1,\ 3\}$$

We can also determine whether a given expression is an element of a set using the function `member`.

```
> member(2,a);
```

$$true$$

```
> member(5,a);
```

$$false$$

```
> a[3];
```

$$3$$

So `member(x,A)` returns the value `true` if x is an element of A and `false` otherwise. Also, the jth element of the set A is `A[j]`.

4.3 Lists

In MAPLE, a *list* takes the form

$$[expr1,\ expr2,\ expr3,\ \ldots,\ exprn].$$

Here order is important.

```
> a:='a':  b:='b':
> L1 := [x,y,z,y]; L2 := [a,b,c];
```

$$L1 := [x,\ y,\ z,\ y]$$
$$L2 := [a,\ b,\ c]$$

```
> L := [op(L1),op(L2)];
```

$$L := [x,\ y,\ z,\ y,\ a,\ b,\ c]$$

```
> L[5];
```

$$a$$

We observe that the lists $L1$ and $L2$ can be concatenated by the command `[op(L1),op(L2)]` and that `L[j]` gives the jth item in the list L. Lists can be created from sequences:

```
> s := seq( i/(i+1), i=1..6);
```

$$s := 1/2,\ 2/3,\ 3/4,\ 4/5,\ 5/6,\ 6/7$$

```
> M := [s];
```

$$M := [1/2,\ 2/3,\ 3/4,\ 4/5,\ 5/6,\ 6/7]$$

```
> M[2..5];
```

$$[2/3,\ 3/4,\ 4/5,\ 5/6]$$

So, `M[i..j]` gives the ith through jth elements of the list M.

4.4 Tables

In MAPLE, a *table* is an array of expressions whose indexing set is not necessarily a set of integers. Sounds bizarre? Let's look at some examples. Tables are created by the `table` function.

```
> T := table([a,b]);
```

$$T := \text{table}([\\ 1 = a\\ 2 = b\\ ])$$

```
> T[2];
```

$$b$$

So, if L is a list, then `table(L)` converts L into a table. The jth element of this table T is given by `T[j]`. Try

```
> S := table([(1)=A,(3)=B+C,(5)=A*B*C]);
> S[3];
> S;
> op(S);
```

For the table S, the indexing set is $\{1, 3, 5\}$ and thus does not necessarily have to be a set of consecutive integers. See `?table` for more bizarre examples. In your session you should have found that `S` did not return the table, but that `op(S)` did.

4.5 Arrays

In MAPLE, an *array* is a special kind of a table. It most resembles a matrix. Let's look at some examples.

```
> A := array(1..2,1..3);
```

$$A := \text{array}(1..2, 1..3, [\ ])$$

```
> op(A);
```

$$\begin{bmatrix} ?_{1,1} & ?_{1,2} & ?_{1,3} \\ ?_{2,1} & ?_{2,2} & ?_{2,3} \end{bmatrix}$$

```
> B := array(1..2,1..2,1..2);
```

$$B := \text{array}(1..2, 1..2, 1..2, [\ ])$$

```
> op(B);
```

$$\text{array}(1..2, 1..2, 1..2, [\\ (1, 1, 1) = ?_{1,1,1}$$

$$(1,1,2) = ?_{1,1,2}$$
$$(1,2,2) = ?_{1,2,2}$$
$$(2,1,1) = ?_{2,1,1}$$
$$(2,1,2) = ?_{2,1,2}$$
$$(2,2,1) = ?_{2,2,1}$$
$$(2,2,2) = ?_{2,2,2}$$

])

We see that the array A corresponds to a 2×3 matrix. The array B corresponds to $2 \times 2 \times 2$ matrix or, if you like, a table with indexing set

$$\{(1,1,1),\ (1,1,2),\ \ldots,\ (2,2,2)\}.$$

We can insert entries into an array by using subscripts (or indices).

```
> C:=array(1..2,1..2):
> C[1,1]:=1:  C[1,2]:=2:  C[2,1]:=3:  C[2,2]:=7:
> op(C);
```

$$\begin{bmatrix} 1 & 2 \\ 3 & 7 \end{bmatrix}$$

```
> print(C);
```

$$\begin{bmatrix} 1 & 2 \\ 3 & 7 \end{bmatrix}$$

Observe that we can print out an array using the `print` command. An alternative method for entering arrray entries is given below.

```
> F:=array(1..2,1..3,[[1,2,3],[5,9,7]]);
```

$$F := \begin{bmatrix} 1 & 2 & 3 \\ 5 & 9 & 7 \end{bmatrix}$$

4.6 Data conversions

The function `type` checks the data type of an object.

```
> A := {1,2,3}:
> s := 1,2,3:
> L := [1,2,3]:
> T := table([1,2,3]):
> M := array(1..3,[1,2,3]):
> type(L,list);
```

true

```
> type(T,set);
```

false

The function `convert` can be used to convert from one data type to the other.

```
> convert(A,list);
```

$$[1, 2, 3]$$

```
> convert(L,set);
```

$$\{1, 2, 3\}$$

The `whattype` function is used find the type of an expression.

```
> whattype(A);
```

$$set$$

```
> whattype(s);
```

$$exprseq$$

```
> whattype(L);
```

$$list$$

```
> whattype(T);
```

$$symbol$$

```
> whattype(op(T));
```

$$table$$

```
> whattype(M);
```

$$symbol$$

```
> whattype(op(M));
```

$$array$$

See `?whattype` for more information.

4.7 Other data types

In this chapter we have seen a small sample of MAPLE's data types. To see a complete list, try

```
> ?type
```

5. CALCULUS

5.1 Defining functions

To enter the function $f(x) = x^2 - 3x + 5$, type

```
> f:= x -> x^2 - 3*x + 5;
```

$$f := x \to x^2 - 3x + 5$$

The arrow symbol is entered by typing the *minus* key, "–" immediately followed by the *greater than* key, ">". We compute $f(2)$.

```
> f(2);
```

$$3$$

Thus, in MAPLE the syntax for creating a function $f(x)$ is `f := x -> expr`, where `expr` is some expression involving x. Functions in more than one variable are defined in the same way.

```
> g := (x,y) -> x*y/(1+x^2+y^2);
```

$$g := (x, y) \to \frac{xy}{1 + x^2 + y^2}$$

We defined the function

$$g(x, y) = \frac{xy}{1 + x^2 + y^2}.$$

Try simplifying $g(\sin t, \cos t)$

```
> g(sin(t),cos(t));
> simplify(%);
```

To convert an expression into a function, we use the `unapply` function.

```
> q := Z^5+3*Z^4-12*Z^3-35*Z^2
    +42*Z+119:
> h := unapply(q,Z);
```

$$h := Z \to Z^5 + 3Z^4 - 12Z^3 - 35Z^2 + 42Z + 119$$

In Sections 3.2 and 3.3 we came across the quintic polynomial q above. Here q is an expression involving Z. To convert q into the function $h(Z)$, we used the command `unapply(q,Z)`. Now we are free to play with the function h.

```
> H := x -> evalf( h(x), 4):
```

$$H := x \to \text{evalf}(\, h(x), 4)$$

```
> X := [seq(evalf(-4+i/10,4),i=0..10)];
```

$$X := [-4., -3.900, -3.800, -3.700, -3.600, \\ -3.500, -3.400, -3.300, -3.200, -3.100, -3.]$$

```
> Y := map(H,X);
```

$$Y := [-97., -73.7, -54.5, -39.0, -26.6, -17.1, \\ -10.4, -5.1, -1.4, .6, 2.]$$

The function $H(x)$ computes $h(x)$ to 4 digits. Then we used `seq` and `map` to produce the lists X and Y, which give a table of x and y values for the function $y = h(x)$.

5.2 Composition of functions

In MAPLE, `@` is the function composition operator. If f and g are functions, then the composition of f and g, $f \circ g(x) = f(g(x))$, is given by `(f@g)(x)`.

```
> (sin@cos)(x);
```

$$\sin(\cos(x));$$

```
> f := x -> x^2:
> g := x -> sqrt(1-x):
> (f@g)(x);
```

$$1 - x$$

```
> (g@f)(x);
```

$$\sqrt{1 - x^2}$$

`@@` gives repeated composition, so that `(f@@2)(x)` gives $f(f(x))$ and `(f@@3)(x)` gives $f(f(f(x)))$. For certain functions known to MAPLE, `f@@(-1)(x)` gives the inverse function $f^{-1}(x)$.

5.3 Summation and product

In MAPLE, the syntax for the sum

$$\sum_{i=1}^{n} f(i) = f(1) + f(2) + \cdots + f(n)$$

is `Sum(f(i),i=1..n)` and `sum(f(i),i=1..n)`.

```
> f := 'f':
> Sum(f(i),i=1..n);
```

$$\sum_{i=1}^{n} f(i)$$

```
> Sum(i^2,i=1..10);
```

$$\sum_{i=1}^{10} i^2$$

```
> sum(i^2,i=1..10);
```

$$385$$

Notice that the difference between `sum` and `Sum` is that in `sum`, the sum is evaluated, but that in `Sum`, it is not. It is recommended that you get into the habit of using `Sum` to first check for typos and then use `value` to evaluate the sum. In our previous session we found

$$\sum_{i=1}^{10} i^2 = 1 + 4 + 9 + \cdots + 100 = 385.$$

This time we will use `Sum` and `value`.

```
> Sum(i^2,i=1..10);
```

$$\sum_{i=1}^{10} i^2$$

```
> value(%);
```

$$385$$

```
> sum(i^2,i=1..n);
```

$$1/3\,(n+1)^3 - 1/2\,(n+1)^2 + 1/6\,n + 1/6$$

```
> factor(%);
```

$$1/6\,n\,(n+1)\,(2n+1)$$

Notice that MAPLE knows certain summation formulas such as

$$\sum_{i=1}^{n} i^2 = \frac{1}{6}n(n+1)(2n+1).$$

In MAPLE, the syntax for the product

$$\prod_{i=1}^{n} f(i) = f(1) \cdot f(2) \cdots f(n)$$

is `Product(f(i),i=1..n)`.

```
> f := 'f': q := 'q':
> Product(f(i),i=1..n);
```

$$\prod_{i=1}^{n} f(i)$$

```
> Product(1-q^i,i=1..5);
```

$$\prod_{i=1}^{5} 1 - q^i$$

```
> value(%);
```

$$(1-q)(1-q^2)(1-q^3)(1-q^4)(1-q^5)$$

```
> expand(%);
```

$$-q^{15} + q^{14} + q^{13} - q^{10} - q^9 - q^8 + q^7 + q^6 + q^5 - q^2 - q + 1$$

As with `sum` and `Sum`, for `product`, the product is evaluated, but with `Product`, it is not. Note that we could have evaluated the product $\prod_{i=1}^{5} 1 - q^i$ using `product(1-q^i,i=1..5)`.

A common problem with `sum` and `product` is the following:

```
> i:=2;
```

$$i := 2$$

```
> sum(i^3,i=1..5);
Error, (in sum) summation variable previously assigned,
second argument evaluates to, 2=1 ..  5
```

The problem occurred in `sum` since i had already been assigned the value 2. There are two ways around this problem. One way is to restore the variable status of i by typing `i := 'i'`. The second way is to replace `i` by `'i'` in the sum.

```
> sum('i'^3,'i'=1..5);
```

$$225$$

5.4 Limits

Naturally, there are two forms of the MAPLE limit function: `Limit` and `limit`. These are analogous to `sum` and `Sum`, etc.

The syntax for computing the limit of $f(x)$ as $x \to a$ is `Limit(f(x), x=a); value(%)`. The `Limit` command displays the limit so that it can be checked for typos and then the `value` command computes the limit. To compute the limit

$$\lim_{x \to 2} \frac{x^2 - 4}{x - 2}$$

we type

```
> Limit((x^2-4)/(x-2),x=2); value(%);
```

$$\lim_{x \to 2} \frac{x^2 - 4}{x - 2}$$

$$4$$

Thus, we see that

$$\lim_{x \to 2} \frac{x^2 - 4}{x - 2} = 4,$$

which can be verified easily with paper and pencil. Alternatively, by typing `limit((x^2-4)/(x-2),x=2)`, we could have found the limit in one step.

Left and right limits can also be calculated as well as limits where x approaches infinity. Try

```
> f:=(x^2-4)/(x^2-5*x+6);
> Limit(f,x=3,right); value(%);
> Limit(f,x=infinity); value(%);
```

5.5 Differentiation

MAPLE can easily find the derivatives of functions of one or several variables. The syntax for differentiating $f(x)$ is `diff(f(x),x)`.

```
> f := sqrt(1 - x^2);
```

$$f := \sqrt{1 - x^2}$$

```
> diff(f,x);
```

$$-\frac{x}{\sqrt{1 - x^2}}$$

```
> g := z -> z^2*exp(z) + sin(log(z)):
> diff(g(x),x);
```

$$2x\, e^x + x^2\, e^x + \frac{\cos(\ln(x))}{x}$$

The second derivative is given by typing `diff(f(x),x,x)`. For the nth derivative, use `diff(f(x),x$n)`. Use MAPLE to show that

$$\frac{d^5 \tan x}{dx^5} = 136\, \tan^2 x + 240\, \tan^4 x$$
$$+ 120\, \tan^6 x + 16.$$

In MAPLE, partial derivatives are computed using `diff`.

```
> z := exp(x*y)*(1+sqrt(x^2+3*y^2-x));
```

$$z := e^{xy}\left(1 + \sqrt{x^2 + 3\,y^2 - x}\right)$$

```
> diff(z,x);
```

$$ye^{xy}\left(1 + \sqrt{x^2 + 3\,y^2 - x}\right) + \frac{e^{xy}\,(2\,x - 1)}{2\,\sqrt{x^2 + 3\,y^2 - x}}$$

```
> normal(diff(z,x,y)-diff(z,y,x));
```

$$0$$

The syntax for $\frac{\partial z}{\partial x}$ is diff(z,x) and for $\frac{\partial^2 z}{\partial y \partial x}$ is diff(z,x,y). For

$$z = e^{xy}\left(1 + \sqrt{x^2 + 3\,y^2 - x}\right)$$

we found that

$$\frac{\partial z}{\partial x} = y e^{xy}\left(1 + \sqrt{x^2 + 3\,y^2 - x}\right) + \frac{e^{xy}\,(2\,x - 1)}{2\,\sqrt{x^2 + 3\,y^2 - x}},$$

and

$$\frac{\partial^2 z}{\partial y \partial x} = \frac{\partial^2 z}{\partial x \partial x}.$$

MAPLE also has the differential operator D. If f is a differentiable function of one variable, then Df is the derivative f'. We calculate $g'(x)$ for our function g above.

```
> g := z -> z^2*exp(z) + sin(z);
```

$$g := z \to z^2\,e^z + \sin(z)$$

```
> D(g);
```

$$z \to 2z\,e^z + z^2 e^z + \cos(z)$$

5.6 Extrema

MAPLE is able to find the minimum and maximum values of certain functions of one or several variables with zero or more constraints. There are three possible approaches: (1) using the built-in functions maximize and minimize, (2) using the extrema function, and (3) using the *simplex* package (for linear functions). Here we will describe (1) and (2). See ?simplex for (3).

The functions maximize and minimize can find the maximum and minimum values of a function of one or several variables. There is also an option for restricting some of the variables to certain intervals. It is advised that this facility be used with care, especially in earlier versions of MAPLE.

We can find the maximum value of the function $f(x)$ using maximize(f(x)). The command maximize(f(x), x=a..b) gives the maximum of the function, with x restricted to the interval $[a, b]$.

```
> maximize(sin(x));
```

$$1$$

```
> maximize(sin(x)+cos(x));
```

$$maximize(\sin(x) + \cos(x))$$

```
> maximize(x^2-5*x+1,x=0..3);
```

$$1$$

```
> maximize(sin(x),x=0..1);
```

$$\sin(1)$$

```
> maximize(sin(x)+cos(x),x=0..1);
```

$$\sqrt{2}$$

```
> maximize(sin(x)+cos(x),x=0..1/2);
```

$$\sin(1/2) + \cos(1/2)$$

MAPLE was able to find the correct maximum value of $\sin x$, but was unable to compute the maximum for the function $\sin x + \cos x$, although it was able to do so correctly when x was restricted to an interval. For $0 \leq x \leq 3$, the maximum value of $x^2 - 5x + 1$ was found to be 1.

Warning: In MAPLE V Release 5 (and earlier versions), the `maximize` function has a different syntax. In these earlier versions, the correct syntax has the form `maximize(f(x), {x},{x=a..b})`. Bugs in earlier versions have been eliminated in MAPLE 6. For instance, in MAPLE V, the call `maximize(sin(x), {x}, {x=0..1})` will return a value of 1 when the correct value for the maximum of $\sin x$ on the interval $[0, 1]$ is $\sin 1$. In MAPLE 6, the correct value is returned.

To find the minimum value of a function, use the command `minimize` whose syntax is analogous to that of `maximize`. MAPLE can also handle functions of more than one variable.

```
> minimize(x^2+y^2);
```

$$0$$

```
> minimize(x^2+y^2,x);
```

$$y^2$$

We found the minimum value of $x^2 + y^2$ to be 0. The function `minimize(x^2 + y^2,x)` found the minimum value of the function $x^2 + y^2$, considered as a function of x with y fixed.

The second method involves using the function `extrema`, which is able to find the minimum and maximum values of algebraic functions of one or several variables, subject to 0 or more constraints. It returns a set of possible maximum and minimum values, with the option of returning a possible set of points where these values occur. The syntax for the function is `extrema(f,{g1,g2, ... ,gn},{x1,x2, ... ,xm},'s')`. Here, f is the function. The constraints are $g_1 = 0$, $g_2 = 0, \ldots, g_n = 0$. The variables are $x_1, x_2, \ldots, x_m$, and s is the unevaluated variable for holding the set of possible points where the extrema occur.

Warning: In MAPLE V Release 5 (and earlier versions), `extrema` is a misc library function, which must be read into our MAPLE session with `readlib(extrema)`.

```
> readlib(extrema):
```

The `readlib` function is obsolete in MAPLE 6 and can be omitted.

```
> f := 2*x^2 + y + y^2;
```

$$f := 2x^2 + y + y^2$$

```
> g := x^2 + y^2 - 1;
```

$$g := x^2 + y^2 - 1$$

```
> extrema(f,{g},{x,y},'s');
```

$$\{0, 9/4\}$$

```
> s;
```

$$\{\{x = 0, y = 1\}, \{x = 0, y = -1\}\},$$
$$\{\{y = 1/2, x = 1/2\mathrm{RootOf}(\_Z^2 - 3)\}\}$$

```
> simplify(subs(s[1],f));
```

$$0$$

```
> simplify(subs(s[2],f));
```

$$2$$

```
> simplify(subs(s[3],f));
```

$$9/4$$

By using the command `extrema(f,{g},{x,y}, 's')`, we found that the extreme values of $f(x, y) = 2x^2 + y + y^2$ (subject to the constraint $x^2 + y^2 = 1$) are 0 and 9/4. The set of possible points where the extrema occured was assigned to the variable s. Using `simplify` and `subs`, we substituted each set of points into f. In this way, we found that the minimum value 0 occurs at $x = 0, y = -1$ and the maximum value 9/4 occurs at $x = \pm\sqrt{3}/2, y = 1/2$.

5.7 Integration

If f is an expression involving x, then the syntax for finding the integral $\int_a^b f(x)\,dx$ is `int(f,x=a..b)`. For the indefinite integral we use `int(f,x)`. There are also the unevaluated forms `Int(f,x=a..b)` and `Int(f,x)`.

```
> Int(x^2/sqrt(1-x^3),x);
```

$$\int \frac{x^2}{\sqrt{1-x^3}}\,dx$$

```
> value(%);
```

$$-2/3\,\sqrt{1-x^3}$$

```
> Int(1/x/sqrt(x^2 - 1),x=1..2/sqrt(3));
```

$$\int_1^{2/\sqrt{3}} \frac{1}{x\sqrt{x^2-1}}\,dx$$

```
> value(%);
```

$$\frac{1}{6}\,\pi$$

MAPLE easily found that

$$\int \frac{x^2}{\sqrt{1-x^3}}\,dx = -\frac{2}{3}\,\sqrt{1-x^3}$$

and
$$\int_1^{2/\sqrt{3}} \frac{1}{x\sqrt{x^2-1}}\,dx = \frac{\pi}{6}.$$

MAPLE can do improper integrals and multiple integrals in the obvious way. Try finding

$$\int_0^{\infty} re^{-r^2}\,dr$$

by typing `int(r*exp(-r^2),r=0..infinity)`. Try evaluating the double integral

$$\int\!\!\int y\,\sin(2x+3y^2)\,dx\,dy$$

by first integrating with respect to x and then with respect to y.

If MAPLE does not know the value of a definite integral, try `evalf`.

```
> Int(sqrt(1+x^6),x=0..1);
```

$$\int_0^1 \sqrt{1+x^6}\,dx$$

```
> value(%);
```

$$\int_0^1 \sqrt{1+x^6}\,dx$$

```
> evalf(%);
```

$$1.064088379$$

Although MAPLE was unable to evaluate the integral, it was able to find the approximation

$$\int_0^1 \sqrt{1+x^6}\,dx \approx 1.064088379.$$

5.7.1 Techniques of integration

MAPLE knows some standard techniques of integration. These are in the *student* package and are loaded with the command `with(student)`.

5.7.1.1 Substitution

In MAPLE, to do integration by substitution, we use the **changevar** command. The syntax is `changevar(f(u)=h(x),`*integral*`,u)` where *integral* is an integral in the variable x, $f(u) = h(x)$ is the substitution, and u is the new variable in the integral.

```
> with(student):
> G:=Int(x^4/sqrt(1-x^10),x);
```

$$\int \frac{x^4}{\sqrt{1-x^{10}}} dx$$

```
> changevar(u=x^5,G,u);
```

$$\int 1/5 \frac{1}{\sqrt{1-u^2}} du$$

```
> G2 := value(%);
```

$$1/5 \arcsin(u)$$

```
> subs(u=x^5,G2);
```

$$1/5 \arcsin(x^5)$$

```
> diff(%,x);
```

$$\frac{x^4}{\sqrt{1-x^{10}}}$$

Using **changevar** with the substitution $u = x^5$, we found

$$\begin{aligned} \int \frac{x^4}{\sqrt{1-x^{10}}}\, dx &= \frac{1}{5}\int \frac{1}{\sqrt{1-u^2}} du \\ &= \sin^{-1} u \\ &= \sin^{-1}(x^5) \end{aligned}$$

```
> G:=Int((3*x^2+1)/sqrt((1-x-x^3)*(1+x+x^3)),x);
```

$$G := \int \frac{3\,x^2+1}{\sqrt{(1-x-x^3)\,(1+x+x^3)}} dx$$

```
> value(G);
```

$$\int \frac{3\,x^2+1}{\sqrt{(1-x-x^3)\,(1+x+x^3)}} dx$$

Although MAPLE was unable to evaluate the integral above, you should be able to help it along by using `changevar` and the substitution $u = x + x^3$.

```
> radsimp(changevar(u=x+x^3,G,u));
```

5.7.1.2 Integration by parts

To do integration by parts, we use the command `intparts`. The syntax is `intparts(`*integral*`, u)` where u is as usual in the formula

$$\int u\,dv = uv - \int v\,du.$$

```
> with(student):
> Int(x*cos(3*x),x);
```

$$\int x \cos 3x\,dx$$

```
> intparts(%,x);
```

$$1/3\,x\,\sin(3x) - \int 1/3\,\sin(3x)\,dx$$

```
> value(%);
```

$$1/3\,x\,\sin(3x) + 1/9\,\cos(3x)$$

Thus MAPLE has helped us by providing the details of the evaluation of the integral by parts:

$$\begin{aligned} \int x \cos 3x\,dx &= 1/3\,x\,\sin 3x - \int 1/3\,\sin 3x\,dx \\ &= 1/3\,x\,\sin 3x + 1/9\,\cos 3x. \end{aligned}$$

5.7.1.3 Partial fractions

The command for finding the partial fraction decomposition of a rational function *ratfunc* (in the variable x) is `convert(`*ratfunc*`,parfrac,x)`. As an example, we use MAPLE to find the integral

$$\int \frac{4x^4 + 9x^3 + 12x^2 + 9x + 4}{(x+1)(x^2+x+1)^2}\,dx.$$

```
> rat := (4*x^4+9*x^3+12*x^2+9*x+4)
    /(x + 1)/(x^2 + x + 1)^2:
> convert(rat,parfrac,x);
```

$$\frac{2}{x+1} + \frac{1+2x}{x^2+x+1} + \frac{1}{(x^2+x+1)^2}$$

```
> int(%,x);
```

$$2\ln(x+1)+\ln(x^2+x+1)+\frac{1}{3}\frac{2x+1}{x^2+x+1}$$
$$+\frac{4}{9}\sqrt{3}\arctan\left(\frac{1}{3}(2x+1)\sqrt{3}\right)$$

5.8 Taylor and series expansions

The command to find the first n terms of the Taylor series expansion for $f(x)$ about the point $x=c$ is `taylor(f(x),x=c,n)`. We compute the first five terms of the Taylor series expansion of $y=(1-x)^{-1/2}$ about $x=0$.

```
> y := 1/sqrt(1-x);
```

$$y := \frac{1}{\sqrt{1-x}}$$

```
> taylor(y,x=0,5);
```

$$1+\frac{1}{2}x+\frac{3}{8}x^2+\frac{5}{16}x^3+\frac{35}{128}x^4+O\left(x^5\right)$$

To find a specific coefficient in a Taylor series expansion, use `coeff`.

```
> J := product(1-x^'i','i'=1..50):
> taylor(J^3,x=0,20);
```

$$1-3x+5x^3-7x^6+9x^{11}-11x^{15}+O(x^{20})$$

```
> coeff(%,x,15);
```

$$-11$$

To convert a *series* into a polynomial, try `convert(`*series*`, polynom)`. Also, see `?series`.

5.9 The *student* package

The *student* package contains many functions to help the calculus student solve problems step-by-step. In Section 5.7.1 we used the *student* package functions `changevar`, `intparts` to do some integration problems. The package includes the following functions:

```
D           Diff         Doubleint       Int        Limit
Lineint     Point        Product         Sum        Tripleint
changevar   combine      completesquare  distance   equate
extrema     integrand    intercept       intparts   isolate
leftbox     leftsum      makeproc        maximize   middlebox
middlesum   midpoint     minimize        powsubs    rightbox
rightsum    showtangent  simpson         slope      summand
trapezoid
```

We give a brief description of the main functions.

`Doubleint`

Calculates double integrals. `Doubleint(f,x,y)` is equivalent to `int(int(f,x), y)` and `Doubleint(f,x=a..b,y=c..d)` is equivalent to `int(int(f,x=a..b), y=c..d)`. Also see Section 10.6.1.

`Lineint`

Calculates line integrals. Suppose a curve $\mathcal{C}$ is parameterized by $x = x(t)$, $y = y(t)$ $(a \le t \le b)$, and $f(x, y)$ is a function defined on $\mathcal{C}$. Let $\vec{r}(t) = x(t)\vec{\imath} + y(t)\vec{\jmath}$. The line integral

$$\int_{\mathcal{C}} f(x, y)\, ds = \int_a^b f(x(t), y(t))\, ||\vec{r}\,'(t)||\, dt$$

is given in MAPLE by `Lineint(f(x,y),x,y,t=a..b)`. Also see Section 10.8.

`Tripleint`

Calculates triple integrals and is analogous to `Doubleint`. Also see Section 10.6.2.

`completesquare`

`completesquare` is used to complete the square.

```
> with(student):
> p := x^2 + 6*x + 13;
```

$$x^2 + 6\,x + 13$$

```
> completesquare(p);
```

$$(x + 3)^2 + 4$$

```
> q := x^2 + 10*x + 2*y^2 + 12*y + 12;
```

$$x^2 + 10\,x + 2\,y^2 + 12\,y + 12$$

```
> completesquare(q);
Error, (in completesquare) unable to choose indeterminate
> completesquare(q,x);
```

$$(x + 5)^2 - 13 + 2\,y^2 + 12\,y$$

```
> completesquare(%,y);
```

$$2\,(y + 3)^2 - 31 + (x + 5)^2$$

We found that

$$x^2 + 6\,x + 13 = (x + 3)^2 + 4,$$
$$x^2 + 10\,x + 2\,y^2 + 12\,y + 12 = 2\,(y + 3)^2 - 31 + (x + 5)^2.$$

distance

Finds the distance between two points in one, two, or three dimensions.

```
> with(student):
> distance(-3,5);
```

$$8$$

```
> distance([1,2],[-3,4]);
```

$$2\sqrt{5}$$

We see that the distance between the two real numbers -3 and 5 is $|-3-5| = 8$ and that the distance between the points $(1,2)$ and $(-3,4)$ is $2\sqrt{5}$.

equate

Generates a set of equations.

```
> with(student):
> equate(x,y);
```

$$\{x = y\}$$

```
> equate([x+y,x-y],[3,-1]);
```

$$\{x - y = -1, x + y = 3\}$$

```
> solve(%);
```

$$\{y = 2, x = 1\}$$

integrand

Extracts the integrand from an inert MAPLE integral.

```
> with(student):
> F := Int(sin(x^3*y),x=0..2*Pi);
```

$$F := \int_0^{2\pi} \sin(x^3 y) dx$$

```
> integrand(F);
```

$$\sin(x^3 y)$$

intercept

Computes the x-intercept as well as the intersection point of two curves.

```
> with(student):
> intercept(y=5*x-3);
```

$$\{y = -3, x = 0\}$$

```
> intercept(y=x^2+3*x-20,y=2*x^2+x-23);
```

$$\{x = -1, y = -22\}, \{x = 3, y = -2\}$$

We see that the x-intercept of the line $y = 5x - 3$ is the point $(0, -3)$ and that the curves $y = x^2 + 3x - 20$, $y = 2x^2 + x - 23$ have two intersection points $(-1, -22)$ and $(3, -2)$.

leftbox

Gives a graphical representation of a certain Riemann sum. The command `leftbox(f(x), x=a..b, n)` graphs $f(x)$ on the interval $[a, b]$ as well as n rectangles whose area approximates the definite integral. The left corner of each rectangle is a point on the graph of $y = f(x)$. We use `leftbox` to give a graphical approximation for the integral $\int_0^{\pi} \sin x^2 \, dx$:

```
> with(student):
> leftbox(sin(x^2), x=0..Pi, 6, shading=green);
```

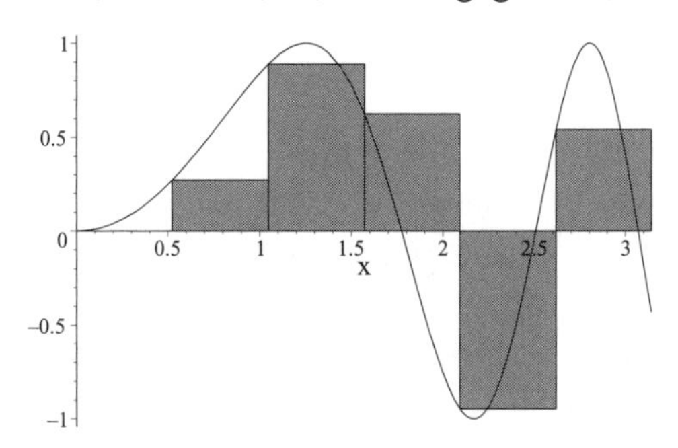

Figure 5.1 Rectangles representing a Riemann sum.

MAPLE's plotting functions are treated in detail in the next chapter. Related functions are `rightbox` and `middlebox`.

leftsum

`leftsum` is the Riemann sum that corresponds to `leftbox`. We compute the Riemann sum, which corresponds to the areas of the rectangles in our previous example.

```
> with(student):
> leftsum(sin(x^2), x=0..Pi, 6);
```

$$1/6\,\pi \sum_{i=0}^{5} \sin(1/36\, i^2 \pi^2)$$

```
> value(%);
```

$$1/6\,\pi \left(\sin(1/36\,\pi^2) + \sin(\frac{1}{9}\,\pi^2) + \sin(\frac{1}{4}\,\pi^2) + \sin(\frac{4}{9}\,\pi^2) + \sin(\frac{25}{36}\,\pi^2) \right)$$

```
> evalf(%);
```

$$0.7212750238$$

```
> evalf(int(sin(x^2),x=0..Pi));
```

$$0.7726517130$$

The required Riemann sum is

$$\frac{\pi}{6}\sum_{i=0}^{5}\sin(i^2\pi^2/36) \approx .7212750238,$$

which is an approximation of the integral

$$\int_0^\pi \sin x^2\,dx = 0.772651713\cdots.$$

Related functions are `rightsum` and `middlesum`.

makeproc

The `makeproc` is used for defining functions and takes three forms. If *expr* is an expression in the variable x, then `makeproc` converts the expression into a function of x.

```
> with(student):
> y := x^2 + x - 3;
```

$$y := x^2 + x - 3$$

```
> f := makeproc(y,x);
```

$$f := x \mapsto x^2 + x - 3$$

```
> f(x);
```

$$x^2 + x - 3$$

We converted the expression $x^2 + x - 3$ into a function of x. Also see `?unapply`. To find the linear function whose graph passes through the two points (a, b), (c, d), use the command `makeproc([a,b],[c,d])`.

```
> with(student):
> f := makeproc([-1,1],[3,7]);
```

$$f := x \mapsto \frac{3}{2}x + \frac{5}{2}$$

```
> f(-1), f(3);
```

$$1, 7$$

We see that $y = \frac{3}{2}x + \frac{5}{2}$ is the line that passes through the two points $(-1, 1)$ and $(3, 7)$. To find the linear function whose graph passes through (a, b) and has slope m, use the command `makeproc([a,b],'slope'=m)`.

```
> with(student):
> f := makeproc([2,5],'slope'=3);
```

$$f := x \mapsto 3\,x - 1$$

```
> f(2);
```

$$5$$

```
> diff(f(x),x);
```

$$3$$

We see that $y = 3x - 1$ is the line with slope 3 that passes through the point $(2, 5)$.

<u>`midpoint`</u>

To find the midpoint of the line segment joining the two points (a, b), (c, d), use the command `midpoint([a,b],[c,d])`.

```
> with(student):
> midpoint([2,3],[5,7]);
```

$$[\frac{7}{2}, 5]$$

We see that the midpoint of the segment joining the points $(2, 3)$, $(5, 7)$ is the point $(7/2, 5)$.

<u>`powsubs`</u>

The `powsubs` function behaves like the `subs` function. See `?powsubs` and `?subs` for more information.

<u>`showtangent(`$f(x)$`, `x`=`a`)`</u>

Produces a graph of the function $y = f(x)$ near $x = a$ together with the tangent that passes through the point $(a, f(a))$. We graph the tangent to the curve $y = \sin x$ at $x = 2\pi/5$ together with the curve.

```
> with(student):
> showtangent(sin(x),x=Pi/4);
```

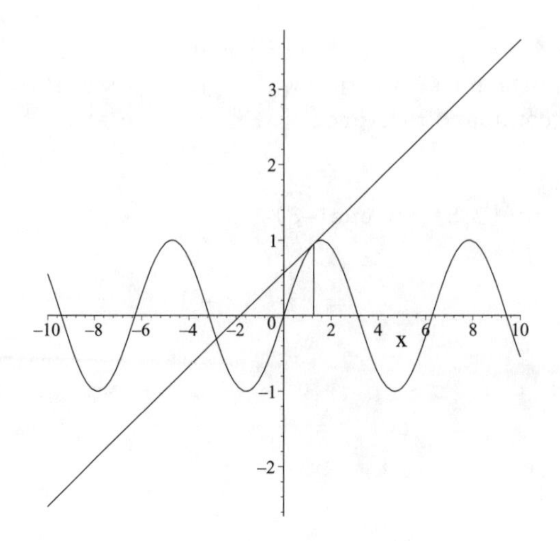

Figure 5.2 The function $y = \sin x$ and the tangent at $x = 2\pi/5$.

simpson

Computes an approximation to a definite integral using Simpson's rule. The call **simpson**($f(x)$,x,n) finds an approximation to the definite integral $\int_a^b f(x)\,dx$ using n subdivisions. We use Simpson's rule with $n = 12$ to find an approximation to $\int_0^1 \frac{1}{\sqrt{1+x^4}}\,dx$:

```
> with(student):
> simpson(1/sqrt(1+x^4),x=0..1,12):
> value(%):
> app := evalf(%);
```

$$app := 0.9270384891$$

```
> xval := evalf(int(1/sqrt(1+x^4),x=0..1));
```

$$xval := 0.9270373385$$

```
> abs(app-xval);
```

$$.11506\,10^{-5}$$

We found that

$$\int_0^1 \frac{1}{\sqrt{1+x^4}}\,dx \approx 0.9270384891,$$

and the error $< 10^{-5}$.

slope

Gives the slope of a line.

```
> with(student):
> slope(y=2*x-5);
```

2

```
> slope(2*y+12=3*x);
Error, (in slope) use slope(y=f(x)), or slope(f(x,y)=g(x,y),y(x))
> slope(2*y+12=3*x,y(x));
```

$$3/2$$

```
> slope([12,5],[3,7]);
```

$$-\frac{2}{9}$$

We found that the slope of the line $y = 2x - 5$ is 2. To find the slope of the line $2y + 12 = 3x$, we need to tell MAPLE that y is the dependent variable. Using the call `slope(2*y+12=3*x,y(x))`, we found the slope to be $3/2$. The call `slope([12,5],[3,7])` gives the slope of the line segment joining the points $(12,5)$ and $(3,7)$.

<u>summand</u>

Gives the summand in a sum.

```
> with(student):
> z3 := Sum(1/n^3,n=1..infinity);
```

$$z3 := \sum_{n=1}^{\infty} \frac{1}{n^3}$$

```
> summand(z3);
```

$$\frac{1}{n^3}$$

<u>trapezoid</u>

Uses the trapezoidal rule to compute an approximation to a definite integral. The call trapezoid($f(x)$,x,n) finds an approximation to the definite integral $\int_a^b f(x)\,dx$ using n subdivisions. We use the trapezoidal rule with $n = 12$ to find an approximation to $\int_0^1 \frac{1}{\sqrt{1+x^4}}\,dx$:

```
> with(student):
> trapezoid(1/sqrt(1+x^4),x=0..1,12):
> value(%):
> app := evalf(%);
```

$$0.9266278484$$

```
> xval := evalf(int(1/sqrt(1+x^4),x=0..1));
```

$$0.9270373385$$

```
> abs(app-xval);
```

$$0.0004094901$$

This time we found that

$$\int_0^1 \frac{1}{\sqrt{1+x^4}}\,dx \approx 0.9266278484,$$

and the error $< 10^{-3}$. The approximation found earlier using Simpson's rule was better.

6. GRAPHICS

MAPLE can plot functions of one variable, planar curves, functions of two variables, and surfaces in three dimensions. It can also handle parametric plots and animations. The two main plotting functions are `plot` and `plot3d`.

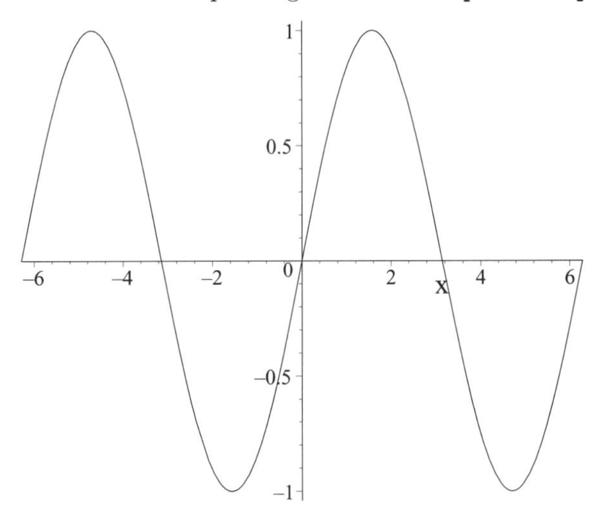

Figure 6.1 MAPLE plot of $y = \sin x$.

6.1 Two-dimensional plotting

The syntax for plotting an expression (or function) in x is `plot(f(x), x=a..b)`. For example, to plot $\sin(x)$ for $-2\pi \leq x \leq 2\pi$, we type

```
> plot(sin(x),x=-2*Pi..2*Pi);
```

The resulting plot appears in Figure 6.1.

Observe that in MAPLE the plot actually appears in the current document. Click on the MAPLE plot with the left mouse button. A rectangle should now border the plot. You will notice eight dots: one in each corner and one at the midpoint of each side. The dots mark positions for resizing the plot. Move the mouse on the dot in the bottom right corner. A little ↘ appears. Try stretching the plot display into a different shape. Notice also that the menu bar and the context bar have changed. The menu bar consists of the File, Edit, View, Format, Style, Legend, Axes, Projection, Animation, Export, Window, and Help menus. The context bar has changed completely. There should be a small window containing a pair of coordinates and nine new buttons. Try clicking on each button to see its effect.

0.53, 0.50	Displays the coordinates of the point under the tracker (i.e., the point clicked).

Render the plot using the usual line style.

Render the plot using the usual point style.

Render the plot using the polygon patch with gridlines style.

Render the plot using the polygon patch style.

Draw the plot axes as an enclosed box.

Draw the plot axes as an exterior frame.

Draw the plot axes in traditional form.

Suppress the drawing of plot axes.

Use the same scale on both axes.

Now click on the plot with the right mouse button. A context menu should appear:

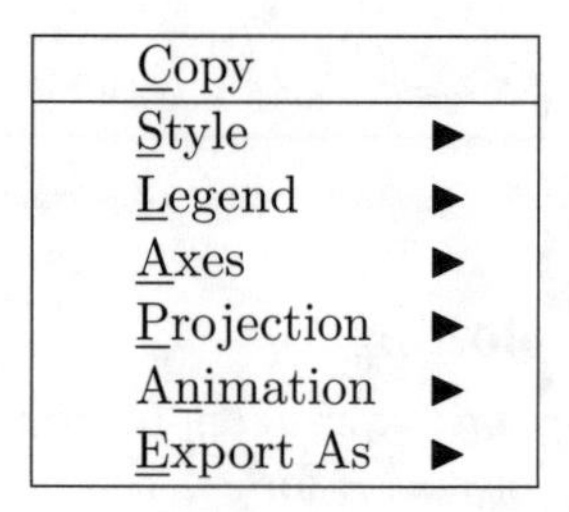

Click in Style. A submenu should appear:

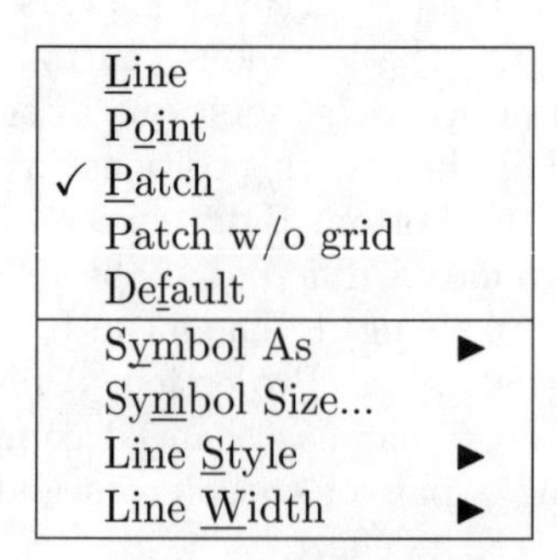

Select Point. The resulting plot is just a set of points interpolating the curve.

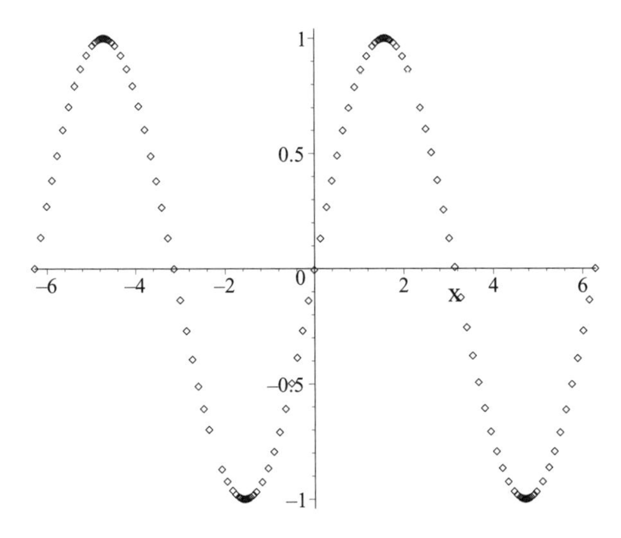

Figure 6.2 MAPLE point-style plot of $y = \sin x$.

Try some of the other selections in the context menu.

6.1.1 Restricting domain and range

Try the plot command `plot(sec(x),x=-Pi.. 2*Pi)`. Notice the "spikes" at $x = -\pi/2$, $\pi/2$, and $3\pi/2$ in your MAPLE plot. These correspond to singularities of $\sec(x)$. We restrict the range to get a more reasonable plot.

```
> plot(sec(x),x=-Pi..2*Pi,y=-5..5);
```

The resulting plot appears in Figure 6.3. Observe the vertical lines in the plot. MAPLE has tried to plot a continuous curve even though the function $\sec x$ has discontinuities at $x = -\pi/2$, $\pi/2$, and $3\pi/2$ in the interval $[-\pi, 2\pi]$. To allow for these discontinuities we can use the `discont` option. Try

```
> plot(sec(x),x=-Pi..2*Pi,y=-5..5,discont=true);
```

So, to plot $y = f(x)$, where $a \le x \le b$, and $c \le y \le d$, in MAPLE we use the command `plot(f(x),x=a..b,y=c..d)`.

6.1.2 Parametric plots

To plot the curve parameterized by

$$x = f(t), \qquad y = g(t), \qquad \text{for } a \le t \le b,$$

we use the command `plot([f(t),g(t),t=a..b])`. The ellipse

$$x^2 + 4y^2 = 1,$$

can be parameterized as

$$x = \cos(t), \quad y = \frac{1}{2}\sin(t), \quad \text{where } 0 \le t \le 2\pi.$$

Try

```
> plot([cos(t),1/2*sin(t),t=0..2*Pi]);
```

This should give you the desired plot.

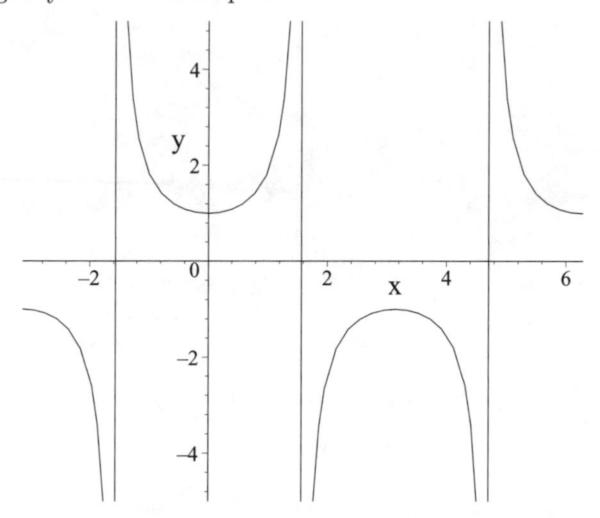

Figure 6.3 MAPLE plot of $y = \sec x$.

6.1.3 Multiple plots

To plot the two functions

$$y = \sqrt{x}, \qquad y = 3\log(x),$$

try

```
> plot([sqrt(x),3*log(x)],x=0..400);
```

The resulting plot is given in Figure 6.4. On the screen, each curve is plotted with a different color. Observe that our plot does not seem to illustrate the expected behavior of the log function near $x = 0$. To get a more accurate plot, we can use the `numpoints` option. Try

```
> plot([sqrt(x),3*log(x)],x=0..400,numpoints=1000);
```

An alternative method for doing multiple plots is to use the `display` function in the *plots* package. Try

```
> with(plots):
> p1:=plot(sqrt(x),x=0..400):
> p2:=plot(3*log(x),x=0..400):
> display(p1,p2);
```

When defining `p1` and `p2`, use a colon unless you want to see all the points MAPLE uses to plot the functions. To see all the functions in the *plots* package, type

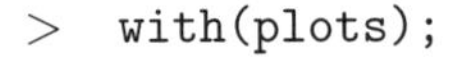

```
> with(plots);
```

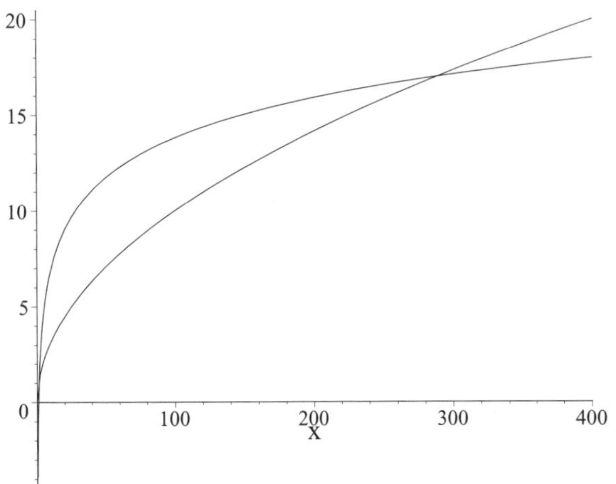

Figure 6.4 MAPLE plot of $y = \sqrt{x}$ and $y = 3\log x$.

6.1.4 Polar plots

To plot polar curves we use the `polarplot` function in the *plots* package. Use the command `polarplot(f(t), t=a..b)` to plot the polar curve $r = f(\theta)$. Try

```
> with(plots):
> polarplot(cos(5*t),t=0..2*Pi);
```

The resulting plot appears in Figure 6.5.

When you try this the first time you will notice the scale on the x-axis is different from that on the y-axis. To make the scales the same, hold the first mouse button on Projection and release on Constrained; or, better still, click on 

.

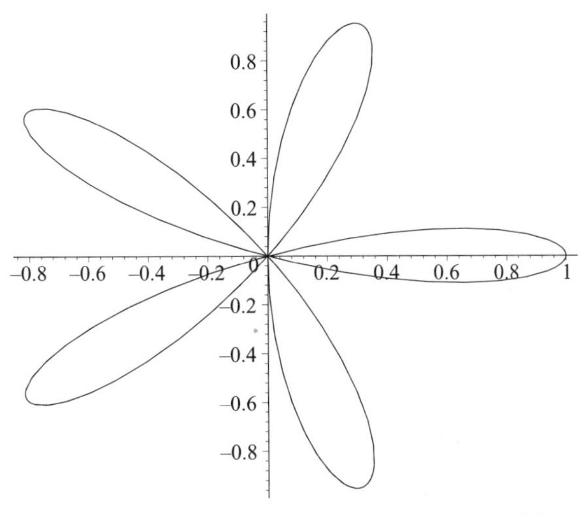

Figure 6.5 MAPLE plot of the polar curve $r = \cos 5\theta$.

We can also plot multiple polar curves. Try

```
> polarplot({cos(5*t),t},t=0..2*Pi);
```

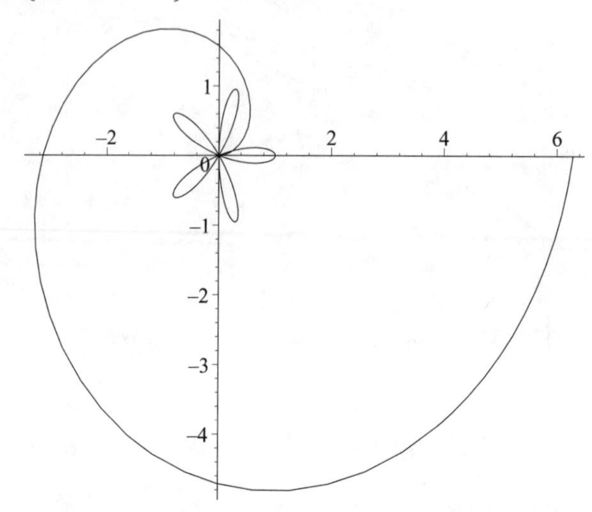

Figure 6.6 MAPLE plot of two polar curves.

You can use `polarplot(L,options)` where L is a list or set. If no range for the angle is specified, the default range $-\pi \leq \theta \leq \pi$ is taken.

There is another way to plot polar curves. Since $x = r\cos\theta$ and $y = r\sin\theta$, the polar curve $r = f(\theta)$ is given parametrically by

$$x = f(\theta)\,\cos\theta, \qquad y = f(\theta)\,\sin\theta.$$

For example, the polar curve $r = \cos 5\theta$ is given parametrically by

$$x = \cos 5\theta\,\cos\theta, \qquad y = \cos 5\theta\,\sin\theta,$$

so try

```
> plot([cos(t)*cos(5*t),sin(t)*sin(5*t),t=0..2*Pi]);
```

You should obtain the same plot.

6.1.5 Plotting implicit functions

In Section 6.1.2 we used a parameterization to plot the curve $x^2 + 4y^2 = 1$. Alternatively, we can plot implicitly defined functions using the `implicitplot` command in the *plots* package. Try

```
> with(plots):
> implicitplot(x^2+4*y^2=1,x=-1..1,y=-1/2..1/2);
```

This should agree with what we obtained before.

6.1.6 Plotting points

In MAPLE, we plot the points

$$(x_1, y_1),\ (x_2, y_2),\ \ldots,\ (x_n, y_n)$$

with the command `plot([[x1,y1],[x2,y2], ... ,[xn,yn]])`. Try

```
> L := [[0,0],[1,1],[2,3],[3,2],[4,-2]]:
> plot(L);
```

The resulting plot is given in Figure 6.7.

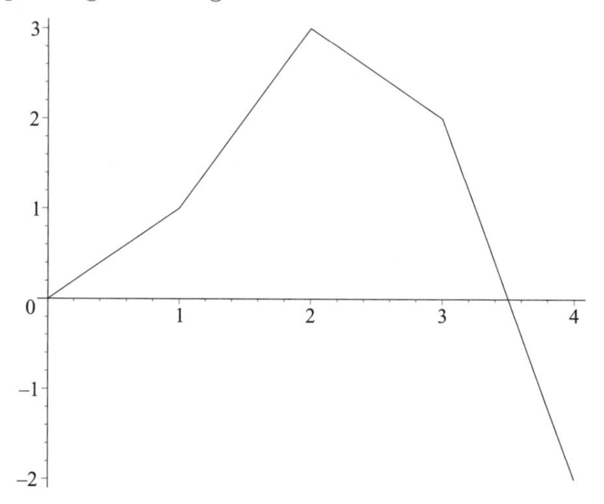

Figure 6.7 MAPLE plot of some data points.

Notice that MAPLE (by default) has drawn lines between the points. To plot the points and nothing but the points, try

```
> plot(L, style=point, symbol=circle);
```

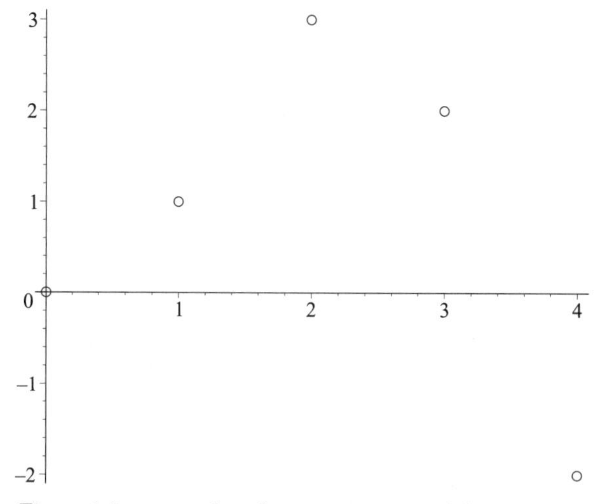

Figure 6.8 MAPLE plot of some unconnected data points.

The points correspond to circles. Try plotting this without the `symbol=circle` option.

6.1.7 Title and text in a plot

To put a title above a plot, we use the option `title`. Try

```
> p1:=plot([sqrt(x),3*log(x)],x=0..400,
   title='The Square Root and log functions'):
> display(p1);
```

To add text to a plot, we use the `texplot` and `display` functions in the *plots* package. Try

```
> p2:=textplot([[360,16,'y=3log(x)'],[130,10,'y=sqrt(x)']]):
> display(p1,p2);
```

`textplot([x1,y1,string])` creates a plot with *string* positioned at (x_1, y_1).

We add a legend to a plot. Try

```
> plot([sqrt(x),3*log(x)],x=0..400,
   title="The Square Root \n and log functions",
   legend=["y=sqrt(x)","y=3log x"]);
```

We can add Greek letters and other symbols to plots using the Symbol font. Below is a table showing Greek letters with corresponding Roman letters.

a	b	c	d	e	f	g	h	i	j	k	l	m	n
α	β	χ	δ	ϵ	ϕ	γ	η	ι	φ	κ	λ	μ	ν

o	p	q	r	s	t	u	v	w	x	y	z
o	π	θ	ρ	σ	τ	υ	ϖ	ω	ξ	ψ	ζ

A	B	C	D	E	F	G	H	I	J	K	L	M	N
A	B	X	Δ	E	Φ	Γ	H	I	ϑ	K	Λ	M	N

O	P	Q	R	S	T	U	V	W	X	Y	Z
O	Π	Θ	P	Σ	T	Y	ς	Ω	Ξ	Ψ	Z

To produce μ at the point $(1, 1)$ in a plot, we try:

```
> with(plots):
> textplot([1,1,'m'], font=[SYMBOL,12]);
```

As an illustration, we will plot two normal curves with means μ and $\mu*$. We need to load the *stats* package so we can plot normal density functions. We will discuss the *stats* package in more detail in Chapter 16.

```
> with(stats):
> with(plots):
> xaxis:=plot([[-5,0],[7,0]]):
> meanl1:=plot([[0,0],[0,0.42]],linestyle=2):
```

```
> meanl2:=plot([[1,0],[1,0.42]],linestyle=2):
> p1:=plot(statevalf[pdf,normald[0,1]](t),t=-5..5):
> p2:=plot(statevalf[pdf,normald[1,1]](t),t=-4..5):
> t1:=textplot([0,-0.02,m],font=[SYMBOL,12],'align=BELOW'):
> t2:=textplot([1,-0.02,"m*"],font=[SYMBOL,12],'align=BELOW'):
> display(xaxis,p1,p2,t1,t2,meanl1,meanl2,
    view=[-5..7,-0.02..0.42], axes=none);
```

The resulting plot appears below in Figure 6.9.

`xaxis` gives a horizontal line corresponding to the x-axis. The two vertical dotted lines are `meanl1` and `meanl2`, indicating the two means μ and $\mu*$. The two normal curves are given by `p1` and `p2`. The *stats* function `statevalf[pdf,normald[`μ`,`σ`]]` computes values of the normal density function with mean μ and standard deviation σ. See Chapter 16 for more details. The symbols μ and $\mu*$ were placed in their correct positions using `textplot`. The `align` option in `textplot` can take the values `ABOVE`, `BELOW`, `RIGHT`, or `LEFT`. See `?plots[textplot]` for more details.

Other keyboard characters give different symbols when using symbol font:

@	$	^	'	|
$\cong$	$\exists$	$\perp$	$\ni$	$\therefore$

Try

```
> textplot([1,1,'@'],font=[SYMBOL,12]);
```

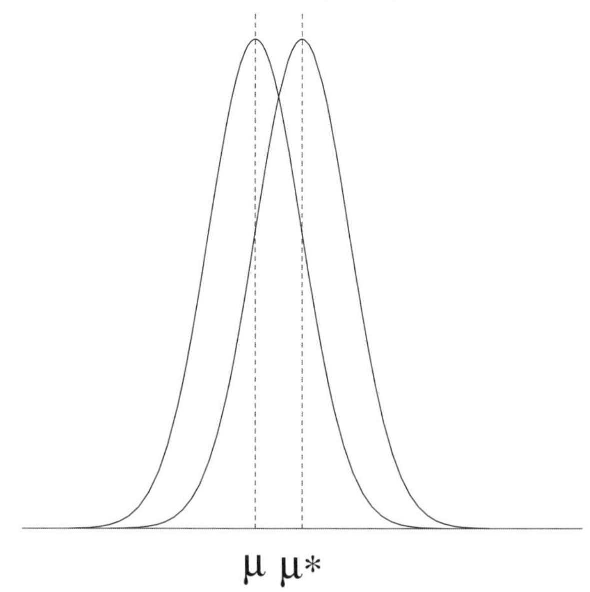

Figure 6.9 Two normal curves.

Other symbols are encoded as character numbers using `convert([`n`], bytes)`. Here n is an integer satisfying $32 \leq n \leq 126$, $161 \leq n \leq 254$. Try

```
> with(plots):
> textplot([0,0,convert([192], bytes)], font=[SYMBOL,12],
  axes=none);
```

ℵ

To view more of these symbols define **chardisplay**:

```
> with(plots):
>   chardisplay:=n -> display(textplot([0,0,convert([n],bytes)],
  font=[SYMBOL,12]),axes=none):
```

Now try `chardisplay(n)` for different values of n.

```
> chardisplay(169);
```

6.1.8 Plotting options

The plotting options are given after the function and ranges in the `plot` command. The following information is taken from the MAPLE help pages. See `?plot[options]`.

<u>`adaptive`</u>

If set to false, disables the use of adaptive plotting.

<u>`axes`</u>

Specifies the type of axes, one of: `FRAME`, `BOXED`, `NORMAL`, and `NONE`.

<u>`axesfont=l`</u>

Font for the labels on the tick marks of the axes, specified in the same manner as `font`.

<u>`color=n`</u>

Allows the user to specify the color of the curves to be plotted. The spelling *colour* may also be used. See `?plot,color` for details.

<u>`coords=name`</u>

Indicates that a parametric plot is in the coordinate system specified by `name`. See `?plot[coords]` for more information about the choices of coordinate system.

<u>`discont=s`</u>

Setting `s` to `true` forces plot to first call the function `discont` to determine the discontinuities of the input and then break the horizontal axis into appropriate intervals where the expression is continuous.

<u>`filled=truefalse`</u>

If the filled option is set to true, the area between the curve and the x-axis is given a solid color. This option is valid only with the following commands: `plot`, `contourplot`, `implicitplot`, `listcontplot`, `polarplot`, and `semilogplot`.

font=l

Font for text objects in the plot; l is a list [family, style, size], where family is one of TIMES, COURIER, HELVETICA, and SYMBOL. For TIMES, style may be one of ROMAN, BOLD, ITALIC, or BOLDITALIC. For HELVETICA and COURIER, style may be omitted or select one of BOLD, OBLIQUE, or BOLDOBLIQUE. SYMBOL does not accept a style option. The final value, size, is the point size to be used. As an example, try font=[HELVETICA,12].

labels=[x,y]

This option specifies labels for the axes. The values of x and y must be strings. The default labels are the names of the variables used in the plotting function.

labeldirections=[x,y]

This option specifies the direction in which labels are printed along the axes. The values of x and y must be HORIZONTAL or VERTICAL. The default direction of any labels is HORIZONTAL.

labelfont=l

Font for the labels on the axes of the plot, specified in the same manner as font.

legend=s

A legend for a plot can be specified by either a string or a list of strings. When more than one curve is being plotted, they must be specified as a list and there must be a legend for each curve.

linestyle=n

Controls the dash pattern used to render lines in the plot. When n=1, the line is solid. For n=2 the style is dot, n=3 gives dash, and n=4 gives dash-dot.

numpoints=n

Specifies the minimum number of points to be generated (the default is n = 50). Note: plot employs an adaptive plotting scheme that automatically does more work when the function values do not lie close to a straight line. Hence plot will often generate more than the minimum number of points.

resolution=n

Sets the horizontal display resolution of the device in pixels (the default is n = 200). The value of n is used to determine when the adaptive plotting scheme terminates. A higher value will result in more function evaluations for non-smooth functions.

sample

Supplies a list of parameter values to be used for the initial sampling of the function(s). When coupled with adaptive=false, this option allows explicit control over the function evaluations performed by plot.

`scaling`

Controls the scaling of the graph. Either `CONSTRAINED` or `UNCONSTRAINED`. Default is `UNCONSTRAINED`. `CONSTRAINED` means the same scale is used on both axes.

`style=s`

The interpolation style must be one of `LINE`, `POINT`, `PATCH`, or `PATCHNOGRID`. The default is `LINE`. `POINT` style plots points only, `LINE` interpolates between the points, `PATCH` uses the patch style for plots containing polygons, and `PATCHNOGRID` is the `PATCH` style without the grid lines.

`symbol=s`

Symbol for points in the plot, `s` is one of `BOX`, `CROSS`, `CIRCLE`, `POINT`, and `DIAMOND`.

`symbolsize=n`

The size (in points) of a symbol used in plotting can be given by a positive integer. This does not affect the symbol `POINT`. The default symbol size is 10.

`thickness=n`

Thickness of lines in the plot; `n` should be 0, 1, 2, or 3. 0 is the default thickness.

`tickmarks=[m,n]`

This option specifies that a reasonable number of points no less than `m` and `n` should be marked along the x-axis and y-axis, respectively. Both `m` and `n` must be either a positive integer or the name *default*. If tickmarks are desired along only one axis, use `xtickmarks` or `ytickmarks` instead.

`title=t`

The title for the plot. `t` must be a character string. The default is no title. You can create multiline titles for standard plots. Use the characters "\n" in the character string to obtain a line break in the title.

`titlefont=l`

Font for the title of the plot, specified in the same manner as `font`.

`view=[xmin..xmax, ymin..ymax]`

This option indicates the minimum and maximum coordinates of the curve to be displayed on the screen. The default is the entire curve.

`xtickmarks=n`

Indicates that a reasonable number of points no less than `n` should be marked along the horizontal axis; `n` must be a positive integer or a list. If `n` is a list, then the list of values is used to mark the axis; the corresponding option `ytickmarks=n` can be used to specify the minimum number of divisions along the vertical axis, or a list of values used to mark the vertical axis.

6.1.9 Saving and printing a plot

There are several ways to save a plot. Any plot that is part of a worksheet will be saved when the worksheet is saved. See Sections 9.2 and 9.3. The `plotsetup` function can be used to save a plot as a file suitable for other drivers. This is done by specifying the `plotdevice` variable. Common settings for `plotdevice` are

`bmp`	Windows BMP file
`cps`	Color Postscript file
`gif`	GIF image file
`ps`	encapsulated Postscript file
`jpeg`	24-bit color JPEG file
`hpgl`	HP GL file

Here is an example.

```
> plotsetup(ps, plotoutput='plot.ps',
    plotoptions='portrait, noborder');
> plot(sin(x),x=-2*Pi..2*Pi);
> interface(plotdevice=inline);
```

In this session, a plot of $y = \sin(x)$ was written to the Postscript file *plot.ps*, in portrait style with no surrounding border. The `interface` function was used so that any future plot will be within the worksheet. Otherwise, if `plotsetup` is not changed, any future plot will overwrite the file *plot.ps*.

A plot may be printed as part of the worksheet using the menu. Alternatively, it can be saved as a file and printed using a graphics driver. For example, try

```
> plotsetup(hpgl, plotoutput='plot.hp',plotoptions='laserjet');
```

when printing a plot with an HP Laserjet printer. For more information, use the help commands `?plotsetup`, `?plot[device]`.

A plot may be also saved using the Export menu. Click on a plot in the worksheet that you want to save and then click on Export. A menu should appear:

Drawing Exchange Format (DXF)...
Encapsulated Postscript (EPS)...
Graphics Interchange Format (GIF)...
JPEG File Interchange Format (JPG)...
Persitence of Vision (POV)...
Windows Bitmap (BMP)...
Windows Metafile (WMF)...

Select your favorite file format. A **Save As** window should appear. Type an appropriate file name in the File name box and click on `Save`.

6.1.10 Other plot functions

We describe briefly the other two-dimensional plotting functions available in the *plots* package. Don't forget to load the *plots* package.

```
> with(plots):
```

complexplot

Suppose $f(t)$ is a complex-valued function, say

$$f(t) = u(t) + i\,v(t),$$

where $u(t)$ and $v(t)$ are real-valued functions. Then the function `complexplot(f(t),t=a..b)` will plot the curve given parametrically by

$$x = u(t), \qquad v(t), \qquad a \leq t \leq b.$$

```
> complexplot(exp(I*x),x=0..2*Pi);
```

conformal

Suppose $f(z)$ is a complex-valued function, then the function `conformal (f(z),z=z1..z2)` will plot the image of a rectangular grid under the mapping $w = f(z)$. The complex numbers `z1` and `z2` determine two corners in the rectangular grid. More details and examples for this function will be given in Section 11.6.

```
> conformal(sin(z),z=-1-I..1+I);
```

The resulting plot appears below in Figure 6.10.

coordplot

The function `coordplot(coord,rangelist,eqns)` plots *graph paper* of the specified `coordinate` system. The available coordinate systems are `bipolar`, `cardioid`, `cartesian`, `cassinian`, `elliptic`, `hyperbolic`, `invcassinian`, `invelliptic`, `logarithmic`, `logcosh`, `maxwell`, `para- bolic`, `polar`, `rose`, and `tangent`. For a description of these coordinate systems see `?coords`. `rangelist` is a list of two coordinate ranges, and `eqns` are optional equations that modify the plot. See `?plots[coord]` for more details.

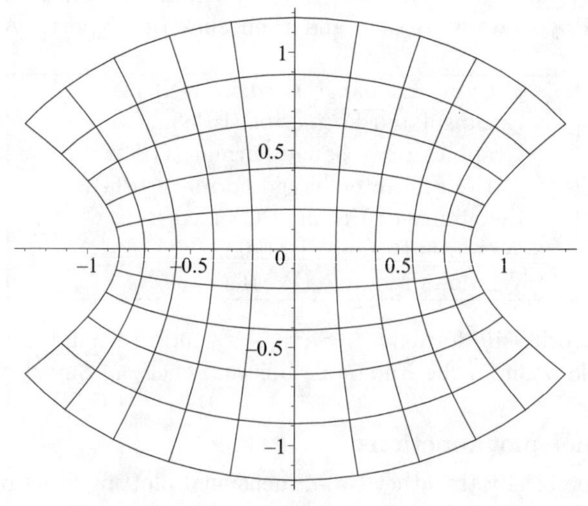

Figure 6.10 The conformal mapping $w = \sin z$.

```
> coordplot(polar,[0..2,0..2*Pi],labelling=true,
    grid=[5,13], view=[-2..2,-2..2],scaling=constrained);
```

The resulting plot appears below in Figure 6.11.

fieldplot

The function `fieldplot([f(x,y),g(x,y),x=a..b,y=c..d)` plots the two-dimensional vector field

$$\vec{F}(x,y) = f(x,y)\,\vec{i} + g(x,y)\,\vec{j},$$

where $a \le x \le b$, and $c \le y \le d$. Let's plot the direction field

$$\vec{F}(x,y) = -y\,\vec{i} + x\,\vec{j},$$

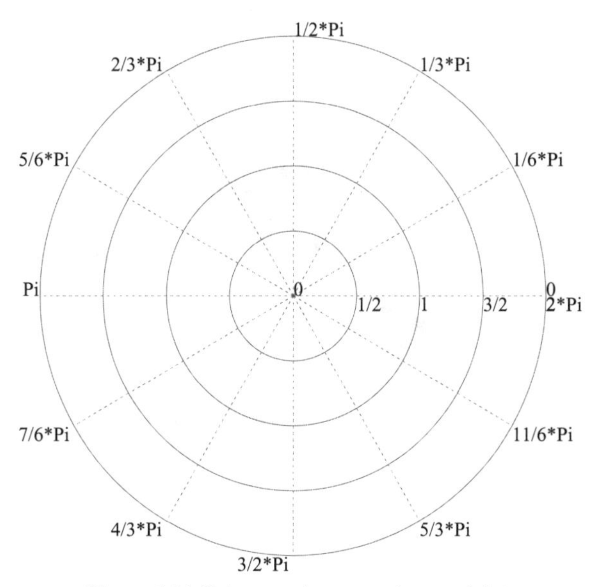

Figure 6.11 Polar graph paper via coordplot.

```
> fieldplot([-y,x],x=-1..1,y=-1..1);
```

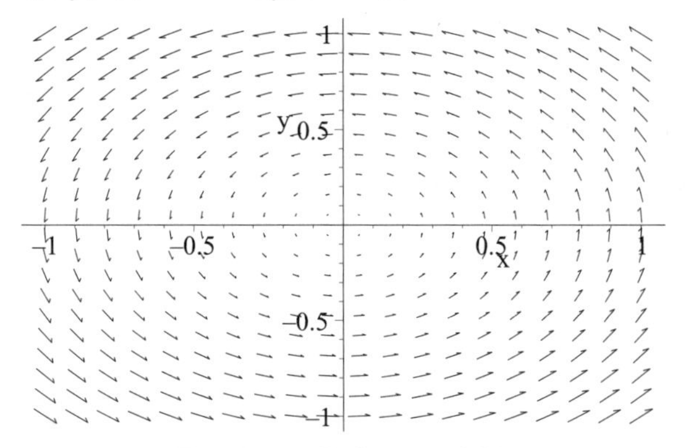

Figure 6.12 2D direction field.

inequal

The function `inequal(ineqs,x=a..b,y=c..d,options)` will plot regions defined by *linear inequalities* in the variables x and y over the specified ranges. We plot the regions specified by the inequalities

$$x - y \leq 0, \quad x + y \leq 1, \quad 5 + 2x \geq y,$$

where $-6 \leq x \leq 3$, and $-6 \leq y \leq 6$. The intersection is colored red and elsewhere is colored yellow.

```
> inequal( { x-y<=0,x+y<=1,5+2*x>=y}, x=-6..3,y=-6..6,
   optionsfeasible=(color=red),optionsexcluded=(color=yellow));
```

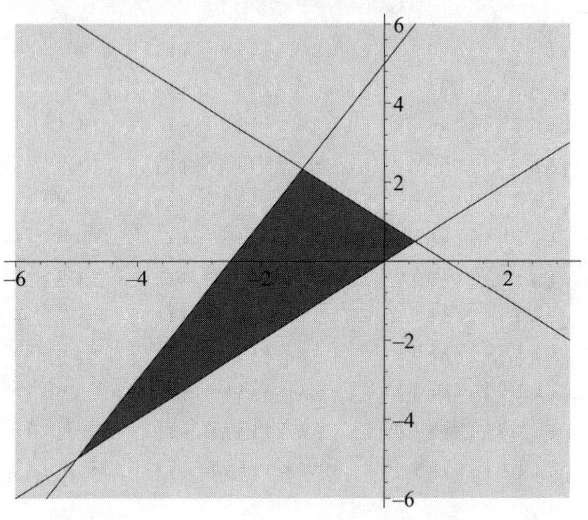

Figure 6.13 Graphing inequalities using inequal.

Warning: This function has some bugs for regions specified with strict inequalities. For example, try

```
> inequal( { x-y<0,x+y<1,5+2*x>y}, x=-6..3,y=-6..6,
optionsfeasible=(color=red),optionsexcluded=(color=blue));
```

logplot

The function `logplot(f(x),x=a..b)` creates a plot of the function $f(x)$ ($a \leq x \leq b$) with a logarithmic scale on the y axis. Try

```
> logplot(tan(x),x=0..1.55);
```

pareto

The `pareto` function plots a Pareto diagram of specified frequencies. For more information see `?plots[pareto]`.

pointplot

The function `pointplot(L)` plots a list or set of points `L`. It is basically equivalent to the command `plot(L,style=point)`.

polygonplot

If L is a list of points, the function polygonplot(L) creates a plot of a polygon whose vertices are these points.

```
> L := [[0,1],[1,1],[1/2,1/2]]:
> polygonplot(L);
```

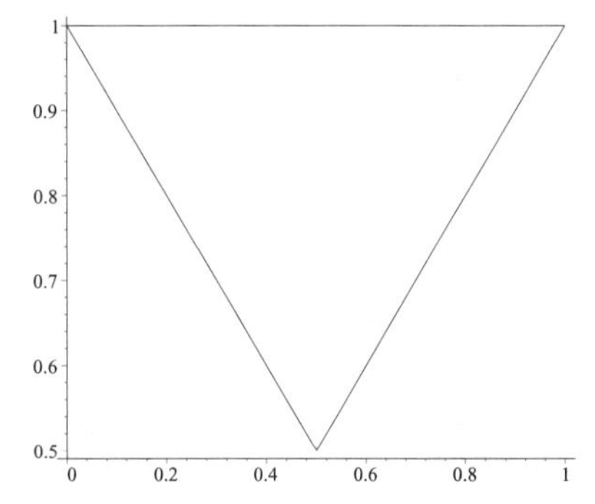

Figure 6.14 A polygon plot of a triangle.

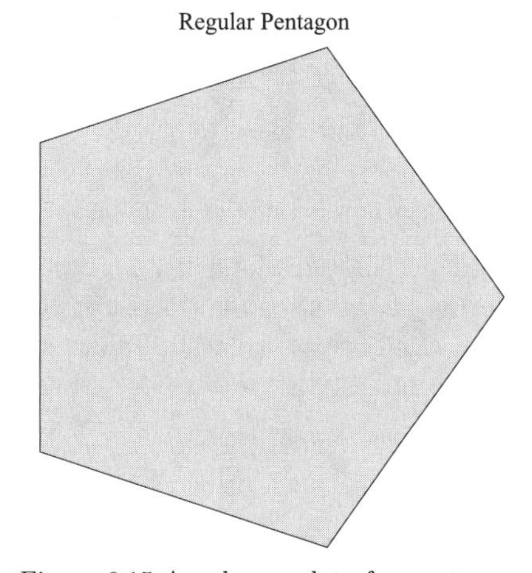

Figure 6.15 A polygon plot of a pentagon.

Observe that we plotted the triangle with vertices $(0, 1)$, $(1, 1)$, $(1/2, 1/2)$. In general, straight lines connect the points of L, and then the last point in L is connected to the first point. We can add color with the color option.

```
> ngon := n -> [seq([ cos(2*Pi*i/n), sin(2*Pi*i/n) ],
    i = 1..n)]:
> polygonplot(ngon(5),scaling=constrained,axes=none,
    title="Regular Pentagon",color=yellow);
```

The resulting plot is given above in Figure 6.15. The function ngon(n) returns n equally spaced points on the unit circle. We plotted a regular polygon by applying the polygonplot function to the list of five points returned by ngon(5). Below we define a function nstar for plotting an n-pointed star using polygonplot.

```
> npt :=(r,i,n) -> [r*cos(2*Pi*i/n),r*sin(2*Pi*i/n)];
```

$$npt := (r, i, n) \mapsto [r\cos(2\,\frac{\pi\, i}{n}), r\sin(2\,\frac{\pi\, i}{n})]$$

```
> shard:=(i,n,col)->polygonplot([npt(1,i,n),npt(2,(2*i+1),
   2*n),npt(1,i+1,n)],color=col):
> nstar:=(n,col)->display(seq(shard(i,n,col),i=1..n),
   scaling=constrained, axes=none):
> nstar(17,blue);
```

Figure 6.16 A polygon plot of a 17 pointed star.

The function `nstar(n,color)` should plot an n-pointed star with the specified color. It is defined in terms of the two functions `npt` and `shard`. `npt(r,i,n)` returns the ith point in a sequence of n equally spaced points on the circle of radius r. `shard(i,n,col)` plots a triangle corresponding to the ith point of the star.

<u>semilogplot</u>

The function `semilogplot(f(x),x=a..b)` creates a plot of the function $f(x)$ $(a \leq x \leq b)$ with a logarithmic scale on the x axis. Try

```
> semilogplot({sqrt(x),log(x)},x=0.1..100);
```

<u>setoptions</u>

This function sets global options for two-dimensional plots. These become default for all subsequent 2D plots in the same MAPLE session. See `?plot[options]` for a list of options.

```
> setoptions(title=`Semilog plot of Sqrt and Log`,
   axes=BOXED);
> semilogplot({sqrt(x),log(x)},x=0.1..100);
```

To remove these options, do

```
> setoptions(title=``, axes=normal);
```

6.2 Three-dimensional plotting

The syntax for plotting an expression (or function) in two variables (say x, y) is `plot3d(f(x,y), x=a..b,y=c..d)`. For example, to plot the function $z = e^{-(x^2+y^2-1)^2}$ for $-2 \leq x, y \leq 2$, we use the command

```
> plot3d(exp(-(x^2 + y^2-1)^2), x=-2..2, y=-2..2);
```

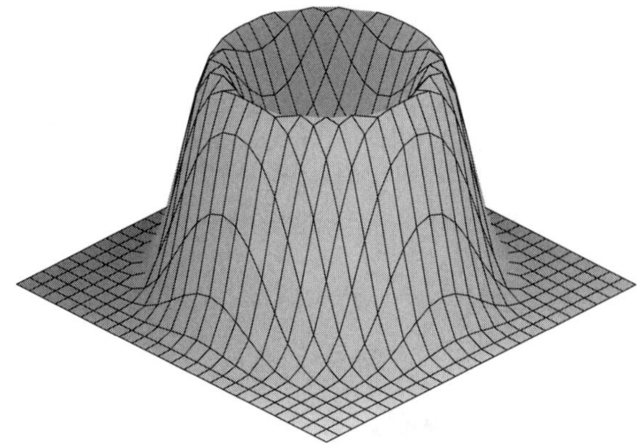

Figure 6.17 A plot of the function $z = e^{-(x^2+y^2-1)^2}$.

Observe (as before with two-dimensional plotting) that the plot appears in the worksheet. Now try clicking on the plot. Notice the appearance of the Style, Colour, Axes, Projection, and Animation menus. The context bar has also changed. There should be a pair of small windows labelled ϑ and ϕ, each containing the number 45. This pair of numbers refers to a point in spherical coordinates and corresponds to the orientation of the plot. There should also be 13 new buttons. Try clicking on each button to see its effect.

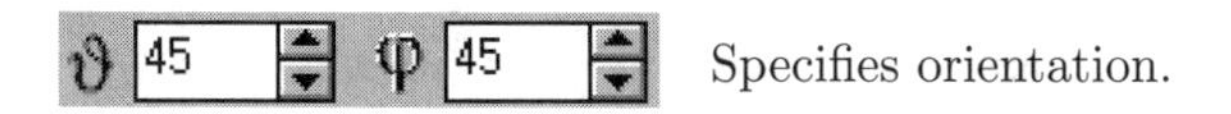

Specifies orientation.

Render the plot using the polygon patch style with gridlines.

Render the plot using the polygon patch style.

Render the plot using the polygon patch and contour style.

Render the plot using the hidden line removal style.

Render the plot using the contour style.

Render the plot using the wireframe style.

Render the plot using the point style.

Draw the plot axes as an enclosed box.

Draw the plot axes as an exterior frame.

Draw the plot axes in traditional form.

Suppress the drawing of plot axes.

Use the same scale on each axis.

Now, hold the first mouse button down on the plot and at the same time move it around. Notice how the plot rotates as you move the mouse, and notice that the value of (ϑ, ϕ) changes. Below in Figure 6.18 is a plot obtained by clicking on and and selecting $(\vartheta, \phi) = (22, 67)$.

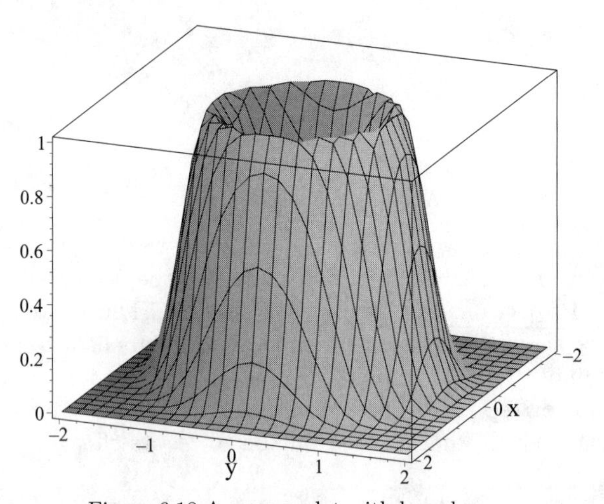

Figure 6.18 A MAPLE plot with boxed axes.

Now, try clicking to see some hidden detail of the plot. You might use the **grid** option to increase the number of contours plotted. Try

```
> plot3d(exp(-(x^2 + y^2)^2), x=-2..2,y=-2..2, grid=[50,50]);
```

This time try clicking the right mouse button on the plot. A context menu should appear:

Copy
Style ▶
Color ▶
Axes ▶
Projection ▶
Animation ▶
Export As ▶

Select Color. Another menu appears.

✓ XYZ
XY
Z
Z (Hue)
Z (Grayscale)
No Coloring
Default Coloring
No Lighting
User Lighting
Light Scheme 1
Light Scheme 2
Light Scheme 3
Light Scheme 4

Select Light Scheme 1. Notice how the coloring of the plot changes. Try out some other selections.

Now let's plot something simpler such as a plane. Remember that the equation of a plane takes the form

$$ax + by + cz = d.$$

To plot such a plane, we solve for z and plot the resulting function of x and y. As an example, we plot the plane

$$2x + 3y + 2z = 6.$$

Solving for z, we find that we must plot the function $f(x, y) = 3 - x - 3y/2$.

```
> plot3d(3 - x - 3*y/2,x=0..3,y=0..2,axes=normal,
    orientation=[20,60], view=[0..4,0..3,0..4]);
```

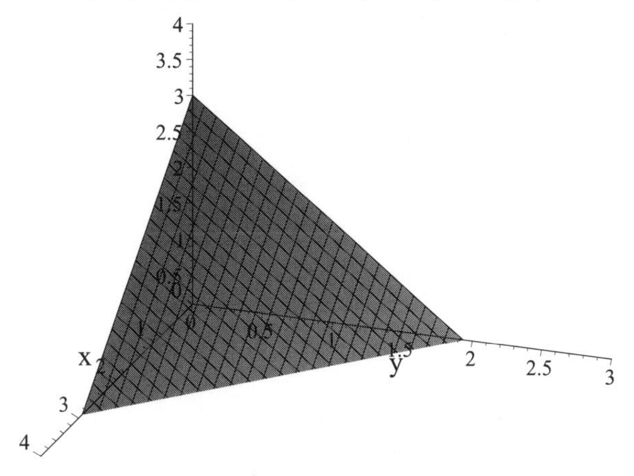

Figure 6.19 Plot of a plane.

The option `axes=normal` gave the usual x-, y- and z-axes. The option `orientation=[20,60]` set $\vartheta = 20$ and $\phi = 60$. The `view` option restricted the range for each variable as $0 \le x \le 4$, $0 \le y \le 3$, $0 \le z \le 4$. This way we were able to plot that portion of the plane that lies in the first octant (i.e., x, y, $z \ge 0$).

6.2.1 Parametric plots

To plot the surface parameterized by

$$x = f(u,v), \quad y = g(u,v), \quad z = h(u,v),$$

where $a \le u \le b$, $c \le v \le d$; use the command `plot3d([f(u,v), g(u,v), h(u,v)], u=a..b, v=c ..d)`. For example, the hyperboloid

$$x^2 + y^2 - z^2 = 1,$$

may be parameterized by

$$x = \sqrt{1+u^2}\cos t, \quad y = \sqrt{1+u^2}\sin t, \quad z = u,$$

where $-\infty < u < \infty$ and $0 \le t \le 2\pi$. Try

```
> plot3d([sqrt(1+u^2)*cos(t),sqrt(1+u^2)*sin(t),u],
   u=-1..1, t=0..2*Pi);
```

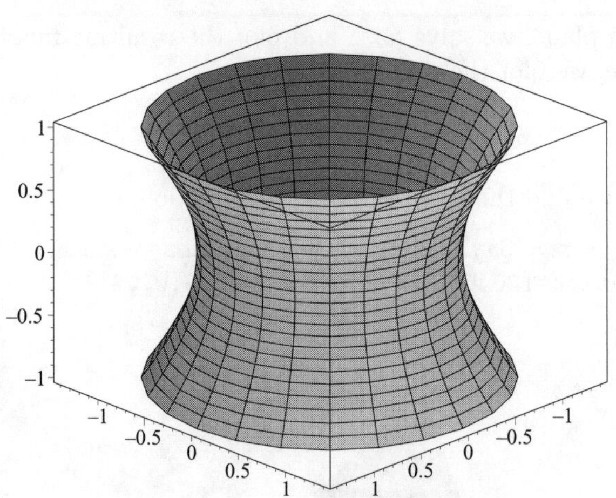

Figure 6.20 MAPLE plot of a hyperboloid.

A plot with $(\vartheta, \phi) = (45, 60)$ is given above in Figure 6.20.

6.2.2 Multiple plots

To plot the two functions

$$z = e^{-x^2-y^2},$$
$$z = x + y + 1,$$

try

```
> plot3d({exp(-x^2-y^2),x+y+1},x=-2..2, y=-1..1);
```

with $(\vartheta, \phi) = (120, 45)$. As with two-dimensional plotting, multiple three-dimensional plots can be produced using the `display` function in the *plots* package. Try

```
> with(plots):
> p1:=plot3d(exp(-x^2-y^2),x=-2..2, y=-1..1):
> p2:=plot3d(x+y+1,x=-2..2,y=-1..1):
> display(p1,p2);
```

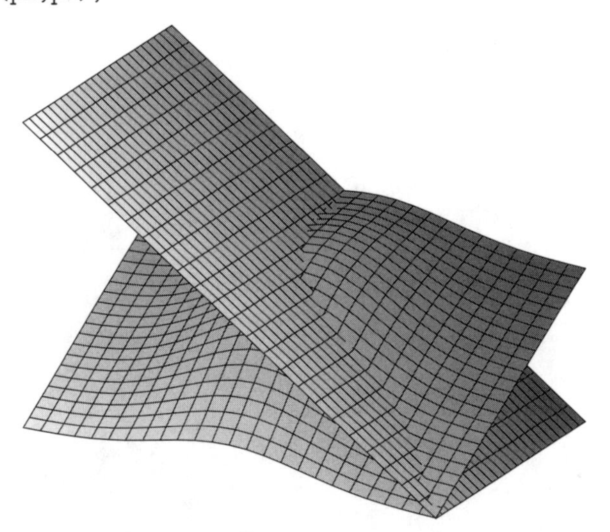

Figure 6.21 Two intersecting surfaces.

6.2.3 Space curves

To plot the space curve

$$x = f(t), \quad y = g(t), \quad z = h(t),$$

where $a \leq t \leq b$, we use the `spacecurve` function in the *plots* package. The command is `spacecurve([f(t),g(t),h(t)],t=a..b)`. We plot the helix

$$x = \cos t, \quad y = \sin t, \quad z = t.$$

Try

```
> with(plots):
> spacecurve([cos(t),sin(t),t],t=0..4*Pi, numpoints=200,
    orientation=[22,60],axes=BOXED);
```

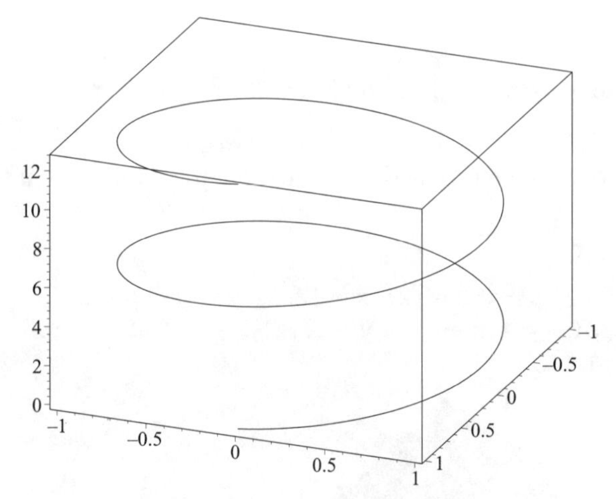

Figure 6.22 MAPLE plot of a helix.

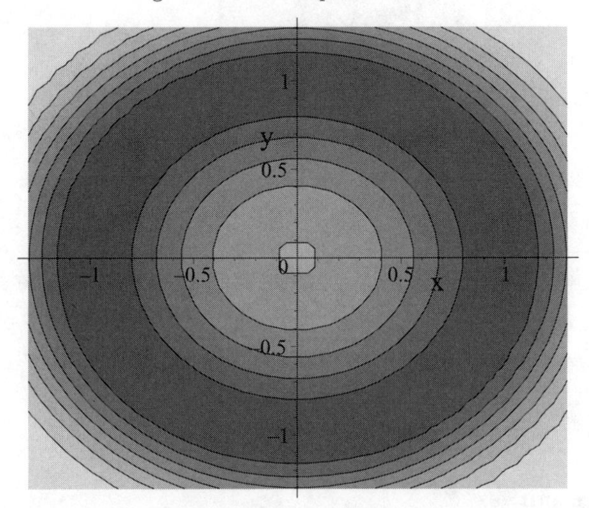

Figure 6.23 A contour plot.

6.2.4 Contour plots

The graph of a function of two variables may be visualized with a two-dimensional contour plot. To produce contour plots, we use the functions `contourplot` and `contourplot3d` in the *plots* package. `Contourplot3d` "paints" the contour plot on the corresponding surface. Try

```
> with(plots):
> contourplot(exp(-(x^2+y^2-1)^2), x=-(1.3)..(1.3),
    y=-(1.3)..(1.3), filled=true, coloring=[blue,red]);
```

The resulting plot is given above in Figure 6.23.

```
> contourplot3d(exp(-(x^2+y^2-1)^2), x=-(1.3)..(1.3),
```

```
y=-(1.3)..(1.3), filled=true, coloring=[blue,red]);
```

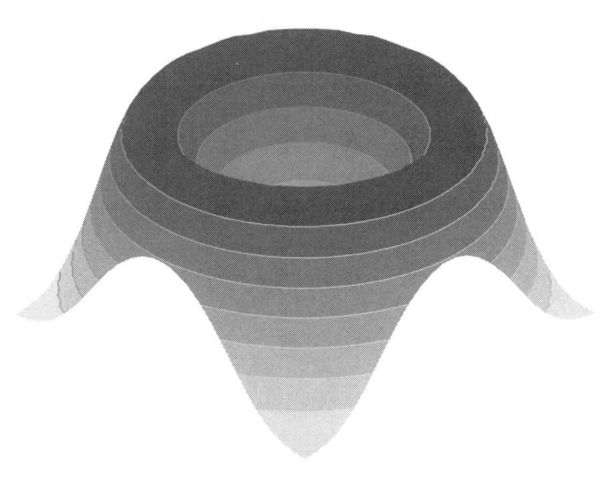

Figure 6.24 A 3D contour plot.

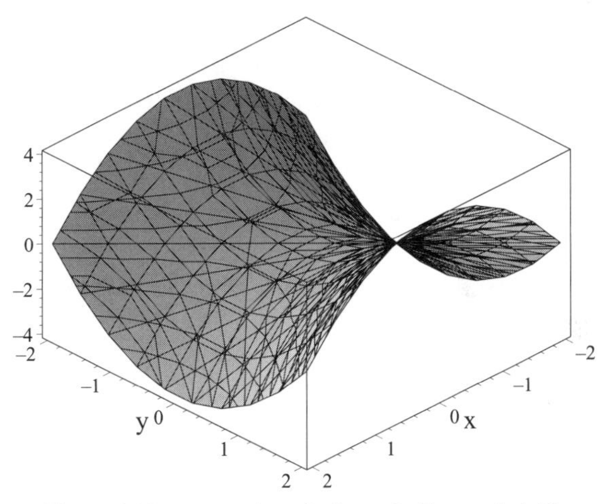

Figure 6.25 MAPLE plot of a hyperbolic paraboloid.

6.2.5 Plotting surfaces defined implicitly

To plot the surface defined implicitly by the equation

$$f(x, y, z) = c,$$

use the command `implicitplot3d(f(x,y,z)=c, x=a..b, y=d..e, z=g..h)` in the *plots* package. For example, to plot the hyperbolic paraboloid

$$y^2 - x^2 = z,$$

try

```
> with(plots):
> implicitplot3d(y^2 - x^2 = z, x=-2..2, y=-2..2,
    z=-4..4);
```

The resulting plot is given above in Figure 6.25.

In Section 6.2.1 we obtained a plot of the surface

$$x^2 + y^2 - z^2 = 1,$$

by using a parameterization. This time, try

```
> implicitplot3d(x^2 + y^2 - z^2 = 1, x=-1..1, y=-1..1,
    z=-1..1);
> implicitplot3d(x^2 + y^2 - z^2 = 1, x=-2..2, y=-2..2,
    z=-1..1);
```

Notice how care must be taken in choosing the range for each variable.

6.2.6 Title and text in a plot

A title or text may be inserted in a three-dimensional plot in the same way it was done in Section 6.1.7 for two-dimensional plots. Try

```
> with(plots):
> p1:=plot3d(exp(-(x^2+y^2-1)^2), x=-2..2,y=-2..2,
font=[TIMES,ROMAN,12],titlefont=[HELVETICA,BOLD,10],
title=`The surface z=exp(-(x^2+y^2-1)^2)`):
> p2:=textplot3d([0,1.1,1,`Circular Rim`], align=RIGHT,
    color=BLUE):
> display(p1,p2);
```

6.2.7 Three-dimensional plotting options

The options `axes`, `font`, `labels`, `labelfont`, `linestyle`, `numpoints`, `scaling`, `symbol`, `thickness`, `title`, `titlefont`, and `view` should work like they did for two-dimensional plotting (see Section 6.1.8). Other options are given below. This information was taken from the MAPLE help pages. See `?plot3d[options]`.

`ambientlight=[r,g,b]`

This option sets the red, green, and blue intensity of the ambient light for user-defined lighting. `r`, `g`, and `b` must be numeric values in the range 0 to 1.

`axes=f`

This option specifies how the axes are to be drawn, where `f` is one of `BOXED`, `NORMAL`, `FRAME`, and `NONE`. The default axis is `NONE`.

axesfont=l

This option defines the font for the labels on the tick marks of the axes, specified in the same manner as `font`.

color=c

This option defines a color value or function, where `c` is a predefined color name in a color function as described in `?plot3d,colorfunc`. See those help pages for details.

contours=n

This option specifies the number of contours or a list of contour values, where `n` is a positive integer or a list of contour values. The default is `n = 10`.

coords=c

This option specifies the coordinate system to be used. The default is the Cartesian system. For other coordinate systems see `?plot3d[coords]`.

filled=true/false

If the filled option is set to true, the region between the surface and the xy-plane is displayed as solid. This option is valid only with the following commands: `plot3d`, `contourplot3d`, and `listcontplot3d`.

grid=[m,n]

This option specifies the dimensions of a rectangular grid on which the points will be generated (equally spaced).

gridstyle=x

This option specifies rectangular or triangular grid; x is either `rectangular` or `triangular`.

labeldirections=[x,y,z]

This option specifies the direction in which labels are printed along the axes. The values of `x`, `y`, and `z` must be `HORIZONTAL` or `VERTICAL`. The default direction of any labels is `HORIZONTAL`.

labelfont=l

This option defines the font for the labels on the axes of the plot, specified in the same manner as `font`.

labels=[x,y,z]

This option specifies labels for the axes. The value of `x`, `y`, and `z` must be a string. The default label is no label.

light=[phi,theta,r,g,b]

This option adds a directed light source from the direction phi, theta in spherical coordinates with red, green, and blue intensities given by r, g, and b, respectively. r, g, and b must be numeric values in the range 0 to 1.

lightmodel=x

This option chooses a predefined light model to illuminate the plot. Valid light models include none, light1, light2, light3, and light4.

numpoints=n

This option specifies the minimum total number of points to be generated (default $625 = 25^2$). Plot3d will use a rectangular grid of dimensions $= \sqrt{n}$.

orientation=[theta,phi]

This option specifies the theta and phi angles of the point in three dimensions from which the plot is to be viewed. The default is at a point that is out perpendicular from the screen (negative z-axis) so that the entire surface can be seen. The point is described in spherical coordinates where theta and phi are angles in degrees, with default 45 degrees in each case.

projection=r

This option specifies the perspective from which the surface is viewed, where r is a real number between 0 and 1. The 1 denotes orthogonal projection, and the 0 denotes wide-angle perspective rendering. r can also be the one of the names, FISHEYE, NORMAL, and ORTHOGONAL, which correspond to the projection values 0, 0.5, and 1, respectively. The default projection is ORTHOGONAL.

scaling=s

This option specifies whether the surface should be scaled so that it fits the screen with axes using a relative or absolute scaling, where s is either UNCONSTRAINED or CONSTRAINED.

shading=s

This option specifies how the surface is colored, where s is one of XYZ, XY, Z, ZGREYSCALE, ZHUE, NONE.

style=s

This specifies how the surface is to be drawn, where s is one of POINT, HIDDEN, PATCH, WIREFRAME, CONTOUR, PATCHNOGRID, PATCHCONTOUR, or LINE. The default style is PATCH for colored surface patch rendering.

tickmarks=[l,n,m]

This option specifies reasonable numbers no less than l; n and m should be marked along the x-axis, y-axis, and z-axis, respectively. Each tick mark value must be either a positive integer or the name DEFAULT.

view=zmin..zmax or [xmin..xmax,ymin..ymax,zmin..zmax]

This option indicates the minimum and maximum coordinates of the surface to be displayed on the screen. The default is the entire surface.

6.2.8 Other three-dimensional plot functions

We describe briefly the other three-dimensional plotting functions available in the *plots* package. Don't forget to load the *plots* package.

```
> with(plots):
```

coordplot3d

The coordplot3d(coord, rangelist, eqns) function plots a graphical representation of most of the three-dimensional coordinate systems currently supported in MAPLE . The available coordinate systems are given below.

```
bipolarcylindrical    bispherical          cardioidal
cardioidcylindrical   casscylindrical      confocalellip
confocalparab         conical              cylindrical
ellcylindrical        ellipsoidal          hypercylindrical
invcasscylindrical    invellcylindrical    invoblspheroidal
invprospheroidal      logcoshcylindrical   logcylindrical
maxwellcylindrical    oblatespheroidal     paraboloidal
paraboloidal2         paracylindrical      prolatespheroidal
rosecylindrical       sixsphere            spherical
tangentcylindrical    tangentsphere        toroidal
```

For a description of these coordinate systems, see ?coords. rangelist is a list of three coordinate ranges, and eqns are optional equations that modify the plot. See ?plots[coord].

```
> coordplot3d(spherical);
```

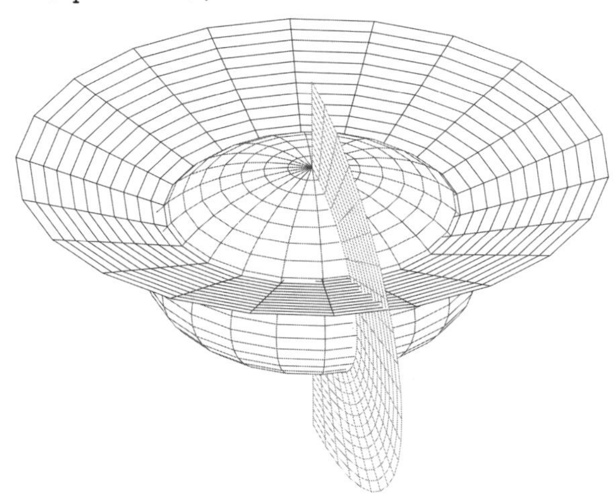

Figure 6.26 Spherical coordinates plot.

Observe the plot (Figure 6.26) of the three surfaces corresponding to the spherical coordinates ρ, ϕ, and θ by setting each to a constant. So the sphere corresponds to ρ, the cone corresponds to ϕ, and the plane corresponds to θ.

<u>cylinderplot</u>

The `cylinderplot(L,r1,r2,options)` function plots a surface in cylindrical coordinates. `L` is an expression for r in term of the two cylindrical coordinate variables z and θ or `L` is a list of three such procedures or expressions. `r1`, `r2` are ranges for the variables. We plot the surface

$$r = z + \cos\theta.$$

```
> cylinderplot(z+cos(theta),theta=0..2*Pi,z=0..1);
```

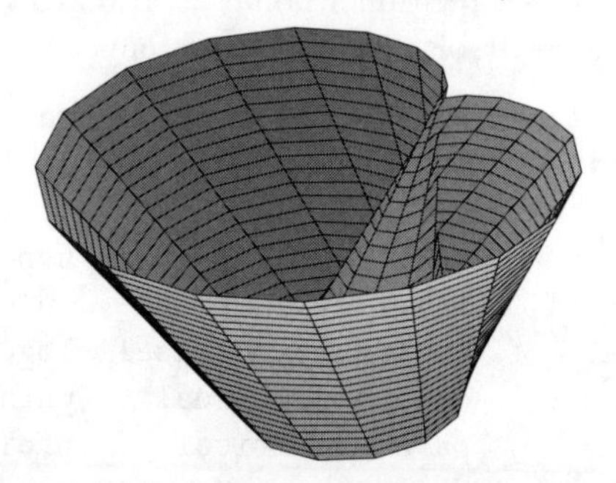

Figure 6.27 A plot using cylindrical coordinates.

<u>fieldplot3d</u>

The function `fieldplot3d` is the three-dimensional analog of `fieldplot`. It plots a three-dimensional vector field. Let's plot the direction field

$$\vec{F}(x,y) = \frac{x\,\vec{i} + y\,\vec{j} + z\,\vec{k}}{\sqrt{x^2+y^2+z^2}},$$

```
> fieldplot3d([x/sqrt(x^2+y^2+z^2),y/sqrt(x^2+y^2+z^2),
   z/sqrt(x^2+y^2+z^2)],x=-1..1,y=-1..1,z=-1..1);
```

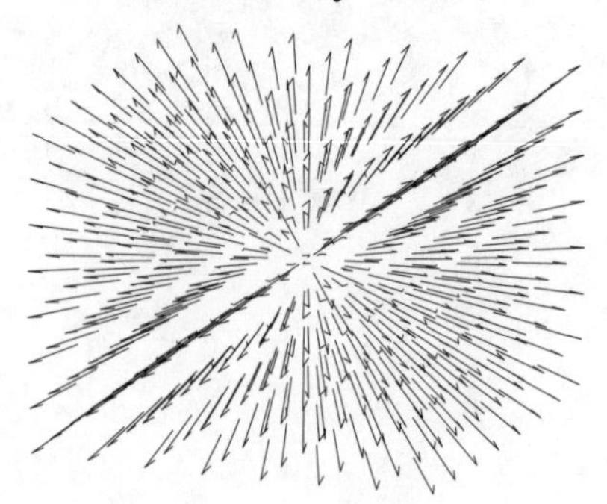

Figure 6.28 A 3D vector field.

polygonplot3d

The polygonplot3d function is used to plot polygons in three dimensions. Let's plot the faces of a square pyramid leaving one face open.

```
> p1:=polygonplot3d([[-1,-1,0],[-1,1,0],[1,1,0],
    [1,-1,0]]):
> p2:=polygonplot3d([[-1,-1,0],[-1,1,0],[0,0,1]]):
> p3:=polygonplot3d([[-1,1,0],[1,1,0],[0,0,1]]):
> p4:=polygonplot3d([[1,1,0],[1,-1,0],[0,0,1]]):
> display(p1,p2,p3,p4);
```

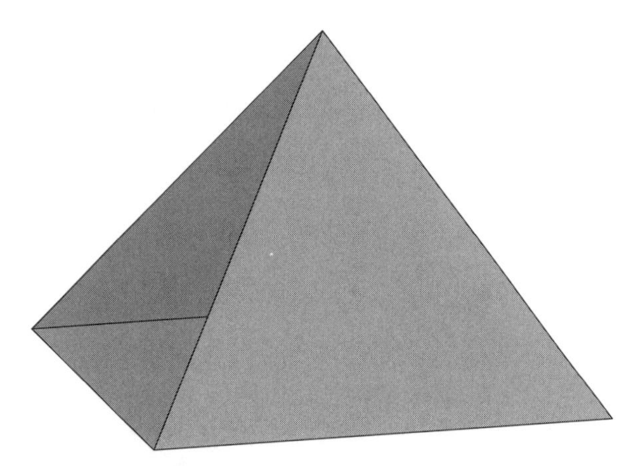

Figure 6.29 A 3D polygon plot of an open pyramid.

polyhedraplot

The polyhedraplot function plots polyhedra at specified points. L is a point or list of points. There are two options specific to this function. The polyscale option controls the size of each polyhedron, and the polytope option specifies the type of polyhedron, such as tetrahedron, octahedron, dodecahedron, etc. To see a complete list of supported polyhedra, try

```
> polyhedra_supported();
```

Let's plot a transparent dodecahedron.

```
> with(plots):
> polyhedraplot([0,0,0],polytype=dodecahedron,
    style=wireframe,scaling=CONSTRAINED,orientation=[71,66]);
```

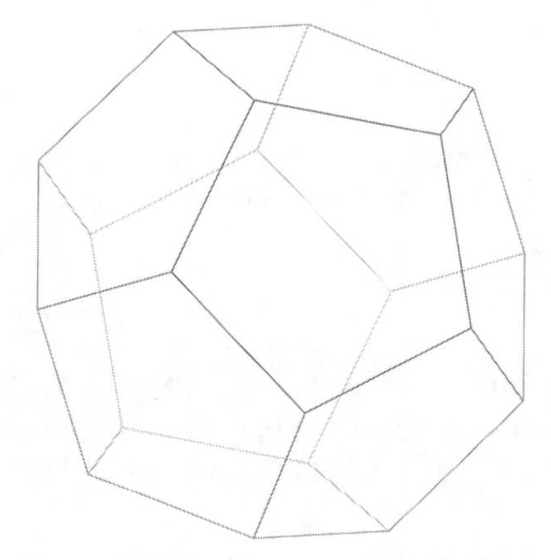

Figure 6.30 A transparent dodecahedron.

This time we plot a solid icosahedron.

```
> polyhedraplot([0,0,0],polytype=icosahedron,
    style=patch,scaling=CONSTRAINED);
```

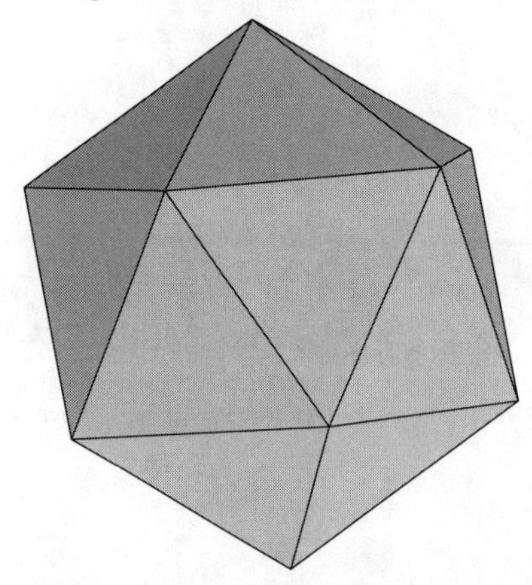

Figure 6.31 A solid icosahedron.

sphereplot

The `sphereplot(L,r1,r2,options)` function plots a surface in spherical coordinates. It is analogous to `cylinderplot`. See `?sphereplot` for more details.

surfdata

The `surfdata` function plots one or more surfaces where a surface is input as a grid of data points of the form [x,y,z]. See `?surfdata` for more details.

tubeplot

The `tubeplot` function basically plots a spacecurve as a tube. Let's use `tubeplot` to plot the helix

$$x = \cos t, \quad y = \sin t, \quad z = t.$$

Try

```
> with(plots):
> spacecurve([cos(t),sin(t),t],t=0..4*Pi, numpoints=200,
    orientation=[22,60],axes=BOXED);
```

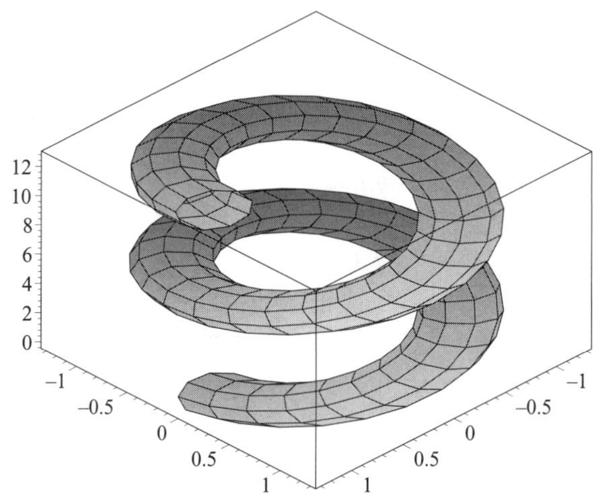

Figure 6.32 Tube plot of a helix.

6.3 Animation

MAPLE is capable of animating two- and three-dimensional plots. The two animation functions are `animate` and `animate3d`. These are in the *plots* package. For fixed t, we consider the function

$$f_t(x) = \frac{1}{1+xt}.$$

We can examine the behavior of this function as t changes using `animate`. Try

```
> with(plots):
> animate(1/(1+x*t),x=0..10,t=0..1, frames=10);
```

A plot of $f_0(x) = 1$ should appear in the worksheet. Now click on the plot. A new context bar should appear containing a window for coordinates and nine new buttons similar to those on a cassette tape player. Try clicking on each button to see its effect.

Stop the animation.

Play the animation.

Move to the next frame.

Set the animation direction to be backward.

Set the animation direction to be forward.

Decrease the speed of the animation.

Increase the speed of the animation.

Set animation to run in single-cycle mode.

Set animation to run in continuous-cycle mode.

Now click on to play the animation. The `frames` option allows you to set the number of separate frames in the animation. To view each frame, click on . Try setting `frame=50`. Now try

```
> animate([Pi/2*sin(t*(u+1)),sin(2*t)*sin(Pi/2*sin(t*u+t)),
  t=-2*Pi..2*Pi], u=0..1,frames=20,numpoints=200,
  color=blue);
```

This time right-click on the plot. You should get the usual context menu for a two-dimensionsal plot. Select Animation. A submenu appears.

Play
Next
Backward
Faster Slower
Continuous

Select Continuous and then Play. This sets the animation in a continuous loop, and has the same effect as pressing and .

The three-dimensional animation command is `animate3d`. The surface

$$x^2 - y^2 = z,$$

may be parameterized by

$$x = r\cos t, \quad y = r\sin t, \quad z = r^2\cos 2t.$$

Try animating a rotation of this surface

```
> with(plots):
> animate3d([r*cos(t+a),r*sin(t+a),r^2*cos(2*t)], r=0..1,
   t=0..2*Pi, a=0..3, frames=10,style=patch,
   title=`The Rotating Saddle`);
```

A little adjusting creates a flying pizza

```
> animate3d([r*cos(t+a),r*sin(t+a),r^2*cos(2*t)+sin(a)],
  r=0..1,t=0..2*Pi, a=0..2*Pi,frames=10,style=patch,
  title=`The Flying Pizza`);
```

Try clicking on to set your pizza in continuous motion.

7. MAPLE PROGRAMMING

MAPLE is a programming language as well as an interactive symbolic calculator. It is possible to solely use MAPLE interactively and not bother with its programming features. However, it is well worth the effort to develop some programming skills. The MAPLE language is much easier to learn than traditional programming languages, and you do not need to be an expert programmer to master it. You will appreciate the real power of MAPLE when you learn some of the basic MAPLE language and use it in combination with its interactive features. If you have gotten this far into the book, you are already familiar with many MAPLE commands, and the step to MAPLE programming is not a big one.

A number of programming exercises are included in this chapter. The answers to all the exercises can be found in the last section.

7.1 The MAPLE procedure

The following is a MAPLE program. Start a MAPLE session and type it in.

```
> f2c := proc(x)
>   evalf(5/9*(x - 32));
> end proc;
```

$$f2c := \textbf{proc}(x)\ \texttt{evalf}(5/9 * x - 160/9)\ \textbf{end proc}$$

Notice that the body of the proc was echoed below it. To avoid this use a colon instead of a semicolon to end the `end proc` statement.

Warning: In MAPLE V Release 5 (and earlier versions), use `end` instead of `end proc`. So in MAPLE V you should enter

```
> f2c := proc(x)
>   evalf(5/9*(x - 32));
> end:
```

Here *proc* is an abbreviation for *procedure*, which is just another name for program. The MAPLE program `f2c` converts degrees from Fahrenheit to centigrade. The program takes one value x within the procedure, and calculates an approximation to $5(x-32)/9$. Since this is the last calculation done within the procedure, the `f2c` procedure returns this approximation. If we call x the degrees in Fahrenheit and y the temperature in centigrade then x and y are related by

$$y = \frac{5}{9}(x-32).$$

This explains the formula within the program. Below are some examples. Try them out.

```
> fc2(32);
```

$$0$$

```
> fc2(100);
```

$$37.77777778$$

```
> f2c(60);
```

$$15.55555556$$

```
> f2c(70);
```

$$21.11111111$$

This means that

$$\begin{aligned} 32^{\circ}F &= 0^{\circ}C, \\ 60^{\circ}F &\approx 15.6^{\circ}C, \\ 70^{\circ}F &\approx 21.1^{\circ}C, \\ 100^{\circ}F &\approx 37.8^{\circ}C. \end{aligned}$$

Exercise 1.

Now, write a MAPLE program called `c2f` that converts the temperature in degrees centigrade to degrees Fahrenheit, and returns the result as a decimal.

```
> c2f :=
```

Check out your program by computing some examples:

```
> c2f(0);
> c2f(37.8);
> c2f(100);
> c2f(f2c(100));
```

Did you get what was expected?

7.1.1 Local and global variables

Variables that you use at the interactive level in MAPLE, that are not within the body of a procedure (or program), are called *global variables.* Variables that are introduced within a procedure and are known to MAPLE only within the procedure are called *local variables.* To illustrate this, we define two nearly identical procedures `g` and `h`. The procedure `g` will be defined using a local variable a. In `h`, the variable a will be a global variable.

```
>  g := proc()
>    local a;
>    a := exp(2);
>    evalf(a);
>  end proc:
```

The empty parentheses () indicate that this procedure requires no input. The procedure `g` computes an approximation to e^2. Remember that in MAPLE, `exp(x)` corresponds to the exponential function e^x. In the procedure the variable a is declared a local variable.

```
> g();
```

$$7.389056099$$

```
> a;
```

$$a$$

Notice that `g()` returned the approximation 7.389056099 for e^2, and notice that the variable a remains a variable (unassigned). Now we define `h`:

```
>   h := proc()
>     global a;
>     a := exp(2);
>     evalf(a);
>   end proc:
```

This procedure is the same as `g` except that now a is a global variable.

```
> h();
```

$$7.389056099$$

```
> a;
```

$$e^2$$

Notice this time that the procedure still returned the approximation 7.389056099, but the variable a has been assigned the value e^2, and this value holds outside the procedure.

Exercise 2.

Write a MAPLE procedure `dist` that computes the distance between two points (x_1, y_1) and (x_2, y_2) using at least one local variable. Your procedure should return an exact answer, so do not use `evalf`.

```
> dist := proc(x1,y1,x2,y2)
```

Check your program:

```
> dist(1,3,13,-4);
```

Did you get $\sqrt{193}$?

7.2 Conditional statements

A conditional statement has the form

```
if    condition  then
        statseq
else
        statseq
end if:
```

Here *statseq* is a sequence of statements separated by semicolons (or colons). Also, notice that the *if statement* is closed by `end if`.

For example,

```
>   x:=1;
```

$$x := 1$$

```
>   if x>0 then
>     y:=x+1
>   else
>     y:=x-1
>   end if:
>   y;
```

$$2$$

This conditional statement means that if $x > 0$, then $y = x + 1$, but if $x \leq 0$, then $y = x - 1$. In the session, $x = 1 > 0$, so $y = x + 1 = 2$.

Warning: In MAPLE V Release 5 (and earlier versions), use `fi` instead of `end if` to end the conditional statement.

The conditional or if statement is used to define functions piecewise. For example, consider the function

$$f(x) = \begin{cases} x^2 & \text{if } x > 1, \\ (1 - x^3) & \text{otherwise.} \end{cases}$$

We illustrate how to define this function (as a proc) in MAPLE:

```
> f := proc(x)
>    if x > 1 then
>        x^2;
>    else
>        (1-x^3);
>    end if;
> end proc;
```

$$f := \textbf{proc}(x) \textbf{ if } 1 < x \textbf{ then } x\hat{\ }2 \textbf{ else } 1 - x\hat{\ }3 \textbf{ end if end proc}$$

Let's try out our function $f(x)$:

```
> f(2);
```

$$4$$

```
> f(-3);
```

$$28$$

Now $1 < 2$, so $f(2) = 2^2 = 4$, and since $-3 \leq 1$, $f(-3) = 1-(-3)^3 = 1+27 = 28$.

7.2.1 Boolean expressions

In the previous section we used the relational operator `<` in our definition of the proc `f`. Other relational operators are given below.

`<`	less than
`>`	greater than
`<=`	less than or equal
`>=`	greater than or equal
`<>`	not equal

We also need the logical operators `and`, `or`, and `not`. Now we are able to define more complicated functions. For instance, consider the function

$$f(x) = \begin{cases} x & \text{if } 0 < x \leq 1, \\ -1 & \text{otherwise.} \end{cases}$$

We can define `f` as a MAPLE proc.

```
> f := proc(x)
>   if 0<x and x<=1 then
>     x;
>   else
>     0;
>   end if;
> end proc:
```

We test our function.

```
> f(-1/2), f(0), f(1/2), f(1), f(3/2);
```

$$-1, -1, 1/2, 1, -1$$

We found $f(-1/2) = f(0) = -1$, $f(1/2) = 1/2$, $f(1) = 1$, and $f(3/2) = -1$, as expected.

Consider the function

$$g(x) = \begin{cases} x^2 - 3x + 2 & \text{if } x > 2, \\ 1 - x^3 & \text{if } 0 < x \leq 2, \\ x^3 & \text{otherwise.} \end{cases}$$

We define a MAPLE proc `g`, which corresponds to this function.

```
> g := proc(x)
>   if x>2 then
```

```
>         x^2-3*x+2;
>       else
>         if 0<x and x<=2 then
>           1 - x^3;
>         else
>           x^3;
>         end if;
>       end if;
>   end proc:
```

We can write this more compactly using `elif`, which means *else if*.

```
>   g := proc(x)
>         if x > 2 then x^2 - 3*x + 2;
>         elif x>0 and x<=2 then 1-x^3;
>         else x^3;
>         end if;
>   end proc:
```

Notice that `elif` is not closed by an `end if`. We can even plot this function. See Figure 7.1.

```
>   plot(g, -1..3, discont=true);
```

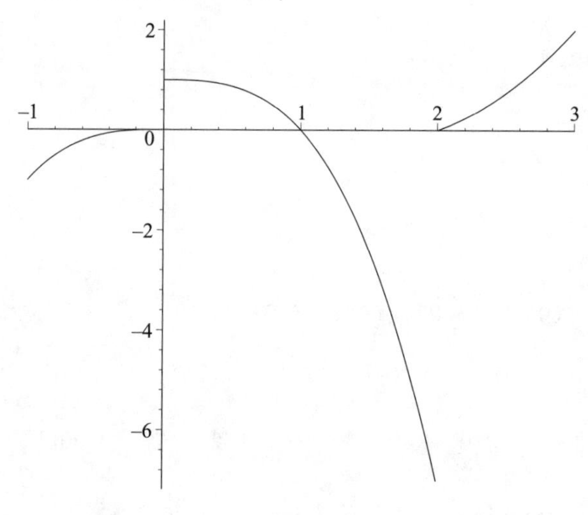

Figure 7.1 MAPLE plot of a proc.

This is correct form for plotting a proc, if it corresponds to a function of one variable. We set `discont=true` since $g(x)$ is a discontinuous function. Look what happens if we try to use `g(x)` instead of `g` in `plot`:

```
>   plot(g(x), x=-1..3, discont=true);
Error, (in g) cannot evaluate boolean:  -x < -2
```

Exercise 3.

Write a MAPLE proc **h** corresponding to the function

$$h(x) = \begin{cases} -1 & \text{if } x \leq 0, \\ x & \text{if } 0 < x \leq 1, \\ 1 & \text{if } 1 < x \leq 3, \\ 0 & \text{otherwise.} \end{cases}$$

Check your function by plotting it on the interval $[-1, 4]$.

7.3 The "for" loop

Loops are used to do the same or similar computation several times. There are two kinds of loops: the "for" loop and the "while" loop. A for loop statement has the form

```
for   var   from   num1   to num2   do
            statseq
end do:
```

For instance, we can print out the numbers from 1 to 5.

```
>    for i from 1 to 5 do
>      print(i);
>    end do:
```

$$1$$
$$2$$
$$3$$
$$4$$
$$5$$

Warning: In MAPLE V Release 5 (and earlier versions), use `od` instead of `end do` to close a for loop.

Now we will write a MAPLE procedure that incorporates a for loop.

```
>  SUM := proc(n)
>    local i, tot;
>    tot := 0;
>    for i from 1 to n do
>      tot := tot + 1;
>    end do;
>    tot;
>  end proc:
```

This procedure computes the sum of the integers from 1 to n. Let's test it out.

```
>   1+2+3+4+5+6+7+8+9+10;
```

$$55$$

```
>   SUM(10);
```

$$55$$

It may be hard for you to see what is going on inside this program. One way to get more information is to change the MAPLE global variable `printlevel`.

```
> printlevel;
```

$$1$$

Observe that the default value of `printlevel` is 1. Let's increase the value of printlevel and see what happens.

```
> printlevel := 20;
```

$$printlevel := 20$$

```
> SUM(10);
(--> enter SUM, args = 10
```

$$tot := 0$$
$$tot := 1$$
$$tot := 3$$
$$tot := 6$$
$$tot := 10$$
$$tot := 15$$
$$tot := 21$$
$$tot := 28$$
$$tot := 36$$
$$tot := 45$$
$$tot := 55$$

$$55$$

```
<-- exit SUM (now at top level) = 55)
```

$$55$$

Now we can see more of what is going on. First, `tot` $= 0$, then `tot` $= 1$, then `tot` $= 1 + 2 = 3$, then `tot` $= 1 + 2 + 3 = 6$, then `tot` $= 1 + 2 + 3 + 4 = 10$, etc.

Statements within a particular procedure are recognized in levels, determined by the nesting of conditional statements or loops and the nesting of procedures. The setting of printlevel has the effect of printing out all statements executed up to the level set. The higher the value of printlevel, the more information that will be displayed.

We can also add print statements when we want to understand a program better. First we reassign printlevel back to its default value:

```
> printlevel := 1;
```

We redefine our proc SUM this time adding some print statements.

```
> SUM := proc(n)
>   local i, tot;
>   tot := 0;
>   for i from 1 to n do
>     tot := tot + 1;
>     print('i=',i,' tot=',tot);
>   end do;
>   tot;
> end proc:
> SUM(10);
```

$$i =, 1, \quad tot =, 1$$
$$i =, 2, \quad tot =, 3$$
$$i =, 3, \quad tot =, 6$$
$$i =, 4, \quad tot =, 10$$
$$i =, 5, \quad tot =, 15$$
$$i =, 6, \quad tot =, 21$$
$$i =, 7, \quad tot =, 28$$
$$i =, 8, \quad tot =, 36$$
$$i =, 9, \quad tot =, 45$$
$$i =, 10, \quad tot =, 55$$

$$55$$

Let's examine the output:

$i = 1$ tot $= 1$
$i = 2$ tot $= 1 + 2 = 3$
$i = 3$ tot $= 1 + 2 + 3 = 6$
$i = 4$ tot $= 1 + 2 + 3 + 4 = 10$
$\vdots$
$i = 10$ tot $= 1 + 2 + 3 + 4 + 5 + 6 + 7 + 8 + 9 + 10 = 55$

Now we can see more of what is going on. When we enter the loop for the first time, tot $= 0$. Every time we cycle through the loop, i increases by 1 and i is added to tot. In this way we sum the integers from 1 to 10.

In the loops we have seen, the variable is incremented by one unit for each cycle. There is a way to increment by different amounts. Examine this example.

```
> for i from 2 by 5 to 24 do
>     print(i);
> end do;
```

$$2$$
$$7$$
$$12$$
$$17$$
$$22$$

This time in each cycle i is incremented by 5 units. So the more general form of the for loop takes the shape:

```
for var from a by b to c do
    statseq
end do;
```

Here *var* is incremented by b units in each cycle of the loop.

Exercise 4.

Modify our proc `SUM` to accept two inputs a, b and return the sum of the integers from a to b, i.e., return

$$a + (a + 1) + (a + 2) + \cdots + (b - 1) + b.$$

```
> SUM := proc(a,b)
```

Check your program by computing

```
> SUM(1,10);
> SUM(15,29);
```

Make sure the output is correct by checking by hand or using a calculator.

Exercise 5.

Write a MAPLE procedure `ODDSUM` that returns the sum of the odd integers from 1 to n, assuming n is odd.

```
> ODDSUM := proc(n)
```

```
> 1+3+5+7+9+11+13+15+17+19;
```

$$100$$

```
> ODDSUM(19);
```

Did you get 100? Now use a for loop to print out a table of `ODDSUM`(k) for k from 1 to 19. Do you see a pattern?

7.4 Type declaration

In MAPLE it is possible to declare the type of input that is acceptable for a given procedure. We illustrate this feature by modifying our SUM function.

```
> SUM := proc(n::posint)
>   local i, tot;
>   tot := 0;
>   for i from 1 to n do
>     tot := tot + 1;
>   end do;
>   tot;
> end proc:
```

Notice how n was declared a **positive integer** by typing `n::posint`. Let's try it out.

```
> SUM(10);
```

$$55$$

```
> SUM(sqrt(2));
Error, SUM expects its 1st argument, n, to be of type
posint, but received 2^(1/2)
```

See how MAPLE has informed us that our input $\sqrt{2}$ was not valid because it was not a positive integer. Some common types are `array`, `complex`, `equation`, `even`, `integer`, `list`, `name`, `negint`, `odd`, `posint`, `prime`, `set`, and `string`. See `?type` for more types.

Exercise 6.

Below is a MAPLE procedure called `find2s`, which writes a given prime as a sum of two squares if such a sum exists, otherwise it prints the statement, "*p is not the sum of two squares.*" The procedure checks whether the input is a prime. Fill in the missing parts. Below you will find some sample output with which to check the procedure.

```
> find2s := proc(p::.......)
>   local ................
>   find := 0;
>     for a from 1 to trunc(sqrt(p/2)) do
>         c := p - a^2:
>         if issqr(c) then
>             print(`p = a^2 + b^2 where a=`,a,......);
>             find := 1:
>         end if;
>     end do;
>     if find=0 then
>         .....................;
>     end if;
> end proc:
```

Sample output

```
> find2s(3);
```

$$p \text{ is not the sum of two squares}$$

```
> find2s(13);
```

$$p = a\hat{}2 + b\hat{}2 \text{ where } a =, 2, \; b =, 3$$

```
> find2s(17);
```

$$p = a\hat{}2 + b\hat{}2 \text{ where } a =, 1, \; b =, 4$$

(i) Explain why in the for loop a goes from 1 to `trunc(sqrt(p/2))`.

(ii) Explain what the purpose of the variable *find* is in the procedure.

(iii) Find all primes less than 100 that are the sum of two squares.

7.5 The "while" loop

A while loop statement takes the form

```
while    condition    do
        statseq
end do:
```

In the `while` loop, MAPLE tests the condition recycling through the loop until the condition fails.

We construct a proc `binpow(n)` that returns the highest power of 2 less than or equal to n.

```
> binpow := proc(n::posint)
>    local x,m:
>    x:=0:
>    m:=n:
>    while m>=1 do
>        m := m/2:
>        x:= x + 1:
>    end do:
>    x - 1;
> end:
```

Let's make sure our function works with some examples.

```
> for n from 1 to 8 do
>    n, binpow(n);
```

```
> end do;
```

$$\begin{matrix} 1, & 0 \\ 2, & 1 \\ 3, & 1 \\ 4, & 2 \\ 5, & 2 \\ 6, & 2 \\ 7, & 2 \\ 8, & 3 \end{matrix}$$

Our function seems to work. In the example, $2^2 \le 6 < 2^3$, so that `binpow(6)=2`. In the for loop, m is repeatedly divided by 2 until $m < 1$. At the same time, x keeps track of the number of divisions, and we see that $x - 1$ gives the correct power of 2.

There is a famous algorithm of Euclid's for computing the gcd of two integers. We write a MAPLE proc `euclid(m,n)`, which implements this algorithm.

```
> euclid := proc(m::posint,n::posint)
>   local a,b,r:
>   a:=m:
>   b:=n:
>   r:=irem(a,b):
>   while r>0 do
>       a:=b:
>       b:=r:
>       r:=irem(a,b):
>   end do:
>   b:
> end proc:
```

There is a built-in MAPLE function `igcd` that also computes the gcd. We check our program with an example.

```
> a:=281474439315457:
> b:=33685115929:
> euclid(a,b);
```

$$256999$$

```
> igcd(a,b);
```

$$256999$$

We computed the gcd of 281474439315457 and 33685115929 two ways as 256999 each time.

Our next exercise is to write a program that will convert an integer (in base 10) to binary, i.e., base 2. Recall that the algorithm involves computing successive divisions by 2 and the remainders. Let's convert 213 to binary as an example.

```
> a:=53; r:=irem(a,2);
```

$$a := 53$$
$$r := 1$$

```
> a:=iquo(a,2); r:=irem(a,2);
```

$$a := 26$$
$$r := 0$$

```
> a:=iquo(a,2); r:=irem(a,2);
```

$$a := 13$$
$$r := 1$$

```
> a:=iquo(a,2); r:=irem(a,2);
```

$$a := 6$$
$$r := 0$$

```
> a:=iquo(a,2); r:=irem(a,2);
```

$$a := 3$$
$$r := 1$$

```
> a:=iquo(a,2); r:=irem(a,2);
```

$$a := 1$$
$$r := 1$$

```
> a:=iquo(a,2); r:=irem(a,2);
```

$$a := 0$$
$$r := 0$$

We obtained the remainders 1, 0, 1, 0, 1, 1, 0. Writing these in reverse order we obtain the binary form of 53:

$$(53)_2 = 0110101 = 110101.$$

MAPLE already has a function that converts to binary, which we use to check our result.

```
> convert(53,binary);
```

$$110101$$

This example should help you get started on the next exercise. Notice that we computed successive divisions by 2 until $a = 0$. This suggests our program will involve a while loop.

Exercise 7.

Write a MAPLE proc `conbin(n)` that converts a positive integer n to binary. Don't forget to check your program with lots of examples.

7.6 Recursive procedures

The Fibonacci numbers are defined as follows:

$$f_0 = 0,$$
$$f_1 = 1,$$

and for $n \geq 2$,

$$f_n = f_{n-1} + f_{n-2}.$$

In this way, we say that the Fibonacci numbers are defined recursively. The Fibonacci sequence is thus

$$f_0 = 0,\ f_1 = 1,\ f_2 = f_1 + f_0 = 1 + 0 = 1,\ f_3 = f_2 + f_1 = 1 + 1 = 2,\ \ldots$$

This is the sequence

$$0,\ 1,\ 1,\ 2,\ 3,\ 5,\ 8,\ 13,\ 21,\ 34,\ 55,\ \ldots$$

Notice that each number in the sequence is the sum of the two previous terms and that this corresponds to the equation

$$f_n = f_{n-1} + f_{n-2}.$$

The following proc computes f_n, for any n:

```
> restart;
> FIB := proc(n::nonnegint)
>   if n<2 then
>       n;
>   else
>       FIB(n-1) + FIB(n-2);
>   end if;
```

```
> end proc:
```

Let's compute the first 16 Fibonacci numbers.

```
> seq(FIB(n), n=0..15);
```

$$0,\ 1,\ 1,\ 2,\ 3,\ 5,\ 8,\ 13,\ 21,\ 34,\ 55,\ 89,\ 144,\ 233,\ 377,\ 610$$

Let's examine our proc FIB. First observe that $f_n = n$ when $n = 0, 1$. This explains the first part of the conditional statement in the proc. FIB is an example of a recursive procedure. It is a procedure that calls itself. For instance, when we enter FIB(5), MAPLE goes through something like

$$\begin{aligned} FIB(5) &= FIB(4) + FIB(3) = (FIB(3) + FIB(2)) + (FIB(2) + 1) \\ &= ((FIB(2) + 1) + (1 + 0)) + ((1 + 0) + 1) \\ &= (((1 + 0) + 1) + (1 + 0)) + ((1 + 0) + 1) \\ &= 5. \end{aligned}$$

The FIB proc is not very efficient. The time command returns the number of cpu seconds taken in executing a given command.

```
> time(FIB(20));
```

$$0.570$$

It took 0.57 seconds to compute f_{20}. This is pretty slow. Many calculations are repeated and forgotten. There is a way to get the FIB proc to remember its results. This is done using the remember option.

```
> FIB := proc(n::nonnegint)
>   option remember;
>   if n<2 then
>       n;
>   else
>       FIB(n-1) + FIB(n-2);
>   end if;
> end proc:
```

Let's see how this new version of FIB performs.

```
> time(FIB(20));
```

$$0.$$

```
> time(FIB(200));
```

$$.010$$

This time f_{20} was computed in no time at all, and it took only 0.01 seconds to compute f_{200}.

Exercise 8.

Assume m and n are nonnegative integers and $0 \le m \le n$. There are two ways to define the binomial coefficients $\binom{n}{m}$. They can be defined in terms of factorials:

$$\binom{n}{m} = \frac{n!}{(n-m)!\,m!}.$$

They can defined recursively as follows:

$$\binom{n}{n} = \binom{n}{0} = 1,$$

and for $n \ge 1$ and $1 \le m \le n-1$,

$$\binom{n}{m} = \binom{n-1}{m-1} + \binom{n-1}{m}.$$

Write a recursive proc `MYBINOM(n,m)` which computes the binomial coefficient $\binom{n}{m}$. Use your program to compute the binomial coefficient $\binom{20}{8}$. Check your result by using a different method.

```
> MYBINOM := proc(n::nonnegint,m::nonnegint)
>    if n >= 1 and m <= n-1 and .....  then
>       ............................  ;
>    else
>       .....  ;
>    end if:
> end proc:
```

7.7 Explicit return

By default, a MAPLE proc will return the last computation encountered in the proc. Quite often this is difficult to arrange. The **return** statement is used to invoke an explicit return. For instance, suppose we want to define the function

$$f(x) = \begin{cases} x & \text{for } x < 0, \\ 1-x & \text{for } 0 \le x \le 1, \\ x^2 & \text{for } x > 1, \end{cases}.$$

but we don't want to bother with **else** and **elif** statements, or with making sure that the last statement executed returns the correct result. This is easy if we use **return** statements.

```
> f := proc(x)
>    if x < 0 then
```

```
>          return x;
>      end if;
>      if x >= 0 and x <= 1 then
>          return 1 - x;
>      end if;
>      if x > 1 then
>          return x^2;
>      end if;
>  end proc:
```

Let's test our function.

```
> f(-0.4), f(0.3), f(2);
```

$$-.4, \quad .7, \quad 4$$

Warning: In MAPLE V Release 5 (and earlier versions), this form of the `return` statement will cause an error. In these earlier versions use `RETURN(x)`.

Exercise 9.

Use `return` statements to write a proc called `GRADE` that returns the usual letter grades. In other words, `GRADE(x)` returns

`A`	if $x \geq 90$,
`B+`	if $85 \leq x < 90$,
`B`	if $80 \leq x < 85$,
`C+`	if $75 \leq x < 80$,
`C`	if $70 \leq x < 75$,
`D`	if $60 \leq x < 70$, and
`E`	if $x < 60$.

7.8 Error statement

An `error` statement is used in a proc to print out an informative error message. The proc `SCORETOT` computes the total of a list of five lab scores.

```
> SCORETOT := proc(L::list)
>   local num, TOT, k;
>   num := nops(L):
>   if num = 5 then
>     TOT := sum(L[k],k=1..5);
>     return TOT;
>   else
>     error "L must have have 5 entries. Your L had %1.",
>     num;
>   end if;
```

```
> end proc:
```

Warning: In MAPLE V Release 5 (and earlier versions), use `ERROR( )` instead of `error`.

```
> L := [48,36.5,24,43,36];
```

$$L := [48, 36.5, 24, 43, 36]$$

```
> SCORETOT(L);
```

$$187.5$$

So the total score in this example is 187.5. Suppose we accidently omitted one of the scores.

```
> L:=[48,36.5,24,43];
```

$$L := [48, 36.5, 24, 43]$$

```
> SCORETOT(L);
Error, (in SCORETOT) L must have have 5 entries.  Your L had
4.
```

Notice how helpful the `error` message was. The syntax of the `error` statement has two forms:

```
error string
error string, listparams
```

Here `string` is the message string, and `listparams` is a list of parameters. In our example above, the message string was `"L must have have 5 entries. Your L had %1."` , and there was one parameter `num`. The percent sign is used to refer to parameters. In our example, `%1` refers to the parameter `num`.

Now we write a new version of `SCORETOT` that (1) prints out a table of lab scores, and (2) returns the total score as a percentage. Since we want a percentage, we need to know the total possible score for each lab. This time we require two input arguments `L` and `M`. `L` is a list of lab scores and `M` is a list of the total possible scores for each lab.

```
> SCORETOT := proc(L::list,M::list)
>   local num, TOT, k, numlabs, numtots, A, j, TOTPOSS,
>   TOTPER;
>   numlabs:=nops(L);
>   numtots:=nops(M):
>   lprint(`LAB GRADE COMPUTATION`);
>   lprint(`ASSUMPTIONS: Overall grade is computed by `);
>   lprint(`summing all lab scores, then dividing by`);
>   lprint(`the total possible and then converting to a`);
```

```
>     lprint(`percentage.`);
>     A:=matrix(numlabs+1,4);
>     A[1,1]:=`LAB`:  A[1,2]:=`Raw Score`:
>     A[1,3]:=`Total Possible`:
>     A[1,4]:=`Score as Percentage`:
>     if numlabs=numtots then
>         for j from 1 to numlabs do
>             A[j+1,1]:=j:  A[j+1,2]:=L[j]:
>             A[j+1,3]:=M[j]:
>             A[j+1,4]:=evalf(L[j]/M[j]*100,4):
>         end do;
>         print(A);
>         TOT:=sum(L[k],k=1..numlabs);
>         TOTPOSS:=sum(M[k],k=1..numlabs):
>         TOTPER:=evalf(TOT/TOTPOSS*100,4);
>         lprint(`Total of lab scores = `,TOT);
>         lprint(`Total possible score = `,TOTPOSS);
>         lprint(`TOTAL score as a percentage = `,TOTPER);
>     else
>         error "numlabs must equal numtots but here"
>         "numlabs=%1 and numtots=%2.",numlabs,numtots;
>     end if;
>  end proc:
```

In this proc the matrix A has four columns. The first column is the LAB number. The second column gives the raw LAB scores, the third column gives the total possible score for each LAB, and the fourth column gives each LAB score as a percentage. For example, suppose there are 3 LABS: LAB 1, 2, 3 are out of 20, 30, and 15 points, respectively. In this case, `M` would be the list `[20,30,15]` — each entry corresponds to a maximum possible score. Suppose a student obtains 18, 20, and 14.5 points on LABS 1, 2, 3, respectively. Then `L` would be the list `[18, 20, 14.5]`. We are now ready to illustrate the new version of `SCORETOT`.

```
>  L:=[18,20,14.5];M:=[20,30,15];
```

$$L := [18, 20, 14.5]$$
$$M := [20, 30, 15]$$

```
>  SCORETOT(L,M);
`LAB GRADE COMPUTATION`
`ASSUMPTIONS: Overall grade is computed by `
`summing all lab scores, then dividing by`
`the total possible and then converting to a`
`percentage.`
```

$$\begin{bmatrix} \mathit{LAB} & \mathit{Raw\ Score} & \mathit{Total\ Possible} & \mathit{Score\ as\ Percentage} \\ 1 & 18 & 20 & 90. \\ 2 & 20 & 30 & 66.67 \\ 3 & 14.5 & 15 & 96.67 \end{bmatrix}$$

```
'Total of lab scores = ', 52.5
'Total possible score = ', 65
'TOTAL score as a percentage = ', 80.77
```

Exercise 10.

Complete the MAPLE proc WSCORETOT(L,M) given below. This proc gives each LAB the same weight by computing each LAB score as a percentage. Here L is a list of scores, and M is a list of maximum possible scores. Your proc WSCORETOT(L,M) should work for any list of scores L and list of totals M. It should return an error message if the number of entries of L does not equal the number of entries of M. It should also return an error message if a student's total is larger than the total possible. The proc should return the total score as a percentage, and each lab should count equally.

```
>  WSCORETOT := proc(L::list,M::list)
>    local num, TOT, k, numlabs, numtots, A, j, TOTPOSS,
>    TOTPER;
>    numlabs:=nops(L);
>    numtots:=nops(M):
>    lprint('LAB GRADE COMPUTATION');
>    lprint('ASSUMPTIONS: Overall grade is computed by ');
>    lprint('........................................');
>    lprint('...........................');
>    A:=matrix(numlabs+1,4);
>    A[1,1]:='LAB':  A[1,2]:='Raw Score':
>    A[1,3]:='Total Possible':
>    A[1,4]:='Score as Percentage':
>    if numlabs=numtots then
>      for j from 1 to numlabs do
>        if ......... then
>          error "Score for LAB %1 = %2.  This is too big"
>          "since the total possible for this lab is ...",
>          .,........;
>        end if;
>        A[j+1,1]:=j:  A[j+1,2]:=L[j]:
>        A[j+1,3]:=M[j]:
>        A[j+1,4]:=evalf(L[j]/M[j]*100,4):
>      end do;
>      print(A);
>      LABSUM:=............................);
>      LBS:=evalf(LABSUM,4);
>      WSCORE:=evalf(..............,4);
>      lprint('The sum of the lab scores');
>      lprint('as percentages = ',...);
>      lprint('Overall score as a percentage = ',......);
>    else
>      error "numlabs must equal numtots but here"
```

```
>       "numlabs=%1 and numtots=%2.",numlabs,numtots;
>     end if;
>   end proc:
```

Check your proc with the following example:

```
> L:=[18,20,14.5];M:=[20,30,15];
```

$$L := [18, 20, 14.5]$$
$$M := [20, 30, 15]$$

```
> WSCORETOT(L,M);
`LAB GRADE COMPUTATION`
`ASSUMPTIONS: Overall grade is computed by `
`converting each lab score to a percentage`
`and then taking the average.`
```

$$\begin{bmatrix} LAB & Raw\ Score & Total\ Possible & Score\ as\ Percentage \\ 1 & 18 & 20 & 90. \\ 2 & 20 & 30 & 66.67 \\ 3 & 14.5 & 15 & 96.67 \end{bmatrix}$$

```
`The sum of the lab scores`
`as percentages = `, 253.3
`Overall score as a percentage = `, 84.43
```

7.9 args and nargs

When defining a MAPLE procedure, it is not necessary to supply the names of input parameters. **args** is a special name for the sequence of input parameters and **nargs** is the special name given to the number of input parameters used.

```
> zol := proc()
>   print("The args of this proc are",args);
>   print("The first arg is",args[1]);
>   print("The number of args is",nargs);
> end proc:
> zol(a,b);
```

```
"The args of this proc are", a, b
"The first arg is", a
"The number of args is", 2
```

```
> zol(b,a,a);
```

```
"The args of this proc are", b, a, a
"The first arg is", b
"The number of args is", 3
```

The proc `zol` prints out its arguments, the first argument, and the number of arguments. When `zol(a,b)` is called, the value of `args` is the sequence of inputs `a,b`, so that the number of arguments is `nargs = 2`. The first argument is `args[1]`. In general, the ith argument is `args[i]`.

In our next example, we define a proc `myevalf` that takes one or two arguments. When one argument is given, it returns an approximation to four digits, otherwise the second argument specifies the number of digits.

```
> myevalf:=proc()
>    if nargs=0 or nargs>2 then
>        error "This proc only takes 1 or 2 arguments."
>        "Number of args supplied = %1", nargs;
>    else
>        if nargs=2 then
>            evalf(args[1],args[2]):
>        else
>            evalf(args[1],4):
>        end if;
>    end if;
> end proc:
> myevalf(Pi);
```

$$3.142$$

```
> myevalf(Pi,10);
```

$$3.141592654$$

Notice that the default for `myevalf` is four digits. So `myevalf(Pi)` returned π to four significant digits.

7.10 Input and output

7.10.1 Formatted output

We have already seen two print commands: `print` and `lprint`. To produce formatted printing we use the `printf` command. This command is similar to the C `printf` command. The syntax of `printf` has the form

`printf(` *format*`,` *expressionSequence*`)`

The *format* tells MAPLE how to write the terms in the *expressionSequence*. The *format* has the following syntax

`%[`*flags*`] [`*width*`] [.`*precision*`] [`*modifiers*`]` *code*

The optional *flags* are `+` (numeric value is output with a leading "+" or "-" sign), `-` (output is left justified), blank space (numeric value is output with a leading "+" or "-" sign), and `0` (numeric value is padded with zeroes). The optional *width* specifies the minimum number of characters to output. The optional *precision* specifies number of digits after the decimal point for numeric point and the maximum number of characters for string output. The optional *modifiers* are `l` or `L`, and `zc` or `Z`. See `?printf` for more information. The *code* indicates the type of object to be printed. The value *code* can be `d` (signed decimal integer), `o` (unsigned octal integer), `x` or `X` (unsigned hexadecimal integer), `e` or `E` (floating point number in scientific notation), `f` (fixed-point number), `g` or `G` (automatic `d`, `e` or `f` format depending on value), `y` or `Y` (IEEE hex dump format), `c` (single character), `s` (character string), `a` or `A` (outputs MAPLE object as a string), `q` or `Q` (used for printing all remaining arguments), `m` (".m" file format), or `%` (verbatim output).

To print an integer we use the `%d` format.

```
> x:=2^12;
```

$$x := 4096$$

```
> printf("%d",x);
96
```

It appears that MAPLE has not printed the correct value. The problem is that the `printf` function does not automatically insert a line break. To insert a line break we use the symbol `\n`.

```
> printf("%d\n",x);
4096
> printf("x=%d\n",x);
x=4096
> printf("x=%10d\n",x);
x=      4096
> printf("x=%010d\n",x);
x=0000004096
```

We use the `%e` format to print a floating-point approximation in scientific notation.

```
> y:=1/x;
```

$$y := \frac{1}{4096}$$

```
> evalf(y);
```

$$.0002441406250$$

```
> printf("y=%e\n",y);
y=2.441406e-04
> printf("y=%20e\n",y);
```

```
y=         2.441406e-04
>  printf("y=%20.9e\n",y);
y=     2.441406250e-04
>  printf("x=%d and y=%e.\n",x,y);
x=4096 and y=2.441406e-04.
>  printf("x=%d and \ny=%e.\n",x,y);
x=4096 and
y=2.441406e-04.
```

To print a MAPLE expression we use the %a format.

```
>  printf("Integral is %a.\n",int(1/t,t));
Integral is ln(t).
```

Observe how MAPLE evaluated the integral and printed its value.

7.10.2 Interactive input

There may be some situations in a program when you would want to prompt the user for input. There are two MAPLE functions for entering input interactively: `readline` and `readstat`. To enter a string we use `readline(terminal)`. More generally the call `readline(filename)` will read a line of text from a file. The following proc `getmethod` will ask you to enter 1 or 2. Type 1 and press the enter key.

```
>  methodget:=proc()
>    local p:
>    printf("There are two available methods:\n");
>    printf("1.  Undetermined coefficients\n");
>    printf("2.  Laplace transforms\n");
>    printf("ENTER 1 or 2:  ");
>    p:=readline(terminal);
>    if p="1" or p="2" then
>        return p;
>    else
>        error "You must enter 1 or 2."
>    end if;
>  end proc:
>  methodget();
There are two available methods:
1.  Undetermined coefficients
2.  Laplace transforms
ENTER 1 or 2:  1
                                        "1"

>  whattype(%);
                                      string
```

Notice that when we entered "1" it was read as a string and not as an integer. To read numerical input or a MAPLE expression, we must use the **readstat** command. The following proc will prompt you for an integer. When you run it, type an integer followed by a semicolon.

```
> numget:=proc()
>   local p:
>   p:=readstat("Enter an integer:  ");
>   if type(p,integer) then
>       return p;
>   else
>       error "You must enter an integer."
>   end if;
> end proc:
> numget();
Enter an integer:  13;
```

$$13$$

We typed in "13;". When using `readstat`, input entered must end with a semicolon or colon.

There is a way to read numerical input or a MAPLE expression using `readline`. This can be done by using the `parse` function to convert the string entered into a MAPLE expression.

```
> s:="sin(x^3+2*x)";
```

"sin(x^3+2*x)"

```
> whattype(%);
```

string

```
> parse(s);
```

$$\sin(x^3 + 2\,x)$$

```
> whattype(%);
```

function

Exercise 11.

Write a MAPLE proc `varget` using `readline` and `parse` that produces the following output:

```
> varget();
ENTER an independent variable:  x
```

$$x$$

```
> varget();
ENTER an independent variable:  12
```

```
Error, (in varget) You must enter a variable.
You entered a integer.
```

7.10.3 Reading commands from a file

Although the editing features of MAPLE are getting better and better with each release, it is usually more convenient and wiser to write MAPLE programs using an editor and save them in ordinary text files. For instance, instead of typing the proc SCORETOT (given in Section 7.8) directly into a worksheet within MAPLE, it would be better to create it using an editor in, say, the file *stot*. The MAPLE read function is used to read a file into a maple session. We give an example for Windows. If this file is in the subdirectory *myprogs* within the *Maple 7* directory, try

```
>    read "c:\\Program Files\\Maple 7\\myprogs\\stot";
```

and then the proc SCORETOT will be ready for use. A variant of this should work on other platforms. For instance, in the UNIX version try

```
>    read stot;
```

if your MAPLE session was started in the same directory.

7.10.4 Reading data from a file

There are several ways to read numerical data from a file. The simplest functions are ImportMatrix and readdata. The ImportMatrix function is used to read columns of numbers. Its syntax has the form

ImportMatrix("filename", delimiter=*string*)

Here filename is name of the file containing the data, and *string* is the character that separates entries on each line of the file. The default value of *string* is the tab character. Sometimes a single space " " is used a delimiter. Suppose we have a file *mydata.txt* containing the following data:

```
0 1.
.175e-1 .905
.125 .649
.353 .353
.649 .125
.905 .175e-1
1.  0
```

We use ImportMatrix to read this data into a MAPLE session. We interpret each line as a two-dimensional data point and plot the result. If you are using the Windows version of MAPLE, then you may have to create the file *mydata.txt* in the c:\"Program Files"\"Maple 7" directory.

```
> M:=ImportMatrix("mydata.txt",delimiter=" ");
```

$$M := \begin{bmatrix} 0 & 1. \\ .0175 & .905 \\ .125 & .649 \\ .353 & .353 \\ .649 & .125 \\ .905 & .0175 \\ 1. & 0 \end{bmatrix}$$

```
> whattype(%);
```

$$Matrix$$

```
> datapts:=[seq([M[i,1],M[i,2]],i=1..7)];
```

$$datapts := [[0, 1.], [.0175, .905], [.125, .649], [.353, .353], [.649, .125], [.905, .0175], [1., 0]]$$

```
> plot(datapts,style=point,symbol=circle,color=black);
```

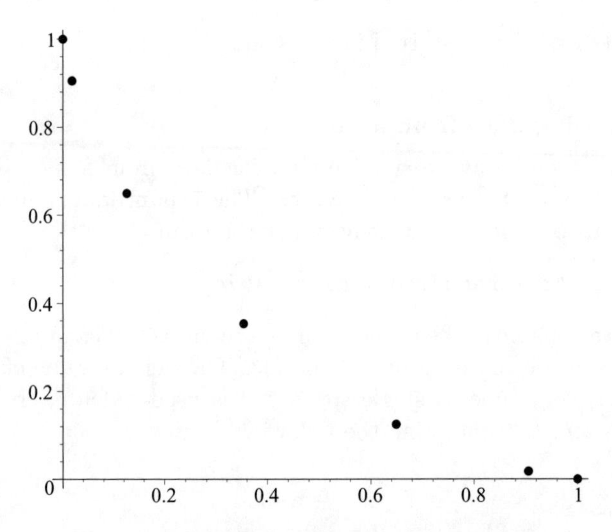

Figure 7.2 MAPLE plot of imported data points.

Alternatively we can use the `readdata` function. Its syntax has the form

`readdata( "filename", numColumns)`

Here `numColumns` is the number of columns of data in the file.

```
> L:=readdata("mydata.txt",2);
```

$$L := [[0, 1.], [.0175, .905], [.125, .649], [.353, .353], [.649, .125], [.905, .0175], [1., 0]]$$

```
> whattype(L);
```

$$list$$

```
> plot(L,style=point);
```

Observe this time that the data is interpreted as a list instead of a matrix. Other functions for reading data are `fscanf`, `scanf`, and `readbytes`. See the help pages for more information.

7.10.5 Writing data to a file

In the previous section we used the `ImportMatrix` function to read numerical data into a MAPLE session. The analogous function for writing data to a file is `ExportMatrix`. We now create a data file, *mydata2.txt*, similar to the one used in the previous section.

```
> M:=Matrix(7,2);
> for i from 0 to 6 do
> x:=sin(Pi*i/12)^3:
> y:=cos(Pi*i/12)^3:
> M[i+1,1]:=evalf(x,3):
> M[i+1,2]:=evalf(y,3):
> end do:
> ExportMatrix("mydata2.txt", M);
```

The call to `ExportMatrix` creates a data file *mydata2.txt* from the entries of the matrix M. Being the default, the tab character is used to separate entries on each line of the file. The syntax of the command has the form

```
ExportMatrix( "filename", matrix)
ExportMatrix( "filename", matrix, delimiter=string)
```

Other functions for writing data are `writedata`, `fprintf`, and `writebytes`. See the help pages for more information.

7.10.6 Writing and saving to a file

In the previous section we used the `ExportMatrix` function to write numerical data to a text file. To write or save more general MAPLE output, we must use other functions. One method is to use the `writeto` command to redirect output from the screen to a file.

```
> restart;
> y:=int(1/x,x);
```

$$y := \ln(x)$$

```
> writeto("myoutput");
> y;
```

```
> writeto(terminal);
> y;
```

$$\ln(x)$$

You should now have a text file *myoutput*, which contains the string `ln(x)`. In the MAPLE session we first computed an integral and assigned its value to `y`. The call `writeto("myoutput")` opened a text file called `myoutput` for writing. The next command entered was written to this file instead of being displayed on the screen. The call `writeto(terminal)` restores output to the screen. The command `appendto` is similar to `writeto` except output is appended to the file if it already exists.

Output saved using `writeto` cannot be reread back into MAPLE. To reuse output one should use the `save` command. The syntax of the command has the form

`save` name$_1$, name$_2$, ... , name$_k$, *filename*

The specified variables will be written to the file specified by *filename*. If the file name ends with the characters `.m` then values are saved in MAPLE's internal format.

```
> x:=5;
```

$$5$$

```
> y:=7;
```

$$7$$

```
> z:=int(1/u,u);
```

$$\ln(u)$$

```
> save x,y,z,"first.m";
> save x,y,part1;
> quit
```

In the session above, all the variables were saved in the MAPLE internal file *first.m*. The values of x and y were saved in the text file *part1*. In another MAPLE session we can restore these variables using the `read` command.

```
> read part1;
```

$$x := 5$$
$$y := 7$$

```
> anames();
```

$$x, y$$

```
> read "first.m";
> z;
```

$$\ln(u)$$

In case you forget the names of the saved variables, use the **anames** function. The call `anames()` will return a list of variables that have been assigned values.

7.11 Generating C and Fortran code

The *codegen* package contains functions for converting MAPLE code to C or Fortran. As an example we use MAPLE to construct a Fortran function `f(x)`, which will return an approximation of the Bessel function $J_0(x)$ for x near 1. First we load the *codegen* package.

```
> restart:
> with(codegen):
```

Next we compute the first few terms of the series expansion of $J_0(x)$ near $x = 1$.

```
> s:=series(BesselJ(0,x),x=1,6):
> p:=convert(s,polynom):
> ep:=evalf(p);
```

$$ep := 1.205248272 - 0.4400505857\,x - 0.1625735505\,(x - 1.0)^2 + 0.05419118346\,(x - 1.0)^3 + 0.009575290409\,(x - 1.0)^4 - 0.002241113982\,(x - 1.0)^5$$

Now we use the **makeproc** function to convert this last expression in x into a proc.

```
> f:=makeproc(ep,[x]);
f := proc(x)
1.205248272 - .4400505857*x - .1625735505*(x - 1.)^2
+ .05419118346*(x - 1.)^3 + .009575290409*(x - 1.)^4
- .002241113982*(x - 1.)^5
end
```

Finally we use the **fortran** function to convert this proc to Fortran code.

```
> fortran(f,optimized);
      real function f(x)
      real x

      real t2
      real t3
      real t7

      t2 = x-1.E0
      t3 = t2**2
      t7 = t3**2
```

```
      f = 0.1205248E1-0.4400506E0*x-0.1625736E0*t3
     #+0.5419118E-1*t3*t2+0.957529E-2*t7-0.2241114E-2*t7*t2
      return
      end
```

Here we have used the optimized option. This has the effect of doing common subexpression optimization. The analogue for producing C code is C.

```
> C(f);
#include <math.h>
double f(x)
double x;
{
  {
    return(0.1205248272E1-0.4400505857*x-0.1625735504*
pow(x-1.0,2.0)+0.5419118346E-1*pow(x-1.0,3.0)
+0.9575290409E-2*pow(x-1.0,4.0)-0.2241113982E
-2*pow(x-1.0,5.0));
  }
}
```

7.12 Viewing built-in MAPLE code

One of the great features of MAPLE is that most of the built-in functions are written in the MAPLE programming language, and the code is accessible to the user. To see how MAPLE defines the Gamma function, try

```
>   interface(verboseproc=2);
```

```
>   op(GAMMA);
```

7.13 The MAPLE interactive debugger

MAPLE has several functions that allow interactive debugging of programs. As an example we reexamine the MAPLE proc euclid from Section 7.5. However, this version has a bug.

```
> euclid := proc(m::posint,n::posint)
>   local a,b,r:
>   a:=m:
>   b:=n:
>   r:=irem(a,b):
>   while r>0 do
>       a:=b:
>       b:=r:
>       r:=irem(a,b):
>   end do:
>   r:
```

```
> end proc:
> euclid(45,51);
```

$$0$$

As you see, this version has a bug. The call `euclid(45,51)` should return a gcd of 3, not 0. The `showstat` command displays a MAPLE procedure with statement numbers.

```
> showstat(euclid);
euclid := proc(m::posint, n::posint)
local a, b, r;
   1   a := m;
   2   b := n;
   3   r := irem(a,b);
   4   while 0 < r do
   5     a := b;
   6     b := r;
   7     r := irem(a,b)
       end do;
   8   r
end proc
```

To invoke the MAPLE debugger we use the `stopat` command.

```
> stopat(euclid);
```

$$[euclid]$$

```
> euclid(45,51);
euclid:
   1*  a := m;
```

DBG>

This `stopat` call inserts a break point when the `euclid` proc is called. The debugger prompt is **DBG>**. At this point MAPLE is waiting for a debugger command. See `?debugger` for available debugger commands. After the DBG prompt we type `next`.

```
DBG> next
45
euclid:
   2   b := n;
```

DBG>

Notice that MAPLE printed out 45, the value of a in statement 1, and then printed out the next statement followed by the DBG prompt. Next we keep using the `next` command until the program is finished.

```
DBG> next
51
euclid:
   3   r := irem(a,b);

DBG> next
45
euclid:
   4   while 0 < r do
          ...
       end do;

DBG> next
0
euclid:
   8   r

DBG> next
                                   0

>
```

Notice that we did not step through each statement in the while loop. To step into a loop we use the `step` command. This time, after statement 4 is printed, we use the `step` command.

```
DBG> next
45
euclid:
   4   while 0 < r do
          ...
       end do;

DBG> step
45
euclid:
   5     a := b;

DBG> step
51
euclid:
   6     b := r;

DBG> step
45
euclid:
   7     r := irem(a,b)
```

```
DBG> step
6
euclid:
   5      a := b;

DBG>
```

The step command allows us to step through each statement in the loop until we exit the loop. Now we could issue another step command or a next command to continue to the next statement. Instead we issue a cont command to continue execution.

```
DBG> cont
```

$$0$$

To turn off the MAPLE debugger, use the unstopwhen function.

```
> unstopwhen();
```

$$[]$$

Exercise 12.

Find the bug in the euclid proc.

7.14 Writing your own packages

There are two methods of writing MAPLE packages. The first method uses a table to save functions in a package. With MAPLE 7, the modern and preferred method is to use modules.

7.14.1 Packages as tables

As an example, we will write a package called *qprod*, which contains three functions: aqprod, etaq, and qbin. The idea is to define a table with the name qprod. The table qprod will contain the three functions. Finally, the table is saved as the file *qprod.m*.

```
> qprod:=table();
> qprod[aqprod]:=proc(a,q,n)
>     local x,i:
>     if type(n,nonnegint) then
>         x:=1:
>         for i from 1 to n do
>             x := x * (1-a*q^(i-1)):
>         end do:
>     else
>         x:=``(a,q)[n];
>     end if:
```

```
>        return x:
>    end proc:
>    qprod[etaq]:=proc(q,i,trunk)
>    # This proc returns (q^i:q^i)_inf up to q^trunk
>        local k,x,z1,z,w:
>        z1:=(i + sqrt( i*i + 24*trunk*i ) )/(6*i):
>        z:=1+trunc( evalf(z1) ):
>        x:=0:
>        for k from -z to z do
>            w:=i*k*(3*k-1)/2:
>            if w<=trunk then
>                x:=x+ q^( w )*(-1)^k:
>            end if:
>        end do;
>        return x:
>    end proc:
>    qprod[qbin]:=proc(q,m,n)
>        if whattype(m)=integer and whattype(n)=integer then
>            if m>=0 and m<=n then
>                return normal(aqprod(q,q,n)/aqprod(q,q,m)
>                /aqprod(q,q,n-m));
>            else
>                return 0;
>            end if:
>        else
>            error "m and n must be integers.";
>        end if:
>    end proc:
>    save( qprod,
>    "c:\\Program Files\\Maple 7\\mylib\\qprod.m");
```

In the last two lines, the table **qprod** was saved to the file

c:\Program Files\Maple 7\mylib\qprod.m

on a Win95 machine. The name of this file should be changed to suit your platform. You must edit MAPLE's initialization file before you can load this package into a MAPLE session. On a Windows machine the initialization file is called *maple.ini* and is contained in the *bin.wnt* subdirectory of the MAPLE 7 directory. Add the following lines to this file:

```
mylib := "c:\\Program Files\\Maple 7/mylib":
libname := libname, mylib:
```

Now when you start a session, MAPLE will know the location of the *mylib* directory. The name of the initialization file is platform-dependent. On a UNIX machine it has the name *.mapleinit*. We try loading our new package into a MAPLE session.

```
> with(qprod);
```

$$[aqprod, etaq, qbin]$$

```
> qbin(q,3,6);
```

$$\left(q^2 - q + 1\right)\left(q^4 + q^3 + q^2 + q + 1\right)\left(q^3 + q^2 + q + 1\right)$$

7.14.2 Modules for packages

A MAPLE module is a collection of MAPLE procs and data that are bound together. A module is similar in construction to a procedure in that some objects are local and global. Objects are made visible outside the module using the `export` declaration. The syntax of a module has the form

```
module()
    local localvars;
    export exportvars;
    global globalvars;
    options optionseq;
    description descriptionstring;
    ⋮
    maple statements and procs
    ⋮
end module
```

As an example we construct a module called **etaproduct**, which contains three procs: **GPmake**, **cuspmake**, and **etaprodcuspord**, for computing orders of cusps of eta-products on the group $\Gamma_0(N)$. Don't worry whether you understand what a cusp or an eta-product is. This example only serves to illustrate the form of a module when it is used to create a MAPLE package. Use an editor to write the following MAPLE code in a text file called *etaprod.txt*.

```
etaproduct := module()
  description "A package for cusps of eta products";
  export GPmake, cuspmake, etaprodcuspord;
  option package;
  GPmake:=proc(etaprod)
    description "This proc finds the GP corresponding"
    " to the given etaproduct.";
    local L1,L1n,GP,i,r,p,t:
    if whattype(etaprod)=`^` then
      L1:=[etaprod]:
    else
      if whattype(etaprod)=function then
```

```
      return [op(etaprod)/tau,1];
    else
      L1:=[op(etaprod)]:
    end if:
  end if:
  L1n:=nops(L1):
  GP:=NULL:
  for i from 1 to L1n do
    r:=degree(L1[i]):
    if r=1 then
      p:=op(L1[i]):
      t:=p/tau:
    else
      p:=op(L1[i]):
      t:=op(p[1])/tau:
    end if:
    GP:=GP,t,r:
  end do:
  return [GP];
end proc:
cuspmake:=proc(N)
  description "Computes a set of inequivalent "
  "cusps for GAMMA_0(N)";
  local S,SoD,c,a,lasta,SSc,lastd,gcN,d,md:
  SoD:=numtheory[divisors](N):
  SoD := SoD minus 1:
  S:=0:
  for c in SoD do
    SSc:=:
    lastd:=c-1:
    gcN:=gcd(c,N/c):
    for d from 1 to lastd do
      md:=modp(d,gcN):
      if gcd(d,c)=1 and member(md,SSc)=false then
        S:= S union d/c:
        SSc:= SSc union md:
      end if:
    end do:
  end do:
  return S:
end proc:
etaprodcuspord:=proc(etaprod,N)
  description "Prints the order at each cusp"
  "from (GAMMA_0(N)) of the given etaproduct.";
  local GP,ngp,S,s,ords,i,t,r,c:
```

```
    GP:=etaproduct:-GPmake(etaprod):
    ngp:=nops(GP):
    S:=etaproduct:-cuspmake(N):
    for s in S do
      ords:=0:
      for i from 1 to (ngp/2) do
        t:=GP[2*i-1]:
        r:=GP[2*i]:
        c:=denom(s):
        ords:=ords+gcd(t,c)^2/t*r/24:
      end do:
      printf(" Order at cusp %a is %a.\n",s,ords);
    end do:
    return :
  end proc:
end module:
```

We have assigned the name `etaproduct` to the module. There are three procs defined inside the module: `GPmake`, `cuspmake`, and `etaprodcuspord`. The declaration

```
export GPmake, cuspmake, etaprodcuspord;
```

tells MAPLE that these three procs are visible outside the module. We call these exports. Also note we have used the *package* option to tell MAPLE we intend to use the `etaproduct` module as a package. The syntax for accessing an export has the form

```
modulename:-functionname
```

We use the `read` function to read in our text file *etaprod.txt* and use the `cuspmake` function to find the cusps of $\Gamma_0(20)$.

```
> read "etaprod.txt";
> etaproduct:-cuspmake(20);
```

$$\left\{0, \frac{1}{2}, \frac{1}{4}, \frac{1}{5}, \frac{1}{10}, \frac{1}{20}\right\}$$

We use `etaprodcuspord` function to print out the order of the eta-product

$$\left(\frac{\eta(5\tau)}{\eta(\tau)}\right)^6$$

at each cusp of $\Gamma_0(20)$.

```
> ep:=(eta(5*tau)/eta(tau))^6;
```

$$\frac{(\eta(5\,\tau))^6}{(\eta(\tau))^6}$$

```
> etaproduct:-etaprodcuspord(ep,20);
 Order at cusp 0 is -1/5.
 Order at cusp 1/2 is -1/5.
 Order at cusp 1/10 is 1.
 Order at cusp 1/4 is -1/5.
 Order at cusp 1/5 is 1.
 Order at cusp 1/20 is 1.
```

We use the `savelib` command to save our module `etaproduct` as a package. This way we will be able to use the package in a later session. We describe the process on a UNIX system. Other systems will be similar, the main difference being the way the files are named. Before we can use `savelib`, the global variables `libname` and `savelibname` should be set correctly. As well, certain MAPLE library files need to be created. In Section 7.14.1 we showed how to set up the MAPLE initialization file *maple.ini* on a Windows machine. On a UNIX machine the MAPLE initialization file is *.mapleinit*. In this file add the name of the directory where you want the package to be saved. Something like the following should be added to the initialization file:

```
mylib := "/home/fac0/frank/maple/mylib":
libname := mylib, libname:
```

Let's start a MAPLE session.

```
> libname;
```

"/home/fac0/frank/maple/mylib", "/opt/maple_6/lib"

We read our text file *etaprod.txt* and then use MAPLE's archiving function `march` to create the necessary MAPLE library files in the `mylib` directory.

```
> read "etaprod.txt";
> march('create', mylib, 100);
```

There should now be two MAPLE library files in the `mylib` directory: `maple.ind` and `maple.lib`. We set the global `savelibname` variable and save the `etaproduct` package using the `savelib` command.

```
> savelibname := mylib;
```

savelibname := /home/fac0/frank/maple/mylib

```
> savelib('etaproduct');
> quit
```

The `etaproduct` package has now been saved to a library file in the `mylib` directory. We start a new MAPLE session and load the package using the `with` command.

```
> with(etaproduct);
```

$$[GPmake, cuspmake, etaprodcuspord]$$

```
> ep:=(eta(5*tau)/eta(tau))^6;
```

$$ep := \frac{(\eta(5\,\tau))^6}{(\eta(\tau))^6}$$

```
> GPmake(ep);
```

$$[5, 6, 1, -6]$$

7.15 Answers to programming exercises

1.

```
> c2f := proc(x)
>   evalf(9/5*x + 32);
> end proc:
```

2.

```
> dist := proc(x1,y1,x2,y2)
>   local s;
>   s := (x1-x2)^2 + (y1-y2)^2;
>   sqrt(s);
> end proc:
```

3.

```
> h:=proc(x)
>   if x>3 then 0;
>       elif x>0 and x<=1 then x;
>       elif x>1 and x<=3 then 1;
>   else
>       -1;
>   end if;
> end proc:
> plot(h, -1..4, discont=true);
```

4.

```
> SUM := proc(a,b)
>   local i, tot;
>   tot := 0;
>   for i from a to b do
>     tot := tot + i;
>   end do;
```

```
>     tot;
>   end proc:
>   SUM(1,10);
```

55

```
>   SUM(15,19);
```

85

5.

```
>   ODDSUM := proc(n)
>     local i, tot;
>     tot := 0;
>     for i from 1 by 2 to n do
>       tot := tot + i;
>     end do;
>     tot;
>   end proc:
>   ODDSUM(19);
```

100

```
>   for k from 1 by 2 to 19 do
>       print(`k=`,k,` ODDSUM=`,ODDSUM(k));
>   end do;
```

k=, 1, ODDSUM=, 1
k=, 3, ODDSUM=, 4
k=, 5, ODDSUM=, 9
k=, 7, ODDSUM=, 16
k=, 9, ODDSUM=, 25
k=, 11, ODDSUM=, 36
k=, 13, ODDSUM=, 49
k=, 15, ODDSUM=, 64
k=, 17, ODDSUM=, 81
k=, 19, ODDSUM=, 100

Do you see the pattern? If N is odd, it seems that

$$1+3+5+\cdots+N=\left(\frac{N+1}{2}\right)^2.$$

6.

```
>   find2s := proc(p::prime)
>     local a,c,find;
>     find := 0;
>     for a from 1 to trunc(sqrt(p/2)) do
```

```
>         c := p - a^2:
>         if issqr(c) then
>             print(`p = a^2 + b^2 where a=`,a,
>             ` b=`,sqrt(c));
>             find := 1:
>         end if;
>     end do;
>     if find=0 then
>         print(`p is not the sum of two squares`);
>     end if;
> end proc:
```

(i) We want $a^2 + b^2 = p$. Since p is prime, $a \geq 1$. We may assume $a \leq b$. Then $2a^2 \leq p$ and $a \leq \sqrt{(p/2)}$. Since a must be an integer, this explains why in the loop a goes from 1 to `trunc(sqrt(p/2))`.

(ii) The value of *find* indicates whether p is the sum of two squares. It is set to 1 if in the loop we find a as the sum of two squares. So $find = 1$ if and only if p is the sum of two squares.

(iii) The primes less than 100 that are the sum of two squares are 2, 5, 13, 17, 29, 37, 41, 53, 61, 73, 89, and 97.

7.

```
> conbin:=proc(n::posint)
>   local m,x,y,r;
>   m:=n:
>   x:=0:
>   y:=1:
>   while m>0 do
>       r:=irem(m,2):
>       m:=iquo(m,2):
>       x:=x + r*y:
>       y:=10*y:
>   end do:
>   x;
> end proc:
```

8.

```
> MYBINOM := proc(n::nonnegint,m::nonnegint)
>   if n >= 1 and m <= n-1 and m >= 1 then
>       MYBINOM(n-1,m-1) + MYBINOM(n-1,m);
>   else
>       1 ;
>   end if:
> end proc:
```

```
> MYBINOM(20,8);
```

$$125970$$

```
> 20!/8!/12!;
```

$$125970$$

We found $\binom{20}{8} = 125970$, and checked our result using the factorial definition. MAPLE has a built-in binomial function. See `?binomial` for more information.

9.

```
> GRADE := proc(x)
>   if x >= 90 then return `A`; end if;
>   if x >= 85 and x < 90 then return `B+`; end if;
>   if x >= 80 and x < 85 then return `B`; end if;
>   if x >= 75 and x < 80 then return `C+`; end if;
>   if x >= 70 and x < 75 then return `C`; end if;
>   if x >= 60 and x < 70 then return `D`; end if;
>   if x < 60 then return `E`; end if;
> end proc:
```

10.

```
> WSCORETOT := proc(L::list,M::list)
>   local num, TOT, k, numlabs, numtots, A, j, TOTPOSS,
>   TOTPER;
>   numlabs:=nops(L);
>   numtots:=nops(M):
>   lprint(`LAB GRADE COMPUTATION`);
>   lprint(`ASSUMPTIONS: Overall grade is computed by `);
>   lprint(`converting each lab score to a percentage`);
>   lprint(`and then taking the average.`);
>   A:=matrix(numlabs+1,4);
>   A[1,1]:=`LAB`:  A[1,2]:=`Raw Score`:
>   A[1,3]:=`Total Possible`:
>   A[1,4]:=`Score as percentage`:
>   if numlabs=numtots then
>     for j from 1 to numlabs do
>       if L[j]>M[j] then
>         error "Score for LAB %1 = %2.  This is too big"
>         "since the total possible for this lab is %3.",
>         j,L[j],M[j];
>       end if;
>       A[j+1,1]:=j:  A[j+1,2]:=L[j]:
>       A[j+1,3]:=M[j]:
```

```
>         A[j+1,4]:=evalf(L[j]/M[j]*100,4):
>       end do;
>       print(A);
>       LABSUM:=sum(L[k]/M[k]*100,k=1..numlabs);
>       LBS:=evalf(LABSUM,4);
>       WSCORE:=evalf(LABSUM/numlabs,4);
>       lprint(`The sum of the lab scores`);
>       lprint(`as percentages = `,LBS);
>       lprint(`Overall score as a percentage = `,WSCORE);
>     else
>       error "numlabs must equal numtots but here"
>       "numlabs=%1 and numtots=%2.",numlabs,numtots;
>     end if;
>   end proc:
```

11.

```
>   varget:=proc()
>     local p,q:
>     printf("ENTER an independent variable:  ");
>     p:=readline(terminal);
>     q:=parse(p);
>     if not type(q,name) then
>         error "You must enter a variable."
>         "You entered a %1.",whattype(q);
>     else
>         return q;
>     end if;
>   end proc:
```

12.

The second to last line of the `euclid` proc should be `b:`, not `r:`.

8. DIFFERENTIAL EQUATIONS

8.1 Solving ordinary differential equations

Remember in MAPLE that there are two ways to code the derivatives $\frac{dy}{dx}, \frac{d^2y}{dx^2}, \frac{d^3y}{dx^3} \ldots$ We can use the diff, or Diff function:

`diff(y(x),x), diff(y(x),x,x), diff(y(x),x,x,x), ...`

or we can use the differential D operator:

`D(y)(x), (D@@2)(y)(x), (D@@3)(y)(x), ...`

To solve the differential equation de involving $Y = y(x)$, we use the command `dsolve(`de`,Y)`.

```
> y:='y':
> Y:=y(x);
```

$$Y := y(x);$$

```
> dY := diff(Y,x);
```

$$\frac{\partial}{\partial x}\, y(x)$$

```
> ddY := diff(%,x);
```

$$\frac{\partial^2}{\partial x^2}\, y(x)$$

```
> de := ddY+5*dY+6*Y = sin(x)*exp(-3*x);
```

$$de := \frac{\partial^2}{\partial x^2}\, y(x) + 5\, \frac{\partial}{\partial x}\, y(x) + 6\, y(x) = \sin(x)\, e^{-3x}$$

```
> ans := dsolve(de, Y);
```

$$ans := y(x) = e^{-3\,x}\_C2 + e^{-2\,x}\_C1 + \frac{1}{2}\,(\cos(x) - \sin(x))\, e^{-3\,x}$$

We found that the general solution to the differential equation

$$y'' + 5y' + 6y = \sin x\, e^{-3x}$$

is

$$y = \frac{1}{2}(\cos(x) - \sin(x))e^{-3x} + c_1 e^{-3x} + c_2 e^{-2x},$$

where c_1 and c_2 are any constants. We use MAPLE to check the solution. There are two methods. First we check the solution by computing derivatives and verifying the differential equation.

```
> sol := rhs(ans);
```

$$sol := 1/2\cos(x)e^{-3x} - 1/2\sin(x)e^{-3x} + \_C1\, e^{-3x} + \_C2\, e^{-2x}$$

```
> dsol:=diff(sol,x);
```

$$dsol := \sin(x)e^{-3x} - 2\cos(x)e^{-3x} - 3\,\_C1\, e^{-3x} - 2\,\_C2\, e^{-2x}$$

```
> ddsol:=diff(sol,x,x);
```

$$ddsol := 7\cos(x)e^{-3x} - \sin(x)e^{-3x} + 9\,\_C1\, e^{-3x} + 4\,\_C2\, e^{-2x}$$

```
> simplify(ddsol + 5*dsol + 6*sol);
```

$$\sin(x)e^{-3x}$$

This confirms the solution found before. A quicker method is to use MAPLE's `odtest` function.

```
> odetest(ans,de);
```

$$0$$

Because 0 was returned, the solution has been verified. See `?odetest` for more information.

Now let's look at our DE again, this time using a context menu.

```
> de;
```

$$\frac{\partial^2}{\partial x^2}\, y(x) + 5\,\frac{\partial}{\partial x}\, y(x) + 6\, y(x) = \sin(x)\, e^{-3x}$$

Right-click on the DE. A context menu should appear:

Copy	
Left-hand Side	
Right-hand Side	
Negate a Relation	
Move to Left	
Move to Right	
Solve D.E.	▶
Add an Initial Condition	▶
Classify the O.D.E.	▶
Conversions	▶

First highlight the DE with the mouse. Click on `Solve D.E.` and then `y(x)`.

```
> y(x) = exp(-3*x)*_C2+exp(-2*x)*_C1
    +1/2*(cos(x)-sin(x))*exp(-3*x);
```

$$y(x) = e^{-3\,x}\_C2 + e^{-2\,x}\_C1 + \frac{1}{2}\,(\cos(x) - \sin(x))\,e^{-3\,x}$$

Now try selecting `Classify the O.D.E` and then `y(x)`.

```
> [[_2nd_order, _linear, _nonhomogeneous]];
```

$$[[\_2nd\_order, \_linear, \_nonhomogeneous]]$$

MAPLE has recognized that the DE is second order, linear, and nonhomogeneous.

8.1.1 Implicit solutions

Quite often it is preferable to find solutions given implicitly. Consider the DE:

$$(3y^2 + e^x)\frac{dy}{dx} + e^x(y+1) + \cos x = 0.$$

```
> DE:=(3*y(x)^2+exp(x))*diff(y(x),x)+exp(x)*(y(x)+1)
    +cos(x) = 0;
```

$$\left(3\,(y(x))^2 + e^x\right)\frac{\partial}{\partial x}\,y(x) + e^x\,(y(x)+1) + \cos(x) = 0$$

```
> dsolve(DE,y(x),implicit);
```

$$e^x y(x) + e^x + \sin(x) + (y(x))^3 + \_C1 = 0$$

The general solution to the DE is given implicitly by the equation

$$e^x y + e^x + \sin x + y^3 = c,$$

where c is any constant. We found this solution by adding the `implicit` option to the `dsolve` function. For other options see `?dsolve`. Without the `implicit` option MAPLE would have returned three slightly horrible explicit solutions. Try it:

```
> dsolve(DE,y(x));
```

8.1.2 Initial conditions

We consider the initial value problem

$$y'' + 5y' + 6y = \sin x\, e^{-3x}, \quad y(0) = -\frac{5}{2},\ y'(0) = 2.$$

We continue our MAPLE session.

```
> de;
```

$$\frac{\partial^2}{\partial x^2}\, y(x) + 5\, \frac{\partial}{\partial x}\, y(x) + 6\, y(x) = \sin(x)\, e^{-3x}$$

```
> dsolve({de,y(0)=-5/2,D(y)(0)=2},Y);
```

$$y(x) = 1/2\, \cos(x) e^{-3\,x} - 1/2\, \sin(x) e^{-3\,x} + 2\, e^{-3\,x} - 5\, e^{-2\,x}$$

The solution to the initial value problem is given by

$$y = \frac{1}{2}\, \cos(x) e^{-3\,x} - \frac{1}{2}\, \sin(x) e^{-3\,x} + 2\, e^{-3\,x} - 5\, e^{-2\,x}.$$

Notice that in MAPLE we entered the initial condition $y'(0) = 2$, using the differential operator `D`. So in MAPLE `D(y)(0)` means $y'(0)$. For a higher order derivative $y^{(k)}$ we use the MAPLE notation `(D@@k)`. To enter an initial condition of the form $y^{(k)}(x_0)$, use the MAPLE notation `(D@@k)(y)(x0)`. For example, let's solve the following third-order initial value problem:

$$y + y' - 3y'' + y''' = e^{-x}(10 - 4x),\ y(0) = 5,\ y'(0) = 6,\ y''(0) = 3.$$

```
> DE := y(x) +D(y)(x) -3*(D@@2)(y)(x)+ (D@@3)(y)(x)
    = exp(-x)*(10-4*x);
```

$$y(x) + D(y)(x) - 3\,(D^{(2)})(y)(x) + (D^{(3)})(y)(x) = e^{-x}\,(10 - 4\,x)$$

```
> dsolve({DE,y(0)=5,D(y)(0)=6,(D@@2)(y)(0)=3},y(x));
```

$$y(x) = e^{-x} x + 5\, e^{x}$$

The solution to the initial value problem is given by

$$y = xe^{-x} + 5e^{x}.$$

The syntax of the `dsolve` function has the form

`dsolve({DE,`*sequence of initial conditions*`},` *dependent variable*)

Here the sequence of initial conditions

$y(x_0) = y_0,\ y(x_0) = y_1,\ y''(x_0) = y_2,\ \ldots$

is coded using the `D` operator as

```
y(x0)=y0, D(y)(x0)=y1, (D@@2)(x0)=y2, ...
```

8.1.3 Systems of differential equations

Systems of differential equations can be solved in an analogous fashion. To solve the initial value problem

$$\begin{aligned} y' + z' &= e^x, \qquad & y(0) &= 8/9 \\ y' - 3z &= x, \qquad & z(0) &= 10/9, \end{aligned}$$

try

```
> de1 := diff(y(x),x) + diff(z(x),x) = exp(x);
> de2 := diff(y(x),x) -3*z(x) = x;
> dsolve({de1,de2,y(0)=8/9,z(0)=10/9},{y(x),z(x)});
```

When solving a system of DEs, the syntax of the `dsolve` command has the form

`dsolve({sysDE,`*sequence of initial conditions*`}, {`*dependent variables*`})`

Here `sysDE` is a sequence of differential equations.

8.2 First-order differential equations

MAPLE knows the standard methods for solving first-order equations: separable equations, linear equations, exact equations, and the method of integrating factors. As well, MAPLE knows the standard first order equations, such as Abel's equation, Bernoulli's equation, Clauraut's equation, Ricatti's equation, Chini's equation, and d'Alembert's equation.

8.2.1 odeadvisor

There are many facilities for solving ODEs in the *DEtools* package. There is a neat function in this package, `odeadvisor`, which analyzes a given ODE and gives advice. Let's see how this function works. Consider the first-order equation

$$\frac{dy}{dx} = \frac{x-3}{y^2}.$$

```
> restart:
> with(DEtools):
> DE := diff(y(x),x) = (x-3)/y(x)^2;
```

$$\frac{\partial}{\partial x}\, y(x) = \frac{x-3}{(y(x))^2}$$

```
> odeadvisor(DE);
```

$$[\_separable]$$

MAPLE is telling us that this DE is separable. Try

```
> odeadvisor(DE,help);
```

This time the `odeadvisor` should bring you to a help page on separable equations. This page contains information on separable equations including the general form of the solution as well as some examples. This should give you some help in completing the problem by hand. You can then check your answer by typing

```
> dsolve(DE, y(x));
```

Thus MAPLE is able to recognize a separable equation. The `odeadvisor` recognizes the following first-order types:

`_separable` — Separable equations
A first-order equation is separable if it is of the form

$$\frac{dy}{dx} = f(x)\,g(y).$$

`_linear` — First order linear equation
A first-order linear equation has the form

$$\frac{dy}{dx} + f(x)\,y = g(x).$$

`_exact` — Exact equation
A first-order equation is exact if it can be written in the form

$$\frac{d}{dx}f(x, y(x)) = 0;$$

i.e., in the form

$$f_x(x, y(x)) + f_y(x, y(x))\,\frac{dy}{dx} = 0.$$

`_homogeneous` — Homogeneous equation
A first-order homogeneous equation has the form

$$\frac{dy}{dx} = f(y/x).$$

See also `?odeadvisor[homogeneousB]`, `?odeadvisor[homogeneousC]`, `?odeadvisor[homogeneousD]`, and `?odeadvisor[homogeneousG]`.

`_quadrature` — Quadrature format
A first-order de is said to be in quadrature format if the right side is a function of x only or a function of y only.

`_rational` — Rational equation
A first-order rational equation has the form

$$\frac{dy}{dx} = \frac{p(x, y)}{q(x, y)},$$

where p and q are polynomials.

`_Bernoulli` — Bernoulli's equation
Bernoulli's equation has the form

$$\frac{dy}{dx} + f(x)\, y = g(x)\, y^n.$$

`_Riccati` — Riccati's equation
Riccati's equation has the form

$$\frac{dy}{dx} = f(x)\, y^2 + g(x)\, y + h(x).$$

Other first-order types are given below:

`Abel`	`Abel2A`	`Abel2C`	`Bernoulli`	`Chini`
`Clairaut`	`dAlembert`	`patterns`	`sym_implicit`	

The `odeadvisor` recognizes the following second-order types:

`Bessel`	`Duffing`	`ellipsoidal`	`elliptic`
`Emden`	`erf`	`exact_linear`	`exact_nonlinear`
`Gegenbauer`	`Halm`	`Hermite`	`Jacobi`
`Lagerstrom`	`Laguerre`	`Lienard`	`Liouville`
`linear_ODEs`	`linear_sym`	`missing`	`Painleve`
`quadrature`	`reducible`	`sym_Fx`	`Titchmarsh`
`Van_der_Pol`			

and it recognizes the following higher order types.

`quadrature`	`missing`	`exact_linear`	`exact_nonlinear`
`reducible`	`linear_ODEs`		

In Section 8.1 we saw how MAPLE's `dsolve` function was used to solve differential equations. The `infolevel` function can be used to supply additional information when using the `dsolve` function. Let's consider the following DE:

$$\frac{dy}{dx} = \frac{(x-3)}{e^y}.$$

```
> with(DEtools):
> DE := diff(y(x),x) = (x-3)/exp(y(x));
```

$$\frac{d}{dx}y(x) = \frac{x-3}{e^{y(x)}}$$

```
> odeadvisor(DE);
```

$$[\_separable]$$

```
> infolevel[dsolve]:=3;
```

$$\texttt{infolevel[dsolve]} := 3$$

```
> dsolve(DE,y(x));
Methods for first order ODEs:
Trying to isolate the derivative dy/dx...
Successful isolation of dy/dx
-> Trying classification methods
trying a quadrature
trying 1st order linear
trying Bernoulli
trying separable
separable successful
```

$$y(x) = \ln(1/2\,x^2 - 3\,x + \_C1)$$

8.2.2 Integrating factors

An integrating factor for a first-order differential equation is a function $\mu(x, y)$ such that when the differential equation is multiplied by $\mu(x, y)$, the resulting equation is exact. The MAPLE function `intfactor` in the *DEtools* package looks for an integrating factor for a given ODE. Consider the DE

$$x^4 + y + x\left(y^2 + \ln(x)\right)\frac{dy}{dx} = 0.$$

This DE is not exact. We use `intfactor` to look for an an integrating factor.

```
> with(DEtools):
> de:=(x^4+y(x))+x*(y(x)^2+ln(x))*diff(y(x),x)=0;
```

$$de := x^4 + y(x) + x\left((y(x))^2 + \ln(x)\right)\frac{d}{dx}y(x) = 0$$

```
> mu := intfactor(de,y(x));
```

$$\mu := \frac{1}{x}$$

The function $\mu(x) = \frac{1}{x}$ is an integrating factor. We multiply both sides of the DE by $\mu(x)$.

```
> de2:=expand(mu*de);
```

$$x^3 + \frac{y(x)}{x} + \left(\frac{d}{dx}y(x)\right)(y(x))^2 + \left(\frac{d}{dx}y(x)\right)\ln(x) = 0$$

```
> odeadvisor(de2);
```

$$[\_exact]$$

We see that the resulting equation is exact. Now we can either use `dsolve` to find the general solution or use the method of exact equations to finish the problem by hand.

8.2.3 Direction fields

The `DEplot` function in the *DEtools* package plots the direction field of a first-order differential equation. We plot the direction field of the differential equations

$$\frac{dy}{dx} = \frac{4x}{y}.$$

```
> restart:
> with(DEtools):
> DE:= diff(y(x),x)=4*x/y(x);
```

$$\frac{\partial}{\partial x}\, y(x) = 4\frac{x}{y(x)}$$

```
> DEplot(DE,y(x),x=-2..2,y=-1..3);
```

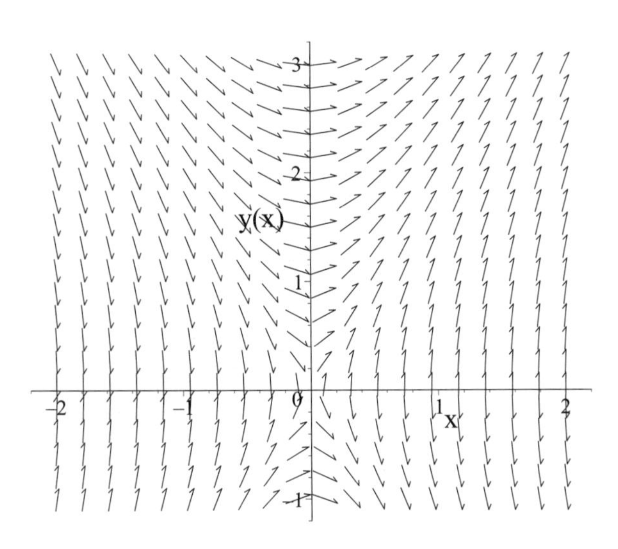

Figure 8.1 Direction field of a first-order DE.

We plotted the direction field in the region given by

$$-2 \leq x \leq 2, \quad -1 \leq y \leq 3.$$

The syntax of the `DEplot` function has the form

```
DEplot(eqn, dvar, ivarrange, dvrange)
```

where deqn is the differential equation, dvar is the dependent variable, ivarrange is the range of the independent variable, and dvrange is the range of the dependent variable.

Continuing with our example, we plot the solution that satisfies the initial condition $y(0) = 2$.

```
> DEplot(DE,y(x),x=-2..2,[[y(0)=2]],y=-1..3);
```

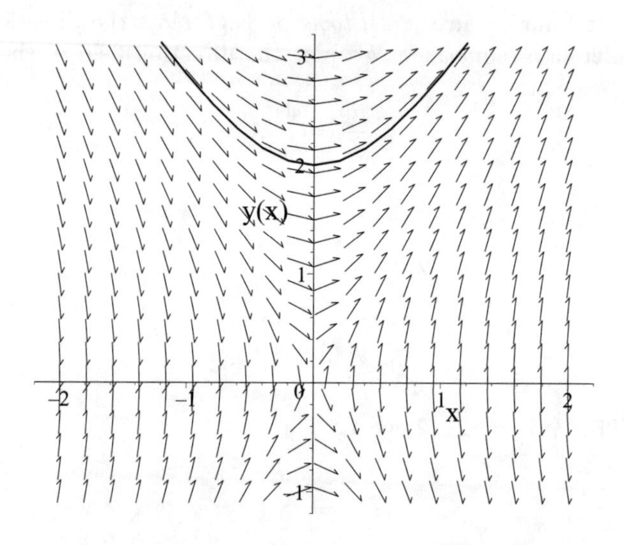

Figure 8.2 Direction field with a solution curve.

We try adding the solution that satisfies $y(1) = 1$.

```
> DEplot(DE,y(x),x=-2..2,[[y(0)=2],[y(1)=1]],y=-1..3);
```

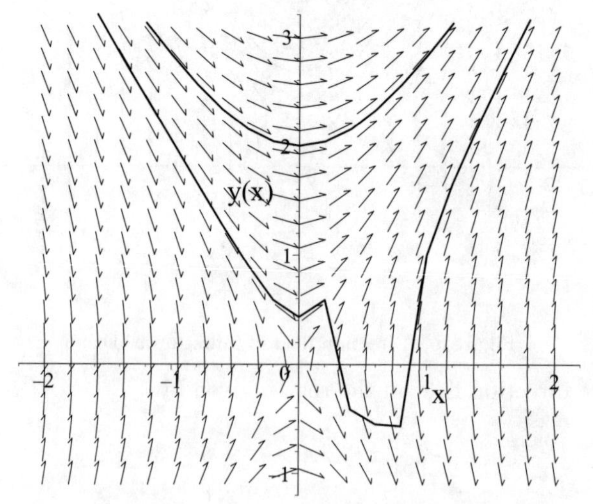

Figure 8.3 Direction field with two solution curves.

It is clear that MAPLE has plotted this second solution incorrectly. Any solution curve that comes near the x-axis must be nearly vertical there. The plotted solution seems fairly accurate for $x > 1$, but otherwise the plot is totally wrong. The problem is that MAPLE has not plotted enough points when y is close to zero. One way to fix this is to choose a small step-size. Try

```
> DEplot(DE,y(x),x=-2..2,[[y(0)=2],[y(1)=1]],y=-1..3,
    stepsize=0.0005);
```

This should give a more accurate plot. But beware! Don't compute blindly without thinking!

When using initial conditions, the syntax of `DEplot` has the form

```
DEplot(eqn, dvar, ivarrange, ICondtions, dvrange)
```

where `IConditions` is a list of initial conditions. Here each initial condition has the form `[y(x0)=y0]`. To plot the solution curves and omit the direction field, use the option `arrows=NONE`. Try

```
> DEplot(DE,y(x),x=-2..2,[[y(0)=2],[y(1)=1]],y=-1..3,
    stepsize=0.0005, arrows=NONE);
```

8.3 Numerical solutions

Numerical solutions to differential equations can be found using the `dsolve` function with the `numeric` option. Consider the initial value problem

$$\frac{dy}{dx} = x^2 + y^3, \quad y(0) = 0.$$

```
> de:=diff(y(x),x)=x^2+y(x)^3:
> IVP:={de,y(0)=0};
```

$$\left\{\frac{d}{dx}y(x) = x^2 + (y(x))^3, y(0) = 0\right\}$$

```
> nsol := dsolve(IVP,y(x),numeric);
```

$$nsol := \textbf{proc}(\texttt{rkf45\_x}) \ \ldots \ \textbf{end proc}$$

Observe that the call to `dsolve` (with the `numeric` option) has returned a MAPLE procedure. The *rkf45* refers to the Frehlberg fourth-fifth order Runge-Kutta method which is MAPLE's default method for numerical solutions. Now let's compute some values of the solution.

```
> for i from 0 to 9 do
>   nsol(i/10.);
```

```
> end do;
```

$$[x = 0., \ y(x) = 0.]$$
$$[x = .1000000000, \ y(x) = .000333333333579307896]$$
$$[x = .2000000000, \ y(x) = .00266666703929017805]$$
$$[x = .3000000000, \ y(x) = .00900002181584397612]$$
$$[x = .4000000000, \ y(x) = .0213337214734835162]$$
$$[x = .5000000000, \ y(x) = .0416702833992176336]$$
$$[x = .6000000000, \ y(x) = .0720224053831288041]$$
$$[x = .7000000000, \ y(x) = .114438119153382908]$$
$$[x = .8000000000, \ y(x) = .171065984044121100]$$
$$[x = .9000000000, \ y(x) = .244303629574327802]$$

The default numerical method is `rkf45`. MAPLE knows many other numerical methods, including the classical Euler methods, the seventh-eighth order Runge-Kutta method, and the Burlirsch-Stoer rational extrapolation method. See `?dsolve,numeric` for more details.

We can plot numerical solutions using the `odeplot` function in the *plots* package.

```
> with(plots):
> odeplot(nsol,[x,y(x)],0..1,title="Numerical solution to
   dy/dx=x^2+y^3, y(0)=0");
```

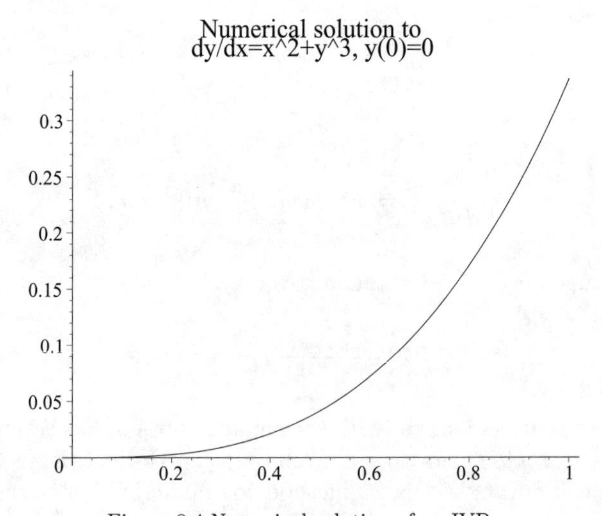

Figure 8.4 Numerical solution of an IVP.

8.4 Second- and higher order linear DEs

8.4.1 Constant coefficients

The `constantcoeffsols` function in the *DEtools* package returns a basis of the space of solutions to a homogeneous linear DE with constant coefficients.

```
> with(DEtools):
> de:=diff(y(x),x,x)+2*diff(y(x),x)+4*y(x)=0;
```

$$\frac{d^2}{dx^2}y(x) + 2\,\frac{d}{dx}y(x) + 4\,y(x) = 0$$

```
> constcoeffsols(de,y(x));
```

$$[e^{-x}\sin(\sqrt{3}x), e^{-x}\cos(\sqrt{3}x)]$$

We see that the functions

$$y_1 = e^{-x}\sin(\sqrt{3}x), \quad y_2 = e^{-x}\cos(\sqrt{3}x)$$

form a basis for the solution space of the homogeneous DE

$$\frac{d^2y}{dx^2} + 2\,\frac{dy}{dx} + 4y = 0.$$

Hence, the general solution is

$$y = c_1e^{-x}\sin\sqrt{3}x + c_2e^{-x}\cos\sqrt{3}x,$$

where c_1, c_2 are any constants.

8.4.2 Variation of parameters

Variation of parameters is a method for computing a particular solution to a nonhomogeneous linear DE, given a basis of solutions for the corresponding homogeneous linear DE. The corresponding MAPLE function is `varparam` from the *DEtools* package. Consider the DE

$$\frac{d^2y}{dx^2} + y = \tan x.$$

Two independent solutions to the corresponding homogeneous equation are

$$y_1 = \cos x \quad y_2 = \sin x.$$

```
> with(DEtools):
> varparam([cos(x),sin(x)],tan(x),x);
```

$$\_C_1\cos(x) + \_C_2\sin(x) - \cos(x)\ln(\sec(x) + \tan(x))$$

We see MAPLE has found that the general solution of our nonhomogeneous DE is

$$y = c_1 \cos x + c_2 \sin x - \cos x \ln(\sec x + \tan x),$$

where c_1, c_2 are any constants. The general syntax of `varparam` is

`varparam(`*sols*`,` *v*`,` *ivar*`)`

where *sols* is a list of independent solutions providing a basis for the corresponding homogeneous equation, v is the function on the right side of the nonhomogeneous equation, and *ivar* is the independent variable. A general solution to the nonhomogeneous equation is returned.

8.4.3 Reduction of order

Reduction of order is a method for reducing the order of a given linear homogeneous DE when a nontrivial solution is known. The `reduceOrder` function in the *DEtools* package does the job. Suppose we know that $y_1 = e^x$ is a solution to the DE

$$y''' + y'' + 3y' - 5y = 0.$$

We can use reduction of order to obtain a second-order equation.

```
> with(DEtools):
> de:=diff(y(x),x,x,x)+diff(y(x),x,x)+3*diff(y(x),x)
   -5*y(x)=0;
```

$$de := \frac{\partial^3}{\partial x^3} y(x) + \frac{\partial^2}{\partial x^2} y(x) + 3\frac{\partial}{\partial x} y(x) - 5y(x) = 0$$

```
> reduceOrder(de,y(x),exp(x));
```

$$\frac{d^2}{dx^2}y(x) + 4\frac{d}{dx}y(x) + 8\,y(x)$$

```
> dsolve(%,y(x));
```

$$y(x) = \_C1\; e^{-2\,x} \sin(2\,x) + \_C2\; e^{-2\,x} \cos(2\,x)$$

We need two further independent solutions, y_2, y_3. Via reduction of order we see that they must satisfy

$$y'' + 4y' + 8y = 0.$$

We can solve this equation to find $y_2 = e^{-2x} \cos 2x$, $y_3 = e^{-2x} \sin 2x$, so that the general solution of our third-order equation is

$$y = c_1 e^x + c_2 e^{-2x} \cos 2x + c_3 e^{-2x} \sin 2x,$$

where c_1, c_2, c_3 are any constants.

8.5 Series solutions

Series solutions to ordinary differential equations can be found using the `dsolve` function with the `type=series` option. Consider the DE

$$(1+x^2)\frac{d^2y}{dx^2} + 3x\frac{dy}{dx} + y = 0.$$

```
> de:=(1+x^2)*diff(y(x),x,x) + 3*x*diff(y(x),x) + y(x)
    = 0;
```

$$de := (1+x^2)\frac{\partial^2}{\partial x^2}\,y(x) + 3\,x\frac{\partial}{\partial x}\,y(x) + y(x) = 0$$

```
> dsolve(de,y(x),type=series);
```

$$y(x) = y(0) + \mathtt{D}(y)(0)x - \frac{1}{2}\,y(0)x^2 - \frac{2}{3}\,\mathtt{D}(y)(0)x^3 + \frac{3}{8}\,y(0)x^4 + \frac{8}{15}\,\mathtt{D}(y)(0)x^5 + O\left(x^6\right)$$

When no initial conditions are given, the series is computed about the origin. By default, the first six terms of the series are returned. Let's compute the solution satisfying the initial condition

$$y(0) = 1, \quad y'(0) = 0.$$

```
> dsolve({de,y(0)=1,D(y)(0)=0},y(x),type=series);
```

$$y(x) = 1 - \frac{1}{2}x^2 + \frac{3}{8}x^4 + O\left(x^6\right)$$

To compute more terms in the series, we change the value of the `Order` environment variable.

```
> Order:=10;
```

$$10$$

```
> dsolve({de,y(0)=1,D(y)(0)=0},y(x),type=series);
```

$$y(x) = (1 - \frac{1}{2}x^2 + \frac{3}{8}x^4 - \frac{5}{16}x^6 + \frac{35}{128}x^8 + O\left(x^{10}\right))$$

Next we solve the IVP

$$(x^2 - 3x)\frac{d^2y}{dx^2} + 2x\frac{dy}{dx} + y = 0, \quad y(1) = 4,\ y'(1) = -2.$$

```
> Order := 6:
> de:=(x^2-3*x)*diff(y(x),x,x) + 2*x*diff(y(x),x) + y(x)
   = 0;
```

$$\left(x^2 - 3\,x\right)\frac{d^2}{dx^2}y(x) + 2\,x\frac{d}{dx}y(x) + y(x) = 0$$

```
> dsol := dsolve({de,y(1)=4,D(y)(1)=-2},y(x),type=series);
```

$$dsol := y(x) = 4 - 2\,(x-1) - \frac{1}{2}\,(x-1)^3 - \frac{13}{80}\,(x-1)^5 + O\left((x-1)^6\right)$$

Because the initial conditions are at $x = 1$, the series returned is about $x = 1$. We can plot this solution by first converting it to a polynomial. See Figure 8.5.

```
> psol:=convert(rhs(dsol), polynom);
```

$$y(x) = 6 - 2\,x - 1/2\,(x-1)^3 - \frac{13}{80}\,(x-1)^5$$

```
> plot(psol,x=0..2);
```

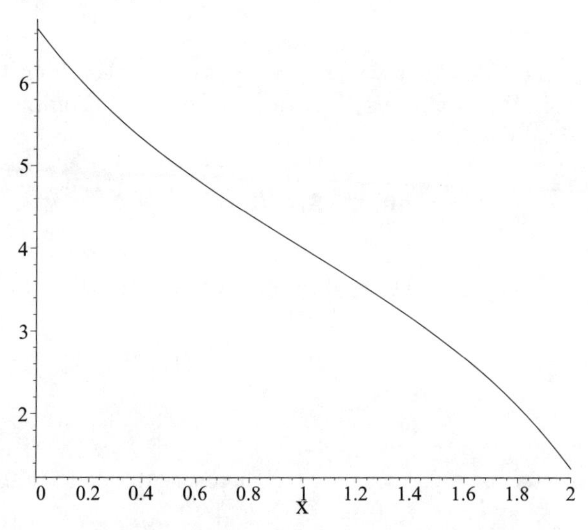

Figure 8.5 Plot of a series solution.

The `dsolve` function uses several methods when trying to find a series solution to an ODE or a system of ODEs. When initial conditions are given, the series is calculated at the given point; otherwise, the series is calculated at the origin, which is assumed to be an ordinary point.

8.5.1 The method of Frobenius

The method of Frobenius is a method for solving a second-order DE of the form

$$(t-t_0)^2 y''(t) + (t-t_0)P(t)y'(t) + Q(t)y(t) = 0,$$

where $P(t)$, and $Q(t)$ are analytic near $t = t_0$. The point t_0 is called a regular singular point. The regular singular points can be found using the `regularsp` in the *DEtools* package. The form of the solution depends on the roots of the indicial equation. This equation can be computed using the `indicialeq` function in the *DEtools* package. As an example, we can consider the hypergeometric equation:

$$t(1-t)y''(t) + (c - (1+a+b)t)y'(t) + aby(t) = 0,$$

where a, b, c are constants.

```
> with(DEtools):
> DE:=t*(1-t)*diff(y(t),t,t)+(c - (1+a+b)*t)*diff(y(t),t)
  -a*b*y(t)=0;
```

$$DE := t\,(1-t)\,\frac{d^2}{dt^2}y(t) + (c - (1+a+b)\,t)\,\frac{d}{dt}y(t) - aby(t) = 0$$

```
> regularsp(DE,t,y(t));
```

$$[0, 1]$$

```
> indicialeq(DE,t,0,y(t));
```

$$t^2 + (-1+c)\,t = 0$$

```
> solve(%,t);
```

$$0, \;\; 1-c$$

The points $t = 0, 1$ are regular singular points of the hypergeometric equation. For the point $t = 0$ the roots of the indicial equation

$$r^2 + (-1+c)r = 0,$$

are

$$r = 0, \;\; 1-c.$$

This means that if c is not an integer, there are two independent solutions of the hypergeometric equation of the form

$$y_1(t) = \sum_{n=0}^{\infty} a_n t^n, \quad y_2(t) = t^{1-c} \sum_{n=0}^{\infty} b_n t^n.$$

The syntax of the `indicialeq` function has the form

`indicialeq(DE,` *independent variable*`,` t_0`,` *dependent variable*`)`

where the point $t = t_0$ is a regular singular point of `DE` with specified independent and dependent variables.

8.6 The Laplace transform

Recall that the Laplace transform, $F(s) = \mathcal{L}\{f\}(s)$, of a function $f(t)$ is defined by

$$F(s) = \int_0^\infty e^{(-s\,t)}\, f(t)\, dt.$$

We need the MAPLE *inttrans* package:

```
> with(inttrans);
```

$$[addtable, fourier, fouriercos, fouriersin, hankel, hilbert, invfourier, invhilbert, invlaplace, invmellin, laplace, mellin, savetable]$$

To find the Laplace transform of a function $f(t)$ we use the command `laplace(f(t),t,s)`. As an example we find the Laplace transform of

$$f(t) = t\, e^{3t}.$$

```
> laplace(t*exp(3*t),t,s);
```

$$\frac{1}{(s-3)^2}$$

To find the inverse Laplace transform, we use the `invlaplace` function. It has the following syntax:

```
invlaplace(F(s), s, t)
```

As an example we find the inverse Laplace transform of the function

$$F(s) = \frac{s-1}{s^2+s+6}.$$

```
> invlaplace((s-1)/(s^2+s+6),s,t);
```

$$-\frac{3}{23}\, e^{-1/2\,t}\sqrt{23}\sin(1/2\,\sqrt{23}t) + e^{-1/2\,t}\cos(1/2\,\sqrt{23}t)$$

We found that

$$\mathcal{L}^{-1}\left\{\frac{s-1}{s^2+s+6}\right\}(t)$$
$$= -\frac{3}{23}\, e^{-1/2\,t}\sqrt{23}\sin(1/2\,\sqrt{23}t) + e^{-1/2\,t}\cos(1/2\,\sqrt{23}t).$$

8.6.1 The Heaviside function

In practice, the Laplace transform is useful for solving initial value problems which involve functions with jump discontinuities. The Heaviside function is the building block of such functions. The Heaviside function $H(t)$, sometimes called the unit step function, is defined by

$$H(t) := \begin{cases} 0 & \text{if } t < 0, \\ 1 & \text{if } t > 0. \end{cases}$$

In MAPLE the Heaviside function is denoted by **Heaviside**(t). Let's plot the Heaviside function on the interval $[-1, 1]$. See Figure 8.6.

```
> plot(Heaviside(t),t=-1..1,discont=true,thickness=3,
   title="The Heaviside Function H(t)");
```

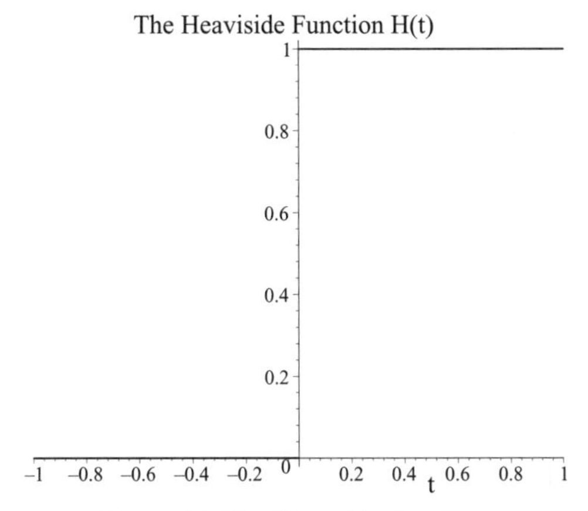

Figure 8.6 The Heaviside function.

To help with solving an initial value problem using Laplace transforms, we enter the following:

```
> restart:
> with(inttrans):
> dy:=diff(y(t),t);
```

$$dy := \frac{\partial}{\partial t} y(t)$$

```
> ddy:=diff(dy,t);
```

$$ddy := \frac{\partial^2}{\partial t^2} y(t)$$

```
> dddy:=diff(ddy,t);
```

$$dddy := \frac{\partial^3}{\partial t^3} y(t)$$

```
> addtable(laplace,y(t),Y(s),t,s);
> laplace(y(t),t,s);
```

$$Y(s)$$

```
> laplace(dy,t,s);
```

$$sY(s) - y(0)$$

```
> laplace(ddy,t,s);
```

$$s\,(sY(s) - y(0)) - \mathtt{D}(y)(0)$$

We have used the `addtable` function to tell MAPLE that we want to denote the Laplace transform of $y(t)$ by $Y(s)$. Also observe that MAPLE correctly returned the Laplace transforms of $y(t)$, $y'(t)$ and $y''(t)$.

We are now ready to solve an IVP using Laplace transforms. Consider the following initial value problem:

$$y''(t) + 4y(t) = f(t), \quad y(0) = 0, \ y'(0) = 0,$$

where

$$f(t) = \begin{cases} 0 & \text{if } 0 \le t < \pi, \\ \sin t & \text{if } \pi \le t < 2\pi, \\ 0 & \text{if } t > 2\pi. \end{cases}$$

We can easily write $f(t)$ in terms of the Heaviside function:

$$f(t) = \sin t\,(H(t-\pi) - H(t-2\pi)).$$

We assign `f` to the right side of the DE:

```
> f := sin(t)*(Heaviside(t-Pi)-Heaviside(t-2*Pi));
```

$$f := \sin(t)\,(\mathit{Heaviside}(t-\pi) - \mathit{Heaviside}(t-2\pi))$$

Next we input the initial conditions:

```
> y(0) := 0;
```

$$y(0) := 0$$

```
> D(y)(0) := 0;
```

$$\mathtt{D}(y)(0) := 0$$

Remember that `D(y)` means the derivative $y'(t)$. We observe that the left side of the DE is `ddy + 4*y(t)`:

```
> ddy + 4*y(t);
```

$$\frac{\partial}{\partial t}\,y(t) + 4\,y(t)$$

Now we take the Laplace transform of both sides of the DE:

```
> laplace(ddy+4*y(t),t,s) = laplace(f,t,s);
```

$$s(sY(s) - y(0)) - D(y)(0) + 4Y(s) = -\frac{e^{(-s\pi)}}{s^2+1} - \frac{e^{(-2s\pi)}}{s^2+1}$$

```
> simplify(%);
```

$$s^2Y(s) + 4\,Y(s) = -\frac{e^{-s\pi}\,(1 + e^{-s\pi})}{s^2+1}$$

Next we solve this equation for $Y(s)$.

```
> solve(%,Y(s));
```

$$-\frac{e^{-s\pi}\,(1 + e^{-s\pi})}{s^4 + 5\,s^2 + 4}$$

```
> S := factor(%);
```

$$S := -\frac{e^{-s\pi}\,(1 + e^{-s\pi})}{(s^2+4)\,(s^2+1)}$$

We see that

$$Y(s) = -\frac{e^{-s\pi}\,(1 + e^{-s\pi})}{(s^2+4)\,(s^2+1)}$$

To find the solution $y(t)$, we find the inverse Laplace transform:

```
> ysol := invlaplace(S,s,t);
```

$$ysol := \frac{1}{6}\,Heaviside(t-\pi)\sin(2\,t) + \frac{1}{3}\,\sin(t)Heaviside(t-\pi) + \frac{1}{6}\,Heaviside(t-2\,\pi)\sin(2\,t) - \frac{1}{3}\,\sin(t)Heaviside(t-2\,\pi)$$

which is our solution $y(t)$.

Of course, we could use `dsolve` to solve this initial value problem.

```
> y:='y':  D(y):='D(y)':
> IC := y(0) = 0, D(y)(0) = 0;
```

$$IC := y(0) = 0, D(y)(0) = 0$$

```
> DE := ddy + 4*y(t) = f:
> dsolve({DE,IC},y(t));
```

$$y(t) = \left(\frac{1}{6}\,Heaviside(t-\pi) + \frac{1}{6}\,Heaviside(t-2\,\pi)\right)\sin(2\,t) + \left(\frac{1}{3}\,Heaviside(t-\pi) - \frac{1}{3}\,Heaviside(t-2\,\pi)\right)\sin(t)$$

```
> ysol2:=rhs(%);
```

$$ysol2 := \left(\frac{1}{6}\,Heaviside(t-\pi) + \frac{1}{6}\,Heaviside(t-2\,\pi)\right)\sin(2\,t) + \left(\frac{1}{3}\,Heaviside(t-\pi) - \frac{1}{3}\,Heaviside(t-2\,\pi)\right)\sin(t)$$

```
> simplify(ysol - ysol2);
```

$$0$$

Happily we found the same solution.

8.6.2 The Dirac delta function

In MAPLE the Dirac delta function $\delta(t)$ is denoted by `Dirac(t)`.

```
> with(inttrans):
> assume(t0>=0):
> laplace(Dirac(t-t0),t,s);
```

$$e^{(-s\,t0\sim)}$$

MAPLE knows the Laplace transform for the delta function:

$$\mathcal{L}\{\delta(t-t_0)\} = \int_0^\infty e^{-st}\delta(t-t_0)\,dt = e^{-st_0},$$

for $t_0 \geq 0$. The delta function is used in IVPs, which involve an impulsive force. We consider the IVP:

$$y''(t) + 2y'(t) + y(t) = e^{-t} + 2\delta(t-1), \quad y(0) = 0,\ y'(0) = -1.$$

```
> restart:
> with(inttrans):
> dy:=diff(y(t),t):
> ddy:=diff(dy,t):
> addtable(laplace,y(t),Y(s),t,s):
> f := exp(-t) + 2*Dirac(t-1):
> y(0) := 0:
> D(y)(0) := -1:
> LS := ddy + 2*dy + y(t);
```

$$LS := \frac{\partial^2}{\partial t^2}\,y(t) + 2\frac{\partial}{\partial t}\,y(t) + y(t)$$

We take the Laplace transform of both sides of the DE:

```
> laplace(LS,t,s) = laplace(f,t,s);
```

$$s^2 Y(s) + 1 + 2\,sY(s) + Y(s) = (1+s)^{-1} + 2\,e^{-s}$$

We solve this equation for $Y(s)$.

```
> solve(%,Y(s));
```

$$\frac{-s + 2\,e^{-s} + 2\,e^{-s}s}{3\,s^2 + s^3 + 3\,s + 1}$$

```
> S := factor(%);
```

$$\frac{-s + 2\,e^{-s} + 2\,e^{-s}s}{(1+s)^3}$$

We found that

$$Y(s) = \mathcal{L}\{y(t)\} = \frac{-s + 2\,e^{-s} + 2\,e^{-s}s}{(1+s)^3}.$$

To find the solution $y(t)$, we find the inverse Laplace transform:

```
> invlaplace(S,s,t);
```

$$\frac{1}{2}\,t^2 e^{-t} - te^{-t} + 2\,\mathit{Heaviside}(t-1)e^{-t+1}t - 2\,\mathit{Heaviside}(t-1)e^{-t+1}$$

We found

$$y(t) = 2(t-1)e^{-t+1}H(t-1) + \frac{1}{2}\,t^2 e^{-t} - te^{-t}.$$

8.7 The *DEtools* package

In this chapter we have already seen many useful functions in the *DEtools* package. In this section we give a brief summary of the remaining functions.

8.7.1 DE plotting functions

`DEplot`

This plots the solution to a DE or the solutions to a system of DEs. For a first order DE it also plots the corresponding direction field. See Section 8.2.3.

`DEplot3d`

Plots the solution curve to a system of DEs. If the system involves two functions $y(t)$ and $z(t)$, the curve parametrized by $(x, y, z) = (t, y(t), z(t))$ is plotted in three dimensions.

`dfieldplot`

Plots the direction field of solutions $y(x)$ to a two-element system involving $x(t)$ and $y(t)$.

`phaseportrait`

Unfortunately this does not produce a true phase portrait. It seems to be just an alias for `DEplot`.

8.7.2 Dynamical systems

`generate_ic`

Generates a set of lists of initial conditions satisfying a given Hamiltonian constraint.

`hamilton_eqs`

Generates a sequence of Hamilton equations for a specified Hamiltonian.

`poincare`

Plots the projection of a Poincaré section of a specified Hamiltonian.

`zoom`

The `zoom` function allows for changing the ranges of the display of a given 2D/3D plot without having to recalculate it, thus saving time and memory resources.

8.7.3 DE manipulation

`DEnormal`

Returns a "normalized" form of a linear differential equation. Here normalized means an equivalent DE where coefficients are polynomials with no common factor.

`autonomous`

Determines whether a given DE or system of DEs is strictly autonomous.

`convertAlg`

Returns a coefficient list form for a DE.

`convertsys`

Converts a system of DEs to a first-order system.

`indicialeq`

Computes the indicial equation of a homogeneous linear DE. See Section 8.5.1.

`reduceOrder`

Implements the method of reduction of order. See Section 8.4.3.

`regularsp`

Computes the regular singular points of a specified linear second-order DE. See Section 8.5.1.

translate

This function takes a linear DE in $y(x)$ and returns a linear DE in $y(x - a)$, where a is a constant. The inverse of `translate` is `untranslate`.

Other functions include

convert_ODEs	Dchangevar	declare	DEnormal
dpolyform	hyperode	varparam	X

8.7.4 Lie symmetry methods

The *DEtools* package contains a subpackage of commands and routines for solving ODEs using integrating factors and Lie group symmetry methods, based on the work of Cheb-Terrab et al., [1] and [2]. The functions available are:

Xchange	Xcommutator	Xgauge	buildsol	buildsym
canoni	convert_ODEs	equinv	eta_k	firint
firtest	gensys	infgen	intfactor	invariants
line_int	muchange	mutest	normalG2	odeadvisor
odepde	redode	reduce_order	remove_RootOf	solve_group
symgen	symtest	transinv		

See `?DEtools,Lie` for more information.

8.7.5 Differential operators

MAPLE has facilities for computing with linear differential operators. Here a differential operator takes the form

$$L = R_0(x) + R_1(x)D + R_2(x)D^2 + \cdots + R_n(x)D^n,$$

where the $R_i(x)$ are rational functions in x and $D = \frac{d}{dx}$. There are several functions in the *DEtools* package for manipulating differential operators. See `diffop` for more information.

DFactor

Factors a specified linear differential operator.

mult

Computes a product of differential operators.

diffop2de

Applies a given differential operator to a function. When applied to an unknown function, the result is one side of a linear differential equation.

de2diffop

Converts a homogeneous linear ODE into a differential operator.

Other related functions include:

DFactorLCLM	GCRD	LCLM
adjoint	eigenring	endomorphism_charpoly
exterior_power	formal_sol	gen_exp
integrate_sols	leftdivision	rightdivision
symmetric_power	symmetric_product	

8.7.6 Closed form solutions

The *DEtools* package contains a collection of functions for finding solutions to DEs using special methods.

`DFactorsols`

Returns a basis of solutions of a linear homogeneous ODE by using `DFactor` to factor the corresponding linear differential operator.

`RiemannPsols`

Returns two independent solutions of a second-order linear homogeneous DE that has three regular singular points.

Other functions include:

abelsol	bernoullisol	chinisol	clairautsol
constcoeffsols	eulersols	exactsol	expsols
genhomosol	hypergeomsols	kovacicsols	liesol
linearsol	matrixDE	MeijerGsols	parametricsol
polysols	ratsols	riccatisol	separablesol

8.7.7 Simplifying DEs and `rifsimp`

There is a collection of functions for simplifying systems of DEs. See `rifsimp,overview` for more information. Functions include:

caseplot	checkrank	initialdata	maxdimsystems
rifread	rifsimp	rtaylor	

9. Linear Algebra

MAPLE can do symbolic and floating point matrix and linear algebra computations. There are two packages: the *linalg* package and the new *LinearAlgebra* package. The new *LinearAlgebra* package is more user-friendly for matrix algebra computations. It is also more efficient for numeric computations, especially with large matrices. The *linalg* package is recommended for more abstract computations. We will concentrate mainly on the *LinearAlgebra* package. Try

```
> ?LinearAlgebra
```

for an introduction to the *LinearAlgebra* package and a list of functions.

9.1 Vectors, Arrays, and Matrices

`Matrix`, `Array`, and `Vector` are the main data types used in the *LinearAlgebra* package. Note that the "M", "A" and "V" are capitalized. The lower-case `matrix`, `array`, and `vector` are used in the `linalg` package. `Matrix` and `Vector` are examples of what MAPLE calls an `rtable`. See `?rtable` for more information.

```
> Matrix(3);
```

$$\begin{bmatrix} 0 & 0 & 0 \\ 0 & 0 & 0 \\ 0 & 0 & 0 \end{bmatrix}$$

```
> Matrix(3,4);
```

$$\begin{bmatrix} 0 & 0 & 0 & 0 \\ 0 & 0 & 0 & 0 \\ 0 & 0 & 0 & 0 \end{bmatrix}$$

```
> Matrix(2,3,[[a,b,c],[d,e,f]]);
```

$$\begin{bmatrix} a & b & c \\ d & e & f \end{bmatrix}$$

```
> Matrix(2,3,[[a,b],[d,e,f]]);
```

$$\begin{bmatrix} a & b & 0 \\ d & e & f \end{bmatrix}$$

```
> Matrix(2,3,[[a,b],[c,d,e,f]]);
Error, (in Matrix) initializer defines more columns
(4) than column dimension parameter specifies (3)
> Matrix(2,3,[a,b,c,d,e,f]);
```

```
Error, (in Matrix) initializer defines more columns
(6) than column dimension parameter specifies (3)
```

The call `Matrix(m,n)` returns an $m \times n$ matrix of zeros. Observe matrix entries are assigned by a list of rows.

```
> W:=Vector(4);
```

$$W := \begin{bmatrix} 0 \\ 0 \\ 0 \\ 0 \end{bmatrix}$$

```
> V:=Vector([x,y,z]);
```

$$V := \begin{bmatrix} x \\ y \\ z \end{bmatrix}$$

The call `Vector(m)` returns an $m \times 1$ column vector of zeros. Observe that vector entries can be assigned using a list.

A fun way to create matrices is to use a function $f(x, y)$ of two variables. The function `Matrix(m,n,f)` produces the $m \times n$ matrix whose (i, j)th entry is $f(i, j)$.

```
> f := (i,j) -> x^(i*j);
```

$$F := (i, j) \mapsto x^{ij}$$

```
> A := Matrix(2,2,f);
```

$$A := \begin{bmatrix} x & x^2 \\ x^2 & x^4 \end{bmatrix}$$

Now try

```
> A := Matrix(4,4,f);
> factor(LinearAlgebra[Determinant](A));
```

The `map` function also works on matrices. Let's form a 5×5 matrix of the integers from 1 to 25.

```
> M:=Matrix(5,(i,j)->5*i+j-5);
```

$$M := \begin{bmatrix} 1 & 2 & 3 & 4 & 5 \\ 6 & 7 & 8 & 9 & 10 \\ 11 & 12 & 13 & 14 & 15 \\ 16 & 17 & 18 & 19 & 20 \\ 21 & 22 & 23 & 24 & 25 \end{bmatrix}$$

Now let's use `map` and `ithprime` to form a table of the first 25 primes:

```
> map(ithprime,M);
```

$$\begin{bmatrix} 2 & 3 & 5 & 7 & 11 \\ 13 & 17 & 19 & 23 & 29 \\ 31 & 37 & 41 & 43 & 47 \\ 53 & 59 & 61 & 67 & 71 \\ 73 & 79 & 83 & 89 & 97 \end{bmatrix}$$

Of course, we could have done this without using `map`. Try

```
> Matrix(5,(i,j)->ithprime(5*i+j-5));
```

Try making a table of the first 100 primes:

```
> Matrix(10,(i,j)->ithprime(10*i+j-10));
```

9.1.1 Matrix and Vector entry assignment

It is easy to access entries in a matrix and reassign them.

```
> A:=Matrix(2,3,[[1,2,3],[5,10,16]]);
```

$$\begin{bmatrix} 1 & 2 & 3 \\ 5 & 10 & 16 \end{bmatrix}$$

```
> A[2,3];
```

$$16$$

The entry in the second row and third column is 16. Let's change it to 15.

```
> A[2,3]:=15;
```

$$15$$

```
> A;
```

$$\begin{bmatrix} 1 & 2 & 3 \\ 5 & 10 & 15 \end{bmatrix}$$

In general, `A[i,j]` refers to the ijth entry of the matrix A (i.e., the entry in the ith row and jth column). It is also possible to access a block of entries.

```
> A := Matrix(4,(i,j)->(i+j));
```

$$A := \begin{bmatrix} 2 & 3 & 4 & 5 \\ 3 & 4 & 5 & 6 \\ 4 & 5 & 6 & 7 \\ 5 & 6 & 7 & 8 \end{bmatrix}$$

```
> A[2..3,2..4];
```

$$\begin{bmatrix} 4 & 5 & 6 \\ 5 & 6 & 7 \end{bmatrix}$$

```
> B := Matrix(2,3,[[0,1,2],[3,4,5]]);
```

$$B := \begin{bmatrix} 0 & 1 & 2 \\ 3 & 4 & 5 \end{bmatrix}$$

```
> A[2..3,2..4]:=B;
```

$$A_{2..3,2..4} := \begin{bmatrix} 0 & 1 & 2 \\ 3 & 4 & 5 \end{bmatrix}$$

```
> A;
```

$$\begin{bmatrix} 2 & 3 & 4 & 5 \\ 3 & 0 & 1 & 2 \\ 4 & 3 & 4 & 5 \\ 5 & 6 & 7 & 8 \end{bmatrix}$$

In general, `A[a..b,c..d]` refers to the submatrix of A from rows a to b, and columns c to d. It is also possible to rearrange rows or columns.

```
> B:=Matrix(3,(i,j)->b[i,j]);
```

$$\begin{bmatrix} b_{1,1} & b_{1,2} & b_{1,3} \\ b_{2,1} & b_{2,2} & b_{2,3} \\ b_{3,1} & b_{3,2} & b_{3,3} \end{bmatrix}$$

```
> B[[3,2,2,1],1..3];
```

$$\begin{bmatrix} b_{3,1} & b_{3,2} & b_{3,3} \\ b_{2,1} & b_{2,2} & b_{2,3} \\ b_{2,1} & b_{2,2} & b_{2,3} \\ b_{1,1} & b_{1,2} & b_{1,3} \end{bmatrix}$$

Observe how we created a generic matrix B. The call `B[[3,2,2,1],1..3]` created a new matrix whose rows are rows 3, 2, 2, and 1 of matrix B. Observe how the second row was repeated. In general, we use the syntax `B[L1,L2]`, where $L1$, $L2$ are either lists or of the form $a..b$. Try

```
> A := Matrix(3,4,[[1,2,3,4],[2,4,6,8],[3,6,9,12]]);
> A[[3,2],[4,3,2]];
```

```
> V := Vector([a,b,c,d]);
> W := V[[3,2]];
```

9.1.2 The Matrix and Vector palettes

The **Matrix** palette contains buttons for entering matrices up to a 4×4. To show the **Matrix** palette: in the menu bar click on View, select `Palettes`, slide to `Matrix Palette` and release. The **Matrix** palette should appear in a separate window. See Figure 9.1 below.

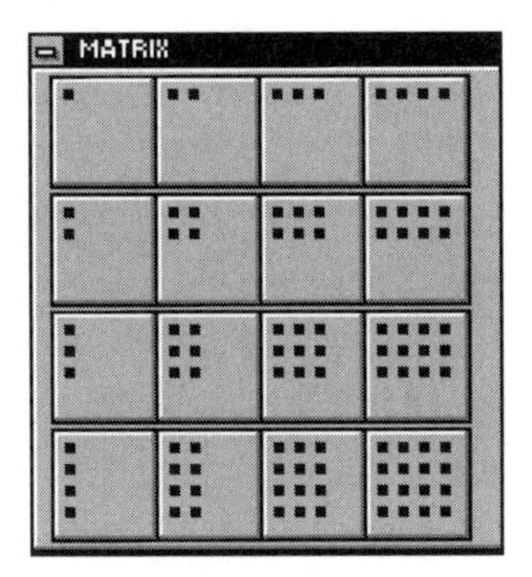

Figure 9.1 The **Matrix** palette.

Let's enter a 2×2 matrix. Click a place in the worksheet where you want to enter the matrix:

```
> |
```

Now click on [2×2 button]. A matrix template should appear in the worksheet:

```
> Matrix([[%?, %?], [%?, %?]]);
```

Type 23:

```
> Matrix([[23, %?], [%?, %?]]);
```

To get to the next entry location, press `Tab`.

```
> Matrix([[23, %?], [%?, %?]]);
```

Type `int(1/x,x=1..2)` and press `Tab`:

```
> Matrix([[23, int(1/x,x=1..2), [%?, %?]]);
```

Type 25 and press `Tab`:

```
> Matrix([[23, int(1/x,x=1..2), [25, %?]]);
```

Finally, type 27 and press `Enter`:

$$\begin{bmatrix} 23 & \ln(2) \\ 25 & 27 \end{bmatrix}$$

The **Vector** palette works in a similar way. In the menu bar, click on View, select `Palettes`, slide to `Vector Palette`, and release. The **Vector** palette should appear in a separate window. See Figure 9.2 below.

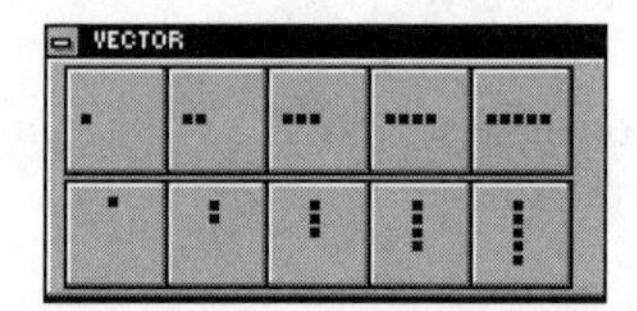

Figure 9.2 The Vector palette.

Let's enter a 3×1 row vector. Click a place in the worksheet where you want to enter the vector:

```
> |
```

Now click on [...]. A vector template should appear in the worksheet:

```
> <%? | %? | %?>;
```

Type `11`:

```
> <11 | %? | %?>;
```

Press `Tab` and type `12`:

```
> <11 | 12 | %?>;
```

Press `Tab`, type `13` and press `Enter`:

```
> <11 | 12 | 13>;
```

$$[11, 12, 13]$$

9.1.3 Matrix operations

MAPLE can do the usual matrix operations of addition, multiplication, scalar multiplication, inverse, transpose, and trace.

Matrix Operation	Mathematical Notation	MAPLE Notation
Addition	$A + B$	`A + B`
Subtraction	$A - B$	`A - B`
Scalar multiplication	$c\,A$	`c*A`
Matrix multiplication	$A\,B$	`A . B` or `Multiply(A,B)`
Matrix power	A^n	`A^n`
Inverse	A^{-1}	`A^(-1)` or `1/A` or `MatrixInverse(A)`
Transpose	A^T	`Transpose(A)`
Trace	$\mathrm{tr}\,A$	`Trace(A)`

We illustrate matrix addition, subtraction and scalar multiplication.

```
> A := Matrix(2,[[1,2],[3,4]]);
```

$$\begin{bmatrix} 1 & 2 \\ 3 & 4 \end{bmatrix}$$

```
> B := Matrix(2,[[-2,3],[-5,1]]);
```

$$\begin{bmatrix} -2 & 3 \\ -5 & 1 \end{bmatrix}$$

```
> A + B;
```

$$\begin{bmatrix} -1 & 5 \\ -2 & 5 \end{bmatrix}$$

```
> A - B;
```

$$\begin{bmatrix} 3 & -1 \\ 8 & 3 \end{bmatrix}$$

```
> 5*A;
```

$$\begin{bmatrix} 5 & 10 \\ 15 & 20 \end{bmatrix}$$

We continue with matrix multiplication, matrix power, and finding an inverse.

```
> A := Matrix(2,[[1,2],[3,4]]):
> B := Matrix(2,[[-2,3],[-5,1]]):
> A . B;
```

$$\begin{bmatrix} -12 & 5 \\ -26 & 13 \end{bmatrix}$$

```
> AI := 1/A;
```

$$\begin{bmatrix} -2 & 1 \\ \frac{3}{2} & -\frac{1}{2} \end{bmatrix}$$

```
> A . AI;
```

$$\begin{bmatrix} 1 & 0 \\ 0 & 1 \end{bmatrix}$$

```
> A^3;
```

$$\begin{bmatrix} 37 & 54 \\ 81 & 118 \end{bmatrix}$$

The functions `Multiply`, `MatrixInverse`, `Transpose`, and `Trace` are part of the *LinearAlgebra* package. Try

```
> with(LinearAlgebra);
```

to see a list of functions in the *LinearAlgebra* package.

```
> with(LinearAlgebra):
> A := Matrix(2,[[1,2],[3,4]]):
> B := Matrix(2,[[-2,3],[-5,1]]):
> Multiply(A , B);
```

$$\begin{bmatrix} -12 & 5 \\ -26 & 13 \end{bmatrix}$$

```
> Multiply(Multiply(A,A),A);
```

$$\begin{bmatrix} 37 & 54 \\ 81 & 118 \end{bmatrix}$$

```
> AI := MatrixInverse(A);
```

$$\begin{bmatrix} -2 & 1 \\ \frac{3}{2} & -\frac{1}{2} \end{bmatrix}$$

```
> Transpose(A);
```

$$\begin{bmatrix} 1 & 3 \\ 2 & 4 \end{bmatrix}$$

```
> Trace(A);
```

$$5$$

Now try the following:

```
> with(LinearAlgebra):
> A:=Matrix(2,3,[[1,2,3],[4,5,6]]);
> B:=Matrix(3,2,[[2,4],[-7,3],[5,1]]);
> C:=Matrix(2,2,[[1,-2],[-3,4]]);
> A . B;
> Multiply(A,B);
> A.B-2*C;
```

Now check your results with pencil and paper. You should have found that

$$AB - 2C = \begin{bmatrix} 1 & 17 \\ 9 & 29 \end{bmatrix}$$

9.1.4 Matrix and vector construction shortcuts

Angled brackets < > are used as a shortcut to construct matrices and vectors. We can construct a column vector:

```
> V := <1,2,3>;
```

$$V := \begin{bmatrix} 1 \\ 2 \\ 3 \end{bmatrix}$$

The construction `<a, b, c, ... >` gives a column vector when a, b, c, ... are scalars. We can construct a row vector:

```
> R := <1|2|3>;
```

$$R := \begin{bmatrix} 1 & 2 & 3 \end{bmatrix}$$

We can construct a matrix from column vectors:

```
> U := <a,b,c>;
```

$$\begin{bmatrix} a \\ b \\ c \end{bmatrix}$$

```
> V := <i,j,k>;
```

$$\begin{bmatrix} i \\ j \\ k \end{bmatrix}$$

```
> W := <x,y,z>;
```

$$\begin{bmatrix} x \\ y \\ z \end{bmatrix}$$

```
> M := <U | V | W>;
```

$$\begin{bmatrix} a & i & x \\ b & j & y \\ c & k & z \end{bmatrix}$$

Similarly, we can build a matrix from row vectors. Try the following:

```
> U := <a|b|c>;
> V := <i|j|k>;
> W := <x|y|z>;
> M := <U , V , W>;
```

Angled brackets can also be used to stack matrices.

```
> A:=Matrix(3,(i,j)->a^i*b^j):
> B:=Matrix(3,(i,j)->b^i*c^j):
> C:=Matrix(3,(i,j)->c^i*a^j):
> A,B,C;
```

$$\begin{bmatrix} ab & ab^2 & ab^3 \\ a^2b & a^2b^2 & a^2b^3 \\ a^3b & a^3b^2 & a^3b^3 \end{bmatrix}, \begin{bmatrix} bc & bc^2 & bc^3 \\ b^2c & b^2c^2 & b^2c^3 \\ b^3c & b^3c^2 & b^3c^3 \end{bmatrix}, \begin{bmatrix} ca & ca^2 & ca^3 \\ c^2a & c^2a^2 & c^2a^3 \\ c^3a & c^3a^2 & c^3a^3 \end{bmatrix}$$

Now we form a new matrix by stacking the matrices A, B, C, to the right of each other:

```
> <A|B|C>;
```

$$\begin{bmatrix} ab & ab^2 & ab^3 & bc & bc^2 & bc^3 & ca & ca^2 & ca^3 \\ a^2b & a^2b^2 & a^2b^3 & b^2c & b^2c^2 & b^2c^3 & c^2a & c^2a^2 & c^2a^3 \\ a^3b & a^3b^2 & a^3b^3 & b^3c & b^3c^2 & b^3c^3 & c^3a & c^3a^2 & c^3a^3 \end{bmatrix}$$

Similarly we can stack A above B:

```
> <A,B>;
```

$$\begin{bmatrix} ab & ab^2 & ab^3 \\ a^2b & a^2b^2 & a^2b^3 \\ a^3b & a^3b^2 & a^3b^3 \\ bc & bc^2 & bc^3 \\ b^2c & b^2c^2 & b^2c^3 \\ b^3c & b^3c^2 & b^3c^3 \end{bmatrix}$$

Now try stacking A, B, and C above each other:

```
> <A,B,C>;
```

9.1.5 Viewing large Matrices and Vectors

Only relatively small matrices and vectors will be displayed on the screen. For instance, a 50×20 matrix of the first 1000 primes is much too big to be displayed on the screen.

```
> M:=Matrix(50,20,(i,j)->ithprime(20*i+j-20));
```

$$M := \begin{bmatrix} 50 \times 20 \ \texttt{Matrix} \\ \texttt{Data Type: anything} \\ \texttt{Storage: rectangular} \\ \texttt{Order: Fortran\_order} \end{bmatrix}$$

Observe that this 50×20 matrix was not displayed on the screen. In its place is a matrix giving the dimensions and some information on `Data Type`, `Storage`, and `Order`. To view entries in this matrix, we can use the context menu, which we will discuss in more detail in the next section. First click the right button of the mouse on the matrix. A menu of options should appear:

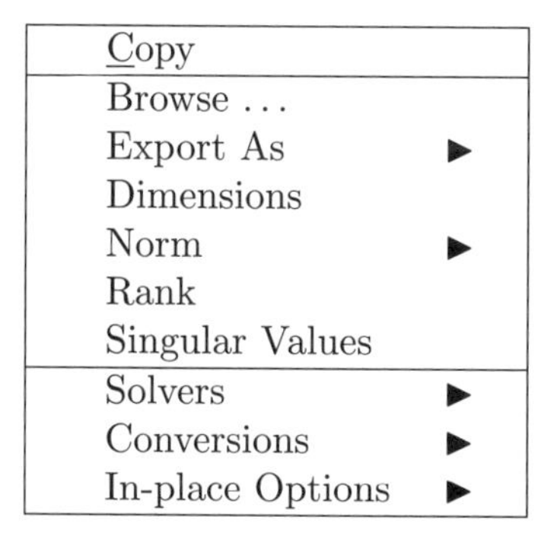

Click on `Browse ...`. **The Structured Data Browser** window should appear. See Figure 9.3 below.

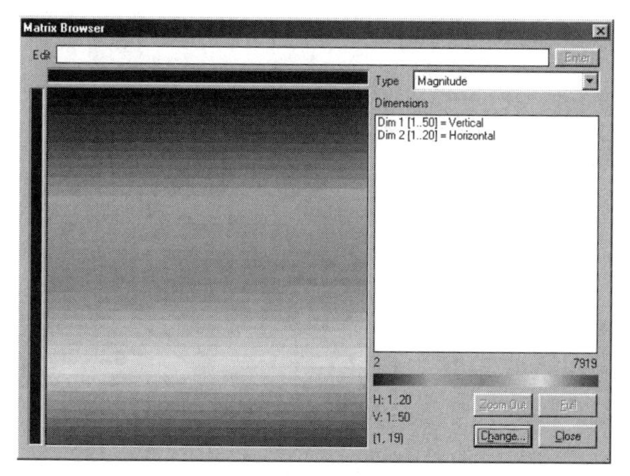

Figure 9.3 The Structured Data Browser.

You will see a view panel consisting of horizontal bars of different colors. Color corresponds to relative magnitude of corresponding entries in the matrix. Colors range from blue (small numbers) to red (large numbers). In the present matrix the bottom right section is the reddest — these corresponding to large prime numbers. You will also see an edit field, a box showing `Dimensions`, and four buttons `Zoom Out`, `Full`, `Change ...` and `Close`. To the left of these buttons you will see three lines of information:

```
H: 1..20
V: 1..50
(20,40)
```

The pair of numbers `(20,40)` correspond with the current position of the mouse in the view panel. Move the mouse around and you will see its value change. You can view a particular section of the matrix by clicking the left button of the mouse in the view panel, holding it down to form a rectangle, and releasing. The view panel should now show an array of numbers corresponding to the subblock of the matrix selected. For instance, if you selected the block corresponding to

```
H: 10..11
V: 26..29
(10,29)
```

you will see a 4×2 array of numbers:

$$\begin{bmatrix} 3643 & 3659 \\ 3821 & 3823 \\ 3989 & 4001 \\ 4139 & 4153 \end{bmatrix}$$

These correspond to the submatrix `M[26..29,10..11]`. Any entry clicked will show up in the edit field. You can change an entry by clicking in the edit field and changing its value. For instance, click on the top left entry, change the number in the edit field to 0 and then press `Enter`. The entry `M[26,10]` has been changed to zero. To check this, press `Close`

```
> M := _rtable[17353556];
```

and try

```
> M[26..29,10..11];
```

$$\begin{bmatrix} 0 & 3659 \\ 3821 & 3823 \\ 3989 & 4001 \\ 4139 & 4153 \end{bmatrix}$$

See `?view,Array` for more information on using **The Structured Data Browser**.

9.2 Matrix context menu

Enter the following matrix:

```
> M:=Matrix(4,(i,j)->2^(i*j));
```

$$M := \begin{bmatrix} 2 & 4 & 8 & 16 \\ 4 & 16 & 64 & 256 \\ 8 & 64 & 512 & 4096 \\ 16 & 256 & 4096 & 65536 \end{bmatrix}$$

Click the right button of the mouse on the matrix. A menu of options should appear:

Copy	
Browse ...	
Export As	▶
Transpose	
Dimensions	
Norm	▶
Rank	
Select Element	▶
Determinant	
Inverse	
Trace	
Characteristic Polynomial	
Eigenvalues	
Eigenvectors	
Singular Values	
Solvers	▶
Conversions	▶
In-place Options	▶

Most of the functions in the menu are self-explanatory. Let's try clicking on `Dimensions`:

```
> R0 := [LinearAlgebra:-Dimensions(_rtable[5673280])];
```

$$R0 := [4, 4]$$

Observe that the result appeared as a MAPLE command in the worksheet together with the output $[4, 4]$, which was assigned the name `R0`. Here `Dimensions` means the number of rows and columns. Now we click on `Determinant`:

```
> R1 := [LinearAlgebra:-Determinant(_rtable[5673280])];
```

$$R1 := 66060288$$

The determinant of our matrix is 66060288. The new MAPLE command and output has appeared just below our matrix and above the previous work. Let's factor the determinant:

```
> ifactor(R1);
```

$$(2)^{20} (3)^2 (7)$$

We saw `Browse` in the previous section.

9.2.1 The Export As submenu

`Export As` has a submenu:

Matlab ...
Matrix Market ...
Tab Delimited ...

In Section 7.10.5 we saw the function `MatrixExport`, the command line version. We can export a matrix in *Matlab*, *Matrix Market*, or *Tab Delimited* format. Let's save our matrix in *Tab Delimited* format. Click on `Tab Delimited ...`. A **Save As** window should appear. Type in a file name such as *matrix.txt* and press `OK`. You should now have a file called *matrix.txt*, which looks something like:

```
2       4       8       16
4       16      64      256
8       64      512     4096
16      256     4096    65536
```

Entries on each row are separated by the tab character.

9.2.2 The Norm submenu

`Norm` has a submenu:

1
Euclidean
infinity
Frobenius

See Section 9.11.2 or `?LinearAlgebra[Norm]` for more information on these matrix norms.

9.2.3 The Select Element submenu

When you select `Select Element` a menu will appear listing the indices of each entry in the matrix. In our example, the list will be `"1,1"`, `"1,2"`, ... , `"4,4"`. Try selecting `"2,3"`.

9.2.4 The Solvers submenu

`Solvers` has a submenu:

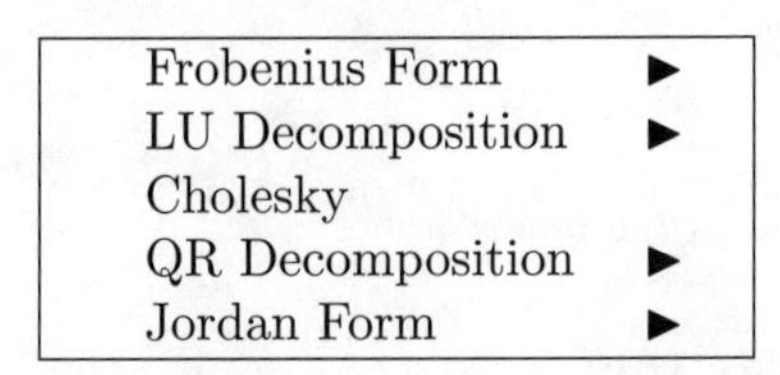

This menu provides various standard matrix decompositions.

`Frobenius Form` has a submenu:

Frobenius Form
Transformation Matrix

See Section 9.15 or `?LinearAlgebra[FrobeniusForm]` for more information on the Frobenius (rational canonical) form.

`LU Decomposition` has a submenu:

Gaussian Elimination
RREF
Fraction Free

See Section 9.4 or `?LinearAlgebra[LUDecomposition]` for more information.

`QR Decomposition` has a submenu:

QR Decomposition
Unitary Factor (Q)
Upper Triangular Factor (R)
Rank

See Section 9.13 or `?LinearAlgebra[QRDecomposition]` for more information.

`Jordan Form` has a submenu:

Jordan Form
Transformation Matrix

See Section 9.10 or `?LinearAlgebra[JordanForm]` for more information.

9.2.5 The Conversions submenu

`Conversions` has a submenu:

Approximate ▶
Maple
LaTeX
C Language
FORTRAN
Data Type ▶

`Approximate` has a submenu:

5
10
20
50
100

This provides floating point approximations to 5, 10, 20, 50, or 100 digits. For example, enter the following matrix:

```
> M:=<<Pi,exp(1)>|<log(2),int(1/(sqrt(1+x^10)),x=0..1)>>;
```

$$M := \begin{bmatrix} \pi & \ln(2) \\ e^1 & \int_0^1 \frac{1}{\sqrt{1+x^{10}}}\,dx \end{bmatrix}$$

Right-click on the matrix, and then select `Conversions`, `Approximate`, `10`:

```
> R0 := evalf(M,10);
```

$$R0 := \begin{bmatrix} 3.141592654 & 0.6931471806 \\ 2.718281828 & 0.9662361773 \end{bmatrix}$$

`Data Type` has a submenu:

Maple Float ▶
Hardware Float
Complex Maple Float ▶
Complex Hardware Float

This provides options for numeric computation.

Both `Maple Float` and `Complex Maple Float` have the submenu:

5
10
20
50
100

9.2.6 The In-place Options submenu

`In-place Options` has a submenu:

Set to Readonly
C Order

Selecting `Set to Readonly` means entries in the matrix cannot be changed. The default order for storing matrices is *Fortran Order* (by columns). Selecting `C Order` changes the storage mode to *C Order* (by rows). Selecting the button again will bring the matrix back to *Fortran Order*.

9.3 Elementary row and column operations

MAPLE can perform all the elementary row and column operations.

Elementary Row Operation	Operational Notation	MAPLE Notation
Swap two rows	$R_i \longleftrightarrow R_j$	`RowOperation(A,[i,j]`
Multiply a row by constant	$R_i \longrightarrow c\,R_i$	`RowOperation(A,i,c)`
Add a multiple of one row to another	$R_j \longrightarrow R_j + c\,R_i$	`RowOperation(A,[j,i],c)`

Let

$$A = \begin{bmatrix} 1 & 1 & 3 & -3 \\ 5 & 5 & 13 & -7 \\ 3 & 1 & 7 & -11 \end{bmatrix}$$

Let's perform the row operation $R_2 \longrightarrow R_2 - 5\,R_1$ (i.e., replace the second row by the sum of the second row and -5 times the first row).

```
> with(LinearAlgebra):
> A:=Matrix([[1, 1, 3, -3],
> [5, 5, 13, -7],
> [3, 1, 7, -11]]);
```

$$A := \begin{bmatrix} 1 & 1 & 3 & -3 \\ 5 & 5 & 13 & -7 \\ 3 & 1 & 7 & -11 \end{bmatrix}$$

```
> RowOperation(A,[2,1],-5);
```

$$\begin{bmatrix} 1 & 1 & 3 & -3 \\ 0 & 0 & -2 & 8 \\ 3 & 1 & 7 & -11 \end{bmatrix}$$

```
> A;
```

$$\begin{bmatrix} 1 & 1 & 3 & -3 \\ 5 & 5 & 13 & -7 \\ 3 & 1 & 7 & -11 \end{bmatrix}$$

Notice that a new matrix was created. The original matrix A did not change. To replace the original matrix we can use the `inplace` option.

```
> RowOperation(A,[2,1],-5,inplace=true);
```

$$\begin{bmatrix} 1 & 1 & 3 & -3 \\ 0 & 0 & -2 & 8 \\ 3 & 1 & 7 & -11 \end{bmatrix}$$

```
> A;
```

$$\begin{bmatrix} 1 & 1 & 3 & -3 \\ 0 & 0 & -2 & 8 \\ 3 & 1 & 7 & -11 \end{bmatrix}$$

The **inplace** option should be used with caution. If the operation fails, the original matrix may become corrupted. Let's restart a MAPLE session and reduce the matrix A to row echelon form using elementary row operations.

```
> restart:
> with(LinearAlgebra):
> A:=Matrix([[1, 1, 3, -3], [5, 5, 13, -7],
    [3, 1, 7, -11]]);
```

$$A := \begin{bmatrix} 1 & 1 & 3 & -3 \\ 5 & 5 & 13 & -7 \\ 3 & 1 & 7 & -11 \end{bmatrix}$$

```
> A1 := RowOperation(A,[2,1],-5);
```

$$A1 := \begin{bmatrix} 1 & 1 & 3 & -3 \\ 0 & 0 & -2 & 8 \\ 3 & 1 & 7 & -11 \end{bmatrix}$$

```
> A2 := RowOperation(A1,[3,1],-3);
```

$$A2 := \begin{bmatrix} 1 & 1 & 3 & -3 \\ 0 & 0 & -2 & 8 \\ 0 & -2 & -2 & -2 \end{bmatrix}$$

```
> A3 := RowOperation(A2,[2,3]);
```

$$A3 := \begin{bmatrix} 1 & 1 & 3 & -3 \\ 0 & -2 & -2 & -2 \\ 0 & 0 & -2 & 8 \end{bmatrix}$$

Now try the following elementary row operations to continue the reduction to reduced row echelon form.

```
> A4:=RowOperation(A3,2,-1/2);
> A5:=RowOperation(A4,3,-1/2);
> A6:=RowOperation(A5,[2,3],-1);
> A7:=RowOperation(A6,[1,3],-3);
> A8:=RowOperation(A7,[1,2],-1);
```

This last matrix should be

$$A8 = \begin{bmatrix} 1 & 0 & 0 & 4 \\ 0 & 1 & 0 & 5 \\ 0 & 0 & 1 & -4 \end{bmatrix},$$

which is in reduced row echelon form.

In MAPLE the elementary column operations are done in a similar fashion. The analogous function is `ColumnOperation`.

9.4 Gaussian elimination

MAPLE can do Gaussian and Gauss-Jordan elimination. We need the `LUDecomposition` function in the *LinearAlgebra* package. In the previous section we reduced a matrix to echelon form using elementary row operations. In this section we check our results. To reduce A to row echelon form we use the command

`LUDecomposition(A,output='U')`

```
> with(LinearAlgebra):
> A:=Matrix([[1, 1, 3, -3], [5, 5, 13, -7],
    [3, 1, 7, -11]]):
> LUDecomposition(A,output='U');
```

$$\begin{bmatrix} 1 & 1 & 3 & -3 \\ 0 & -2 & -2 & -2 \\ 0 & 0 & -2 & 8 \end{bmatrix}$$

This confirms our earlier computation using row operations. To reduce A to reduced row echelon form, we use the command

`LUDecomposition(A,output='R')`

```
> LUDecomposition(A,output='R');
```

$$\begin{bmatrix} 1 & 0 & 0 & 4 \\ 0 & 1 & 0 & 5 \\ 0 & 0 & 1 & -4 \end{bmatrix}$$

This should agree with the matrix `A8` obtained in the previous section. We will discuss the `LUDecomposition` function in more detail later in Section 9.14.

9.5 Inverses, determinants, minors, and the adjoint

To find the inverse of a matrix and its determinant, we use the functions `MatrixInverse` and `Determinant` in the *LinearAlgebra* package.

```
> with(LinearAlgebra):
> A:= Matrix([[1,1,3],[5,5,13],[3,1,7]]);
```

$$A := \begin{bmatrix} 1 & 1 & 3 \\ 5 & 5 & 13 \\ 3 & 1 & 7 \end{bmatrix}$$

```
> Determinant(A);
```

$$-4$$

```
> B := MatrixInverse(A);
```

$$B := \begin{bmatrix} -\frac{11}{2} & 1 & \frac{1}{2} \\ -1 & \frac{1}{2} & -\frac{1}{2} \\ \frac{5}{2} & -\frac{1}{2} & 0 \end{bmatrix}$$

We first found that $\det(A) = -4 \neq 0$, so that A is invertible, then found that

$$A^{-1} = \begin{bmatrix} \frac{-11}{2} & 1 & \frac{1}{2} \\ -1 & \frac{1}{2} & \frac{-1}{2} \\ \frac{5}{2} & \frac{-1}{2} & 0 \end{bmatrix}.$$

Now check your answer.

```
> B.A;
```

Did you get the identity matrix?

To compute the adjoint of a matrix we use the `Adjoint` function.

```
> with(LinearAlgebra):
> A := Matrix([[1,1,3],[5,5,13],[3,1,7]]):
> C := Adjoint(A);
```

$$C := \begin{bmatrix} 22 & -4 & -2 \\ 4 & -2 & 2 \\ -10 & 2 & 0 \end{bmatrix}$$

We found that

$$\operatorname{adj} A = \begin{bmatrix} 22 & -4 & -2 \\ 4 & -2 & 2 \\ -10 & 2 & 0 \end{bmatrix}.$$

Now check your answer:

```
> C.A;
```

Did you get a diagonal matrix?

The function `Minor(A,i,j)` returns the (i, j)th minor of the matrix A (i.e., the matrix obtained by deleting the ith row and jth column). Let's compute the $(2, 3)$th minor of our matrix A.

```
> with(LinearAlgebra):
> A := Matrix([[1,1,3],[5,5,13],[3,1,7]]):
> Minor(A,2,3);
```

$$\begin{bmatrix} 1 & 1 \\ 3 & 1 \end{bmatrix}$$

9.6 Special matrices and vectors

9.6.1 Band matrix

A *band* matrix is a matrix that is constant along each diagonal in a band. The syntax of the `BandMatrix` function has the form

`BandMatrix(`L`,`n`,`r`,`c`)`

L is a list of scalars, which are constants to appear along diagonals. n is the number of subdiagonals, r is the number of rows, and c is the number of columns.

```
> with(LinearAlgebra):
> BandMatrix([a,b,c,d],1,5,7);
```

$$\begin{bmatrix} b & c & d & 0 & 0 & 0 & 0 \\ a & b & c & d & 0 & 0 & 0 \\ 0 & a & b & c & d & 0 & 0 \\ 0 & 0 & a & b & c & d & 0 \\ 0 & 0 & 0 & a & b & c & d \end{bmatrix}$$

`BandMatrix([a,b,c,d],1,5,7)` produced a 5×7 matrix with one diagonal below the main diagonal. Try

```
> BandMatrix([a,b,c,d],2,5,7);
```

You should obtain a band matrix with two subdiagonals.

9.6.2 Constant matrices and vectors

The syntax of the `ConstantMatrix` function has the form

`ConstantMatrix(`s`, `r`, `c`)`

s is the constant, r is the number of rows, and c is the number of columns.

```
> with(LinearAlgebra):
> ConstantMatrix(-3,2,4);
```

$$\begin{bmatrix} -3 & -3 & -3 & -3 \\ -3 & -3 & -3 & -3 \end{bmatrix}$$

The syntax of the `ConstantVector` function has the form

```
ConstantVector(s, d)
ConstantVector[row](s, d)
```

s is the constant, and d is the number of entries in the vector. The argument `[row]` is optional and produces a row vector.

```
> with(LinearAlgebra):
> ConstantVector(11,2);
```

$$\begin{bmatrix} 11 \\ 11 \end{bmatrix}$$

```
> ConstantVector[row](11,3);
```

$$[11, \quad 11, \quad 11]$$

9.6.3 Diagonal matrices

The syntax of the `DiagonalMatrix` function has the form

```
DiagonalMatrix(V, r, c)
DiagonalMatrix(V, n)
```

V is a list of numbers or a list of matrices to be inserted along the diagonal. r is the number of rows and c is the number of columns. n is the number of rows for a square matrix.

```
> with(LinearAlgebra):
> DiagonalMatrix([a,b,c],3);
```

$$\begin{bmatrix} a & 0 & 0 \\ 0 & b & 0 \\ 0 & 0 & c \end{bmatrix}$$

Try these:

```
> DiagonalMatrix([a,b,c],4);
> DiagonalMatrix([<<a,c>|<b,d>>,e,
```

```
<<f,i,l>|<g,j,m>|<h,k,n>>]);
```

9.6.4 Givens rotation matrices

A *Givens rotation* matrix is a rotation in a plane determined by two coordinates. The syntax of the `GivensRotationMatrix` function has the form

`GivensRotationMatrix(`V`, `i`, `j`)`

V is a list of n numbers; i, j are positive integers corresponding to the coordinates of the plane being rotated. An $n \times n$ matrix is returned. For more details see `GivensRotationMatrix`. Try

```
> with(LinearAlgebra):
> V := <3,4,5>;
> GivensRotationMatrix(V, 1, 2);
```

9.6.5 Hankel matrices

A *Hankel* matrix is a symmetric $A = (a_{i,j})$, where $a_{i,j}$ is a function of $i + j$; i.e., constant on the diagonals $i + j = c$. The syntax of the `HankelMatrix` function has the form

`HankelMatrix(`L`)`
`HankelMatrix(`L`, `n`)`

L is a list of $2n - 1$ scalars that will appear on the the diagonals.

```
> with(LinearAlgebra):
> HankelMatrix([a,b,c,d,e]);
```

$$\begin{bmatrix} a & b & c \\ b & c & d \\ c & d & e \end{bmatrix}$$

Try

```
> HankelMatrix([a,b,c,d,e,f,g,h,i,j,k]);
> HankelMatrix([a,b,c,d,e,f,g,h,i,j,k],4);
```

9.6.6 Hilbert matrices

A generalized *Hilbert* matrix has the form $(1/(i + j - x))$. The syntax of the `HilbertMatrix` function has the form

`HankelMatrix(`n`)`
`HankelMatrix(`r`, `c`)`
`HankelMatrix(`r`, `c`, `x`)`

r is the number of rows and c is the number of columns. n is the number of rows for a square matrix. If x is not specified, it is assumed to be 1.

```
> with(LinearAlgebra):
> HilbertMatrix(3);
```

$$\begin{bmatrix} 1 & \frac{1}{2} & \frac{1}{3} \\ \frac{1}{2} & \frac{1}{3} & \frac{1}{4} \\ \frac{1}{3} & \frac{1}{4} & \frac{1}{5} \end{bmatrix}$$

Try

```
> HilbertMatrix(3,x);
> HilbertMatrix(4,5,x);
```

9.6.7 Householder matrices

A *Householder* matrix corresponds to reflection in a fixed hyperplane. If V is a vector, then `HouseholderMatrix(V)` returns the matrix of the transformation, which is the reflection in the hyperplane orthogonal to the vector V. See `?HouseholderMatrix` for more information.

```
> with(LinearAlgebra):
> V:=Vector([1,2,2]);
```

$$\begin{bmatrix} 1 \\ 2 \\ 2 \end{bmatrix}$$

```
> M := HouseholderMatrix(V);
```

$$\begin{bmatrix} \frac{7}{9} & -\frac{4}{9} & -\frac{4}{9} \\ -\frac{4}{9} & \frac{1}{9} & -\frac{8}{9} \\ -\frac{4}{9} & -\frac{8}{9} & \frac{1}{9} \end{bmatrix}$$

Now compute

```
> M.V;
```

Is this what you expected?

9.6.8 Identity matrix

The function `IdentityMatrix(n)` returns the $n \times n$ identity matrix.

```
> with(LinearAlgebra):
> IdentityMatrix(3);
```

$$\begin{bmatrix} 1 & 0 & 0 \\ 0 & 1 & 0 \\ 0 & 0 & 1 \end{bmatrix}$$

Try

```
> IdentityMatrix(4);
> IdentityMatrix(4,6);
```

9.6.9 Jordan block matrices

A *Jordan block* matrix is a square matrix of the form

$$\begin{bmatrix} \lambda & 1 & 0 & \cdots & 0 & 0 \\ 0 & \lambda & 1 & \cdots & 0 & 0 \\ 0 & 0 & \ddots & \ddots & 0 & 0 \\ \vdots & \vdots & & \lambda & 1 & 0 \\ 0 & 0 & \cdots & 0 & \lambda & 1 \\ 0 & 0 & \cdots & 0 & 0 & \lambda \end{bmatrix}$$

The command `JordanBlockMatrix([[`λ`,`n`]])` returns an $n \times n$ Jordan block matrix with eigenvalue λ. A Jordan matrix is a matrix with Jordan blocks along the diagonal and zeros elsewhere. The corresponding MAPLE command takes the form

`JordanBlockMatrix([[`λ_1`,`n_1`],[`λ_2`,`n_2`], ... ,[`λ_r`,`n_r`]])`

```
> with(LinearAlgebra):
> JordanBlockMatrix([[-1,2],[5,3]]);
```

$$\begin{bmatrix} -1 & 1 & 0 & 0 & 0 \\ 0 & -1 & 0 & 0 & 0 \\ 0 & 0 & 5 & 1 & 0 \\ 0 & 0 & 0 & 5 & 1 \\ 0 & 0 & 0 & 0 & 5 \end{bmatrix}$$

The matrix above has two Jordan blocks: one 2×2 block with eigenvalue $\lambda = -1$, and one 3×3 block with eigenvalue $\lambda = 5$. Try

```
> JordanBlockMatrix([[2,2],[3,1],[4,3],[1,2]]);
> JordanBlockMatrix([[2,2],[3,1],[4,3],[1,2]],10);
```

9.6.10 Random matrices and vectors

MAPLE can produce random matrices and vectors with both integral and numeric entries. The relevant functions are `RandomMatrix` and `RandomVector`. The syntax of `RandomMatrix` has the form

`RandomMatrix(`n`)`
`RandomMatrix(`r`,`c`)`

```
RandomMatrix(r,c,generator=a..b)
RandomMatrix(r,c,density=p,generator=a..b)
RandomMatrix(r,c,density=p,generator=a..b,outopts)
```

Here n is the size of square matrix, r is the number of rows, c is the number of columns, $a..b$ is a range of integers or floating-point numbers, and $0 \leq p \leq 1$.

RandomMatrix(r,c) will return a random $r \times c$ matrix with integer entries from the set $\{-99, -98, \ldots, 98, 99\}$.

```
> with(LinearAlgebra):
> RandomMatrix(2,3);
```

$$\begin{bmatrix} -50 & 62 & -71 \\ 30 & -79 & 28 \end{bmatrix}$$

To specify the range for each entry we use the option generator=a..b. This range can take integral for floating-point values. Try

```
> RandomMatrix(2,3,generator=0..9);
> RandomMatrix(6,generator=-10.0..20.0);
```

When the option density=p is used, the probability that an entry is assigned is p. To generate a 20×3 matrix with lots of zeros try

```
> RandomMatrix(20,3, density=0.1, generator=0..1.0);
```

In this matrix there was a probability of 0.9 that an entry remained a zero. We can use outputoptions to assign the shape of the resulting matrix. To generate a random upper triangular matrix, try

```
> RandomMatrix(4,generator=1..9,
    outputoptions=[shape=triangular[upper]]);
```

We use RandomMatrix to construct a procedure RandUniMat, which returns a random unimodular matrix (i.e., a matrix with integral entries and determinant ± 1). First we need a function rand1, which returns a random ± 1 value.

```
> rand1 := 2*rand(0..1)-1:
```

The following procedure RandUniUpMat returns a random unimodular upper triangular matrix.

```
> RandUniUpMat := proc(n::posint,a::integer,b::integer)
>    local M1,i:
>    M1:=LinearAlgebra[RandomMatrix](n,generator=a..b,
          outputoptions=[shape=triangular[upper]]);
>    for i from 1 to n do
```

```
>         M1[i,i]:=rand1():
>     end do:
>   return M1;
>   end proc;
>   RandUniUpMat(3,-5,5);
```

$$\begin{bmatrix} 1 & -3 & -2 \\ 0 & -1 & 4 \\ 0 & 0 & 1 \end{bmatrix}$$

Now we use `RandUniUpMat` to construct our procedure `RandUniMat`.

```
>   RandUniMat := proc(n::posint,a::integer,b::integer)
>     local M1, M2:
>     M1 := RandUniUpMat(n,a,b):
>     M2 := RandUniUpMat(n,a,b):
>     M1.LinearAlgebra[Transpose](M2);
>   end proc;
```

Let's look at an example.

```
>   with(LinearAlgebra):
>   M:=RandUniMat(3,10,50);
```

$$M := \begin{bmatrix} 3493 & 1296 & -36 \\ 1172 & 963 & -26 \\ 47 & 37 & -1 \end{bmatrix}$$

```
>   Determinant(M);
```

$$-1$$

```
>   MatrixInverse(M);
```

$$\begin{bmatrix} 1 & 36 & -972 \\ 50 & 1801 & -48626 \\ 1897 & 68329 & -1844847 \end{bmatrix}$$

What do you notice about M^{-1}?

The syntax for the `RandomVector` function is analogous to that of the `RandomMatrix` function. Try

```
>   with(LinearAlgebra):
>   RandomVector(4);
>   RandomVector(6,generator=0..1.0);
>   RandomVector[row](6,generator=0..1.0);
```

```
> RandomVector[row](6,generator=(2*rand(0..1)-1));
```

9.6.11 Toeplitz matrices

A square matrix that is constant along diagonals is called a *Toeplitz* matrix. The syntax of the `ToeplitzMatrix` function has the form

```
ToeplitzMatrix(L)
ToeplitzMatrix(L,n)
ToeplitzMatrix(L,n,symmetric)
```

Here L is a list of scalars, and n is the size of the square matrix. Using `symmetric` option produces a symmetric matrix.

```
> with(LinearAlgebra):
> ToeplitzMatrix([a,b,c,d,e]);
```

$$\begin{bmatrix} c & b & a \\ d & c & b \\ e & d & c \end{bmatrix}$$

Try

```
> ToeplitzMatrix([a,b,c,d,e,f,g]);
> ToeplitzMatrix([a,b,c,d,e,f,g],3);
> ToeplitzMatrix([a,b,c,d,e,f,g],symmetric);
```

9.6.12 Vandermonde matrices

A *Vandermonde* matrix is a square matrix of the form (x_i^{j-1}). If L is a list of scalars, then `VandermondeMatrix(L)` will return a Vandermonde matrix whose first column has entries from L.

```
> with(LinearAlgebra):
> VandermondeMatrix([a,b,c]);
```

$$\begin{bmatrix} 1 & a & a^2 \\ 1 & b & b^2 \\ 1 & c & c^2 \end{bmatrix}$$

```
> Determinant(%);
```

$$bc^2 - b^2c + ca^2 - ac^2 + ab^2 - ba^2$$

```
> factor(%);
```

$$-(-b+a)(c-b)(c-a)$$

Observe that

$$\begin{vmatrix} 1 & a & a^2 \\ 1 & b & b^2 \\ 1 & c & c^2 \end{vmatrix} = (a-b)(b-c)(a-c).$$

Try

```
> V := VandermondeMatrix([a,b,c,d,e]);
> factor(Determinant(V));
> VandermondeMatrix([a,b,c,d,e],3,4);
```

Can you see a pattern in the factorization of the determinant of a Vandemonde matrix?

9.6.13 Zero matrices and vectors

`ZeroMatrix(m,n)` returns an $m \times n$ zero matrix.

```
> with(LinearAlgebra):
> ZeroMatrix(2,3);
```

$$\begin{bmatrix} 0 & 0 & 0 \\ 0 & 0 & 0 \end{bmatrix}$$

Try

```
> with(LinearAlgebra):
> ZeroMatrix(4);
```

`ZeroVector(`n`)` returns a zero column vector of dimension n. Try

```
> ZeroVector(4);
> ZeroVector[row](4);
```

9.7 Systems of linear equations

Consider the following system of linear equations:

$$\begin{aligned} 10\,x - 27\,y + z + r + 2\,s - 11\,t &= 1 \\ 20\,x - 62\,y + 29\,z + 20\,r + 11\,s - 16\,t &= 1 \\ -x - 8\,y + 36\,z + 24\,r + 9\,s + 9\,t &= 1 \\ -8\,x + 27\,y - 19\,z - 13\,r - 6\,s + 5\,t &= -5 \end{aligned}$$

We enter this system into MAPLE as the list `EqList` and call the list of variables `Vars`.

```
> with(LinearAlgebra):
> EqList:= [10*x-27*y+z+r+2*s-11*t = 1,
```

```
> 20*x-62*y+29*z+20*r+11*s-16*t = 1,
> -x-8*y+36*z+24*r +9*s+9*t = 1,
> -8*x+27*y-19*z-13*r-6*s+5*t = -5];
```

$$EqList := [10\,x - 27\,y + z + r + 2\,s - 11\,t = 1,\; 20\,x - 62\,y + 29\,z + 20\,r + 11\,s - 16\,t = 1,\; -x - 8\,y + 36\,z + 24\,r + 9\,s + 9\,t = 1,\; -8\,x + 27\,y - 19\,z - 13\,r - 6\,s + 5\,t = -5]$$

```
> Vars:=[x,y,z,r,s,t];
```

$$Vars := [x, y, z, r, s, t]$$

We can use the `GenerateMatrix` function to write our system as a matrix equation.

```
> (A,b) := GenerateMatrix(EqList,Vars);
```

$$A, b := \begin{bmatrix} 10 & -27 & 1 & 1 & 2 & -11 \\ 20 & -62 & 29 & 20 & 11 & -16 \\ -1 & -8 & 36 & 24 & 9 & 9 \\ -8 & 27 & -19 & -13 & -6 & 5 \end{bmatrix}, \begin{bmatrix} 1 \\ 1 \\ 1 \\ -5 \end{bmatrix}$$

This means our linear system can be written as a matrix equation

$$A\,\vec{v} = \vec{b},$$

where

$$A = \begin{bmatrix} 10 & -27 & 1 & 1 & 2 & -11 \\ 20 & -62 & 29 & 20 & 11 & -16 \\ -1 & -8 & 36 & 24 & 9 & 9 \\ -8 & 27 & -19 & -13 & -6 & 5 \end{bmatrix}, \quad \vec{v} = \begin{bmatrix} x \\ y \\ z \\ r \\ s \\ t \end{bmatrix}, \quad \vec{b} = \begin{bmatrix} 1 \\ 1 \\ 1 \\ -5 \end{bmatrix}.$$

We can write the system as an augmented matrix using the `augmented = true` option.

```
> AM := GenerateMatrix(EqList,Vars,augmented=true);
```

$$\begin{bmatrix} 10 & -27 & 1 & 1 & 2 & -11 & 1 \\ 20 & -62 & 29 & 20 & 11 & -16 & 1 \\ -1 & -8 & 36 & 24 & 9 & 9 & 1 \\ -8 & 27 & -19 & -13 & -6 & 5 & -5 \end{bmatrix}$$

Now we use the `LinearSolve` function to solve this linear system.

```
> LinearSolve(AM);
```

$$\begin{bmatrix} 215 + 19\,\_t3_2 - 54\,\_t3_3 - 39\,\_t3_4 \\ \_t3_2 \\ \_t3_3 \\ \_t3_4 \\ -145 - 10\,\_t3_2 + 33\,\_t3_3 + 24\,\_t3_4 \\ 169 + 13\,\_t3_2 - 43\,\_t3_3 - 31\,\_t3_4 \end{bmatrix}$$

We see that there are infinitely many solutions with three free parameters $\_t2$, $\_t3$, $\_t4$. We can assign a name to the free parameters using the `free` option.

```
> SOL := LinearSolve(AM, free='w');
```

$$SOL := \begin{bmatrix} 215 + 19\,w_2 - 54\,w_3 - 39\,w_4 \\ w_2 \\ w_3 \\ w_4 \\ -145 - 10\,w_2 + 33\,w_3 + 24\,w_4 \\ 169 + 13\,w_2 - 43\,w_3 - 31\,w_4 \end{bmatrix}$$

We see that the general solution to the system is given by

$$\begin{aligned} x &= 215 + 19\,w_2 - 54\,w_3 - 39\,w_4, \\ y &= w_2, \\ z &= w_3, \\ r &= w_4, \\ s &= -145 - 10\,w_2 + 33\,w_3 + 24\,w_4, \\ t &= 169 + 13\,w_2 - 43\,w_3 - 31\,w_4, \end{aligned}$$

where w_2, w_3, w_4 are any real numbers. We can easily check the solution.

```
> A . SOL = b;
```

$$\begin{bmatrix} 1 \\ 1 \\ 1 \\ -5 \end{bmatrix} = \begin{bmatrix} 1 \\ 1 \\ 1 \\ -5 \end{bmatrix}$$

There are other forms for the `LinearSolve` function. To solve the linear system

$$A\,\vec{v} = \vec{b},$$

try `LinearSolve(A, b)`.

```
> SOL := LinearSolve(A, b, free='w');
```

The method used for solving the system can also be specified. Methods include Cholesky, LU, QR factorization. See `?LinearSolve` for more information. The function `BackwardSubstitute` is used to perform back substitution on a linear system in upper triangular form. Consider the linear system with augmented matrix

$$\left[\begin{array}{cccc:c} 1 & 1 & 0 & -3 & 4 \\ 0 & 0 & 1 & 0 & 2 \\ 0 & 0 & 0 & 1 & -5 \\ 0 & 0 & 0 & 0 & 0 \end{array}\right]$$

```
> with(LinearAlgebra):
> ATM := <<1,0,0,0>|<1,0,0,0>|<0,1,0,0>|<-3,0,1,0>
    |<4,2,-5,0>>;
```

$$\begin{bmatrix} 1 & 1 & 0 & -3 & 4 \\ 0 & 0 & 1 & 0 & 2 \\ 0 & 0 & 0 & 1 & -5 \\ 0 & 0 & 0 & 0 & 0 \end{bmatrix}$$

```
> V := BackwardSubstitute(ATM);
```

$$V := \begin{bmatrix} -11 - \_t3_1 \\ \_t3_1 \\ 2 \\ -5 \end{bmatrix}$$

We see that the linear system has infinitely many solutions with one free parameter. We check the solution. First we select the coefficient matrix A and the last column $\vec{b}$.

```
> A := ATM[1..-1,1..-2];
```

$$\begin{bmatrix} 1 & 1 & 0 & -3 \\ 0 & 0 & 1 & 0 \\ 0 & 0 & 0 & 1 \\ 0 & 0 & 0 & 0 \end{bmatrix}$$

```
> b := ATM[1..-1,-1];
```

$$\begin{bmatrix} 4 \\ 2 \\ -5 \\ 0 \end{bmatrix}$$

Observe in `ATM[1..-1,1..-2]` that the -1 refers to the last row and the -2 refers to the second-to-last column. This way we can easily select the coefficient matrix A. Now we are ready to check the solution.

```
> A . V = b;
```

$$\begin{bmatrix} 4 \\ 2 \\ -5 \\ 0 \end{bmatrix} = \begin{bmatrix} 4 \\ 2 \\ -5 \\ 0 \end{bmatrix}$$

We can also perform back substitution using the `method` option of `LinearSolve`. Try

```
> V := LinearSolve(ATM, method='subs');
```

We conclude this section with the general syntax of the `LinearSolve` function:

```
LinearSolve(AM)
LinearSolve(A,b)
LinearSolve(AM,free=name)
LinearSolve(AM,method=method)
```

Here `AM` is an augmented matrix; `A,b` correspond to a matrix equation $A\vec{x} = \vec{b}$; *name* is the name used for the free parameters; and the available methods are `'none'`, `'solve'`, `'subs'`, `'Cholesky'`, `'LU'`, `'QR'`, or `'SparseLU'`. See `?LinearSolve` for more options and information.

9.8 Row space, column space, and nullspace

Let

$$A = \begin{bmatrix} 1 & 4 & -10 & 3 & -3 \\ 10 & 41 & -102 & 30 & -31 \\ -9 & -19 & 56 & -27 & 10 \end{bmatrix}.$$

We can use MAPLE to find the rank of A and to find bases for the row space, column space, and null space. The relevant MAPLE functions are `Rank`, `RowSpace`, `ColumnSpace`, and `Nullspace`.

```
> with(LinearAlgebra):
> A:=Matrix(3,5,[[1,4,-10,3,-3],
> [10,41,-102, 30,-31],
> [-9,-19,56,-27,10]]);
```

$$A := \begin{bmatrix} 1 & 4 & -10 & 3 & -3 \\ 10 & 41 & -102 & 30 & -31 \\ -9 & -19 & 56 & -27 & 10 \end{bmatrix}$$

```
> Rank(A);
```

$$2$$

```
> RowSpace(A);
```

$$[[1 \quad 0 \quad -2 \quad 3 \quad 1],[0 \quad 1 \quad -2 \quad 0 \quad -1]]$$

```
> ColumnSpace(A);
```

$$[\begin{bmatrix} 1 \\ 0 \\ -179 \end{bmatrix}, \begin{bmatrix} 0 \\ 1 \\ 17 \end{bmatrix}]$$

```
> NSA := NullSpace(A);
```

$$NSA := \left\{ \begin{bmatrix} 2 \\ 2 \\ 1 \\ 0 \\ 0 \end{bmatrix}, \begin{bmatrix} -1 \\ 1 \\ 0 \\ 0 \\ 1 \end{bmatrix}, \begin{bmatrix} -3 \\ 0 \\ 0 \\ 1 \\ 0 \end{bmatrix} \right\}$$

We see that

$$\operatorname{rank} A = 2.$$

The vectors

$$\vec{u_1} = (1, 0, -2, 3, 1),$$
$$\vec{u_2} = (0, 1, -2, 0, -1),$$

form a basis for the row space. The vectors

$$\vec{v_1} = \begin{bmatrix} 1 \\ 0 \\ -179 \end{bmatrix}, \begin{bmatrix} 0 \\ 1 \\ 17 \end{bmatrix},$$

form a basis for the column space and the vectors

$$\vec{w_1} = \begin{bmatrix} 2 \\ 2 \\ 1 \\ 0 \\ 0 \end{bmatrix}, \quad \vec{w_2} = \begin{bmatrix} -1 \\ 1 \\ 0 \\ 0 \\ 1 \end{bmatrix}, \quad \vec{w_3} = \begin{bmatrix} -3 \\ 0 \\ 0 \\ 1 \\ 0 \end{bmatrix},$$

form a basis for the nullspace. Now we check that the vectors $\vec{w_1}$, $\vec{w_2}$, $\vec{w_3}$, are in the nullspace.

```
> W:=<NSA[1] | NSA[2] | NSA[3]>;
```

$$\begin{bmatrix} 2 & -1 & -3 \\ 2 & 1 & 0 \\ 1 & 0 & 0 \\ 0 & 0 & 1 \\ 0 & 1 & 0 \end{bmatrix}$$

```
> A . W;
```

$$\begin{bmatrix} 0 & 0 & 0 \\ 0 & 0 & 0 \\ 0 & 0 & 0 \end{bmatrix}$$

Is this what you expected?

9.9 Eigenvectors and diagonalization

Let

$$A = \begin{bmatrix} 177 & 77 & -28 \\ -546 & -236 & 84 \\ -364 & -154 & 51 \end{bmatrix}$$

We use `Eigenvalues` to find the eigenvalues of A.

```
> with(LinearAlgebra):
> A:=<<177,-546,-364>|<77,-236,-154>|<-28,84,51>>;
```

$$\begin{bmatrix} 177 & 77 & -28 \\ -546 & -236 & 84 \\ -364 & -154 & 51 \end{bmatrix}$$

```
> Eigenvalues(A);
```

$$\begin{bmatrix} 2 \\ -5 \\ -5 \end{bmatrix}$$

We see that A has two eigenvalues $\lambda = 2$ and $\lambda = -5$ (multiplicity 2). Now, let's find a basis for each eigenspace using `Eigenvectors`.

```
> Eigenvectors(A);
```

$$\begin{bmatrix} 2 \\ -5 \\ -5 \end{bmatrix}, \begin{bmatrix} 1 & 1 & 0 \\ -3 & 0 & 1 \\ -2 & 13/2 & 11/4 \end{bmatrix}$$

Observe that a vector and a matrix were returned. The vector contains the eigenvalues of A, and the columns of the matrix are the corresponding eigenvectors. We see that the eigenspace corresponding to $\lambda = 2$ is one dimensional and that $\{[1, -3, -2]^T\}$ is a basis. For $\lambda = -5$, the eigenspace is two dimensional and a basis is $\{[1, 0, 13/2]^T, [0, 1, 11/4]^T\}$. Hence, we have found three independent eigenvectors and A is diagonalizable. So, we let

$$P = \begin{bmatrix} 1 & 2 & 0 \\ -3 & 0 & 4 \\ -2 & 13 & 11 \end{bmatrix}$$

Then $P^{-1} A P$ should be a diagonal matrix. Try

```
> (EG,P):=Eigenvectors(A);
> MatrixInverse(P).A.P;
```

Did you get a diagonal matrix? Alternatively, we can use `JordanForm` to diagonalize A. Try

```
> JordanForm(A);
> JordanForm(A,output='Q');
```

This time the matrix P might be different (since it is not unique). See the next section for more information on Jordan form.

MAPLE can also compute eigenvalues and eigenvectors for complex matrices and matrices with floating point entries. Try

```
> with(LinearAlgebra):
> A := Matrix(2,2,[[1.0,2.0],[3.0,4.0]]);
> Eigenvalues(A);
> Eigenvectors(A);
> B := Matrix(2,2,[[1+10*I,-8*I],[12*I, 1-10*I]]);
> Eigenvalues(B);
> Eigenvectors(B);
> P := JordanForm(B,output='Q');
```

9.10 Jordan form

We used the function `JordanForm` in the previous section. In general, `JordanForm` gives the Jordan canonical form of a square matrix. Try

```
> with(LinearAlgebra):
> C := Matrix(4,4,[[10,10,-14,15],[0,3,0,0],
> [8,1,-13,8],[1,-8,-2,-4]]);
> Q := JordanForm(C,output='Q');
> MatrixInverse(Q).C.Q;
```

The syntax of the JordanForm function has the form

```
JordanForm(A)
JordanForm(C,output='Q')
```

The first form gives the Jordan canonical form of the matrix A. The second form returns a matrix Q such that $Q^{-1}CQ$ is in Jordan form.

9.11 Inner products, and Vector and matrix norms

9.11.1 The dot product and bilinear forms

The DotProduct function gives the usual dot product on $\mathbb{R}^n$ (or the usual inner product on $\mathbb{C}^n$).

```
> with(LinearAlgebra):
> V:=Vector([seq(v[i],i=1..4)]);
```

$$\begin{bmatrix} v_1 \\ v_2 \\ v_3 \\ v_4 \end{bmatrix}$$

```
> W:=Vector([seq(w[i],i=1..4)]);
```

$$\begin{bmatrix} w_1 \\ w_2 \\ w_3 \\ w_4 \end{bmatrix}$$

```
> DotProduct(V,W);
```

$$\overline{v_1}w_1 + \overline{v_2}w_2 + \overline{v_3}w_3 + \overline{v_4}w_4$$

```
> DotProduct(V,W,conjugate=false);
```

$$v_1w_1 + v_2w_2 + v_3w_3 + v_4w_4$$

```
> DotProduct(<1,2,3>,<3,2,1>);
```

$$10$$

If A is a positive definite $n \times n$ matrix, then

$$\langle \vec{x}, \vec{y} \rangle = \vec{x}^T A \vec{y} \quad (\vec{x}, \vec{y} \in \mathbb{R}^n),$$

defines an inner product on $\mathbb{R}^n$. The function `BilinearForm` is used to construct this inner product. First let's construct a 3×3 positive definite matrix.

```
> with(LinearAlgebra):
> L:=Matrix(3,3,[[1,0,0],[1,1,0],[-3,-2,1]]):
> DG:=DiagonalMatrix([1,2,3]):
> A:=L.DG.Transpose(L);
```

$$\begin{bmatrix} 1 & 1 & -3 \\ 1 & 3 & -7 \\ -3 & -7 & 20 \end{bmatrix}$$

Can you see why A is positive definite? In any case, we can check for positive definiteness using the `IsDefinite` function.

```
> IsDefinite(A);
```

$$true$$

For two column vectors $\vec{u}$, $\vec{v}$, `BilinearForm(`$\vec{u}$`, `$\vec{v}$`, A)` computes the inner product $\vec{u}^T A \vec{v}$.

```
> BilinearForm(<1,1,1>,<1,3,1>,A);
```

$$0$$

We see that the vectors $(1,1,1)^T$ and $(1,3,1)^T$ are orthogonal with respect to the given inner product.

9.11.2 Vector norms

For a column vector $\vec{v} = (v_1, \ldots, v_n)^T$, `VectorNorm(`$\vec{v}$`, p)` gives the usual p-norm

$$||\vec{v}||_p = (|v_1|^p + |v_2|^p + \cdots + |v_n|^p)^{1/p},$$

so that `VectorNorm(`$\vec{v}$`, 2)` gives the usual Euclidean norm.

```
> with(LinearAlgebra):
> V:=Vector([seq(v[i],i=1..4)]);
```

$$\begin{bmatrix} v_1 \\ v_2 \\ v_3 \\ v_4 \end{bmatrix}$$

```
> VectorNorm(V,3);
```

$$\left(|v_1|^3 + |v_2|^3 + |v_3|^3 + |v_4|^3\right)^{1/3}$$

```
> VectorNorm(V,2);
```

$$\sqrt{|v_1|^2 + |v_2|^2 + |v_3|^2 + |v_4|^2}$$

`VectorNorm(`$\vec{v}$`)` or `VectorNorm(`$\vec{v}$`, infinity)` gives the usual infinity-norm

$$||\vec{v}||_\infty = \max_{1\le i\le n} |v_i|\,.$$

```
> VectorNorm(V);
```

$$\max(|v_1|, |v_2|, |v_3|, |v_4|)$$

```
> VectorNorm(<-4,3,-5>,infinity);
```

$$5$$

9.11.3 Matrix norms

For an $m \times n$ matrxix $A = (a_{i,j})$, the Frobenius norm is defined by

$$||A||_F = \left(\sum_{j=1}^{n}\sum_{i=1}^{m} a_{i,j}^2\right)^{1/2}.$$

In MAPLE, this is given by `MatrixNorm(A,Frobenuius)`.

```
> with(LinearAlgebra):
> A:=Matrix(2,2,[seq([a[i,1],a[i,2]],i=1..2)]);
```

$$\begin{bmatrix} a_{1,1} & a_{1,2} \\ a_{2,1} & a_{2,2} \end{bmatrix}$$

```
> MatrixNorm(A,Frobenius);
```

$$\sqrt{|a_{1,1}|^2 + |a_{2,1}|^2 + |a_{1,2}|^2 + |a_{2,2}|^2}$$

```
> B:=<<1,2>|<3,4>>;
```

$$\begin{bmatrix} 1 & 3 \\ 2 & 4 \end{bmatrix}$$

```
> MatrixNorm(B,Frobenius);
```

$$\sqrt{30}$$

For any p-norm, the matrix norm $||A||_p$ is defined by

$$||A||_p = \max_{\vec{x}\neq\vec{0}} \frac{||A\vec{x}||_p}{||\vec{x}||_p}.$$

This norm is implemented in MAPLE for p=1, 2 or ∞. It is given by `MatrixNorm(A,p)` where p is 1, 2 or `infinity`. Try

```
> with(LinearAlgebra):
> A:=<<177,-546,-364>|<77,-236,-154>|<-28,84,51>>;
> MatrixNorm(A,1);
> MatrixNorm(A,2);
> MatrixNorm(A,infinity);
```

9.12 Least squares problems

Let A be an $m \times n$ matrix with $m > n$, and suppose $\vec{b} \in \mathbb{R}^m$. The system

$$A\vec{x} = \vec{b}$$

has a least squares solution $\vec{x_s}$ if the vector $\vec{x_s}$ minimizes

$$||A\vec{x} - \vec{b}||.$$

Here $||\ ||$ is the usual Euclidean 2-norm. When A has full rank, this problem has a unique solution. In MAPLE it is given by `LeastSquares(A,`$\vec{b}$`)`. As an example, we consider the problem of fitting a line to some data points:

x	0.70	0.76	0.37	0.82	0.29	0.56	0.42	0.47
y	0.035	0.025	−0.18	0.045	−0.16	−0.058	−0.11	−0.085

This corresponds to solving a least squares problem. We form the matrix A and the vector $\vec{b}$.

```
> X:=[0.70, 0.76, 0.37, 0.82, 0.29, 0.56, 0.42, 0.47];
```

$$X := [0.70, 0.76, 0.37, 0.82, 0.29, 0.56, 0.42, 0.47]$$

```
> Y:=[0.035, 0.025,-0.18, 0.045,-0.16,-0.058,-0.11,
  -0.085];
```

$$Y := [0.035, 0.025, -0.18, 0.045, -0.16, -0.058, -0.11, -0.085]$$

```
> A:=Matrix([seq([1,X[k]],k=1..8)]);
> b:=Vector([seq(Y[k],k=1..8)]);
```

A is an 8×2 matrix whose first column is a string of 1's and whose second column consists of the x-values of the data points. The vector $\vec{b}$ corresponds to the y-values of the data points. Now we solve the corresponding least squares problem.

```
> with(LinearAlgebra):
> c := LeastSquares(A,b);
```

$$c := \begin{bmatrix} -0.3043067376729589600 \\ 0.443383576624982178 \end{bmatrix}$$

The components of the least squares solution give the line of least squares fit. So here the line of best fit has the equation

$$y = -0.3043 + 0.4433x.$$

We plot the data points together with the line of least squares fit.

```
> pts := [seq([X[k],Y[k]],k=1..8)];
```

$$pts := [[0.70, 0.035], [0.76, 0.025], [0.37, -0.18], [0.82, 0.045], [0.29, -0.16], [0.56, -0.058], [0.42, -0.11], [0.47, -0.085]]$$

```
> bestline := c[1] + c[2]*x;
```

$$-0.3043067376729589600 + 0.443383576624982178\,x$$

```
> with(plots):
> PL1 := plot(pts,style=point,symbol=circle):
> PL2 := plot(bestline,x=0..1):
> display(PL1,PL2);
```

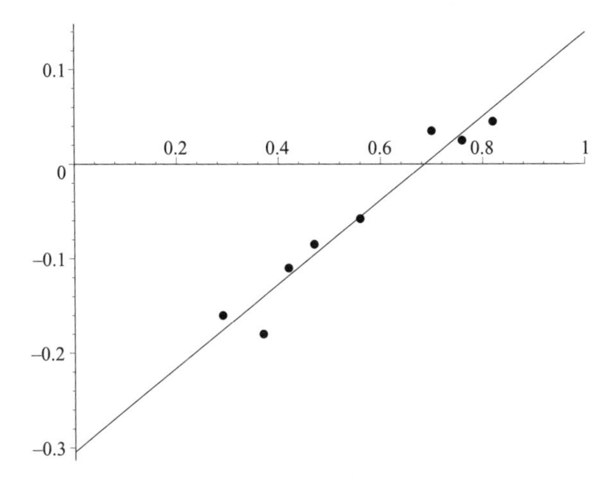

Figure 9.4 Line of best least squares fit.

9.13 QR-factorization and the Gram-Schmidt process

If A is an $m \times n$ matrix with rank n, then A can be factored as

$$A = Q\,R$$

where Q is an $m \times n$ matrix of orthonormal columns and R is an invertible upper triangular matrix. The function `QRDecomposition` computes the QR-factorization. We compute the QR-factorization of the matrix

$$A = \begin{bmatrix} 1 & 12 \\ 2 & 9 \end{bmatrix}.$$

```
> with(LinearAlgebra):
> A:=<<1,2>|<12,9>>;
```

$$A := \begin{bmatrix} 1 & 12 \\ 2 & 9 \end{bmatrix}$$

```
> (Q,R):=QRDecomposition(A);
```

$$Q, R := \begin{bmatrix} \frac{1}{5}\sqrt{5} & \frac{2}{5}\sqrt{5} \\ \frac{2}{5}\sqrt{5} & -\frac{1}{5}\sqrt{5} \end{bmatrix} \begin{bmatrix} \sqrt{5} & 6\sqrt{5} \\ 0 & 3\sqrt{5} \end{bmatrix}$$

```
> Q.R;
```

$$\begin{bmatrix} 1 & 12 \\ 2 & 9 \end{bmatrix}$$

We see that the QR-factorization of A is given by

$$A = \begin{bmatrix} \frac{1}{5}\sqrt{5} & \frac{2}{5}\sqrt{5} \\ \frac{2}{5}\sqrt{5} & -\frac{1}{5}\sqrt{5} \end{bmatrix} \begin{bmatrix} \sqrt{5} & 6\sqrt{5} \\ 0 & 3\sqrt{5} \end{bmatrix}.$$

We check that the columns of A are orthonormal:

```
> Transpose(Q).Q;
```

$$\begin{bmatrix} 1 & 0 \\ 0 & 1 \end{bmatrix}$$

We see $Q^T Q = I$, so the columns are orthonormal.

The Gram-Schmidt process is an algorithm for converting a basis into an orthonormal basis. We can use QR-factorization to compute orthonormal bases. As an example, let's compute an orthonormal basis for the space spanned by the vectors

$$\vec{v}_1 = \begin{bmatrix} 33 \\ -12 \\ -12 \\ -12 \end{bmatrix}, \vec{v}_2 = \begin{bmatrix} 3 \\ 6 \\ -20 \\ -20 \end{bmatrix}, \vec{v}_3 = \begin{bmatrix} 21 \\ 29 \\ 3 \\ 68 \end{bmatrix}$$

We form the matrix whose columns are these three vectors:

```
> V1:=<33,-12,-12,-12>:
> V2:=<3,6,-20,-20>:
> V3:=<21,29,3,68>:
> A := <V1 | V2 | V3>;
```

$$A := \begin{bmatrix} 33 & 3 & 21 \\ -12 & 6 & 29 \\ -12 & -20 & 3 \\ -12 & -20 & 68 \end{bmatrix}$$

We compute the QR-factorization.

```
> with(LinearAlgebra):
> (Q,R) := QRDecomposition(A);
```

$$Q, R := \begin{bmatrix} \frac{11}{13} & -\frac{4}{13} & \frac{4}{13} \\ -\frac{4}{13} & \frac{5}{13} & \frac{8}{13} \\ -\frac{4}{13} & -\frac{8}{13} & -\frac{5}{13} \\ -\frac{4}{13} & -\frac{8}{13} & \frac{8}{13} \end{bmatrix} \begin{bmatrix} 39 & 13 & -13 \\ 0 & 26 & -39 \\ 0 & 0 & 65 \end{bmatrix}$$

The columns of Q give the required orthonormal basis. So the vectors

$$\vec{q}_1 = \begin{bmatrix} \frac{11}{13} \\ -\frac{4}{13} \\ -\frac{4}{13} \\ -\frac{4}{13} \end{bmatrix}, \vec{q}_2 = \begin{bmatrix} -\frac{4}{13} \\ \frac{5}{13} \\ -\frac{8}{13} \\ -\frac{8}{13} \end{bmatrix}, \vec{q}_3 = \begin{bmatrix} \frac{4}{13} \\ \frac{8}{13} \\ -\frac{5}{13} \\ \frac{8}{13} \end{bmatrix}$$

form an orthonormal basis for the vectors spanned by $\vec{v}_1$, $\vec{v}_2$, $\vec{v}_3$. We check that the vectors are orthonormal:

```
> Transpose(Q).Q;
```

$$\begin{bmatrix} 1 & 0 & 0 \\ 0 & 1 & 0 \\ 0 & 0 & 1 \end{bmatrix}$$

Finally, we check that the vectors $\vec{q}_1$, $\vec{q}_2$, $\vec{q}_3$, span the same subspace:

```
> M := <A | Q>;
```

$$M := \begin{bmatrix} 33 & 3 & 21 & \frac{11}{13} & -\frac{4}{13} & \frac{4}{13} \\ -12 & 6 & 29 & -\frac{4}{13} & \frac{5}{13} & \frac{8}{13} \\ -12 & -20 & 3 & -\frac{4}{13} & -\frac{8}{13} & -\frac{5}{13} \\ -12 & -20 & 68 & -\frac{4}{13} & -\frac{8}{13} & \frac{8}{13} \end{bmatrix}$$

```
> Rank(A);
```

$$3$$

```
> Rank(M);
```

$$3$$

Do you see why this implies that the vectors $\vec{q}_1$, $\vec{q}_2$, $\vec{q}_3$, span the same subspace?

Alternatively, we can compute orthogonal bases directly using the function `GramSchmidt`. Try

```
> with(LinearAlgebra):
> V1 := <33,-12,-12,-12>;
> V2 := <3,6,-20,-20>;
> V3 := <21,29,3,68>;
> OBAS := GramSchmidt([V1,V2,V3]);
```

This should give an orthogonal basis for space spanned by the three vectors $\vec{v}_1$, $\vec{v}_2$, $\vec{v}_3$. Check the orthogonality:

```
> M := convert(OBAS, Matrix);
> Transpose(M).M;
```

Did you get a diagonal matrix? To obtain an orthonormal basis we use the `normalized` option in the `GramSchmidt` function:

```
> with(LinearAlgebra):
> V1 := <33,-12,-12,-12>;
> V2 := <3,6,-20,-20>;
> V3 := <21,29,3,68>;
> ONBAS := GramSchmidt([V1,V2,V3],normalized);
```

Did you get the same answer as obtained using the `QRDecomposition` function?

9.14 LU-factorization

In Section 9.4 we used the `LUDecomposition` function to do Gaussian and Gauss-Jordan elimination. LU decomposition is a method for factoring a square matrix as a product of a lower and upper triangular matrix. This is possible if Gaussian elimination can produce an upper triangular matrix without row-swaps. Let's compute the LU factorization of the matrix

$$A = \begin{bmatrix} 2 & 1 & -3 \\ -4 & 4 & 8 \\ -6 & 9 & 3 \end{bmatrix}.$$

```
> with(LinearAlgebra):
> A:=Matrix(3,3,[[2,1,-3],[-4,4,8],[-6,9,3]]);
```

$$\begin{bmatrix} 2 & 1 & -3 \\ -4 & 4 & 8 \\ -6 & 9 & 3 \end{bmatrix}$$

```
> LUDecomposition(A);
```

$$\begin{bmatrix} 1 & 0 & 0 \\ 0 & 1 & 0 \\ 0 & 0 & 1 \end{bmatrix} \begin{bmatrix} 1 & 0 & 0 \\ -2 & 1 & 0 \\ -3 & 2 & 1 \end{bmatrix} \begin{bmatrix} 2 & 1 & -3 \\ 0 & 6 & 2 \\ 0 & 0 & -10 \end{bmatrix}$$

If A is a square matrix, in general `LUDecomposition(`A`)` returns a triple P, L, U, where P is a permutation matrix (i.e., rows are a permutation of the identity matrix I), L is a lower triangular matrix, and U is an upper triangular matrix. Here $P = I$, so the LU factorization exists, and

$$A = \begin{bmatrix} 1 & 0 & 0 \\ -2 & 1 & 0 \\ -3 & 2 & 1 \end{bmatrix} \begin{bmatrix} 2 & 1 & -3 \\ 0 & 6 & 2 \\ 0 & 0 & -10 \end{bmatrix}.$$

A does not have to be a square matrix. Try

```
> with(LinearAlgebra):
> A:=Matrix([[1, 1, 3, -3], [5, 5, 13, -7],
    [3, 1, 7, -11]]);
> LUDecomposition(A);
> (P,L,U):=LUDecomposition(A);
> P.L.U;
```

The matrix U should be the row echelon form of A. Now try

```
> LUDecomposition(A,method=RREF);
> (P,L,U,R):=LUDecomposition(A,method=RREF);
> P.L.U.R;
```

`RREF` stands for row reduced echelon form. The matrix R should be the unique row reduced echelon form of A, so the matrices P, L, U, R were produced using Gauss-Jordan elimination. Also, try

```
> LUDecomposition(A,output=['P','L','U1','R']);
> LUDecomposition(A,output=['R']);
> LUDecomposition(A,output=['P','L']);
```

If A is a real symmteric positive definite matrix, then it can be factored

$$A = L\,L^T,$$

where L is a lower triangular matrix with positive diagonal elements. This is called the Cholesky decomposition of A. Let's compute the Cholesky decomposition of the matrix

$$A = \begin{bmatrix} 1 & -3 & -5 \\ -3 & 13 & 27 \\ -5 & 27 & 62 \end{bmatrix}.$$

We use the command `LUDecomposition(A,method=Cholesky)`. But first let's check that the matrix is positive definite.

```
> with(LinearAlgebra):
> A:=Matrix(3,3,[[1,-3,-5],[-3,13,27],[-5,27,62]]);
```

$$\begin{bmatrix} 1 & -3 & -5 \\ -3 & 13 & 27 \\ -5 & 27 & 62 \end{bmatrix}$$

```
> IsDefinite(A);
```

$$true$$

Now we are ready to compute the decomposition.

```
> L:=LUDecomposition(A,method=Cholesky);
```

$$\begin{bmatrix} 1 & 0 & 0 \\ -3 & 2 & 0 \\ -5 & 6 & 1 \end{bmatrix}$$

Finally, check your answer.

```
> L.Transpose(L);
```

9.15 Other *LinearAlgebra* functions

In this chapter we have already seen many useful functions in the *LinearAlgebra* package. In this section we give a brief summary of the remaining functions.

<u>Add</u>

Add(A, B, c_1, c_2) computes the linear combination $c_1A + c_2B$, provided A and B are both matrices or both vectors with the same dimensions.

`Basis`

Returns a basis for the space spanned by a given set or list of vectors.

`BezoutMatrix`

Computes the Bezout matrix of two polynomials. It is used in the computation of resultants.

`CharacteristicPolynomial`

`CharacteristicPolynomial(`A`, `λ`)` computes the characteristic polynomial $\det(A - \lambda I)$ of a square matrix A.

`Column`

Selects a column or columns of a matrix A.

`ColumnDimension`

Returns the number of columns in a matrix.

`CompanionMatrix`

Returns the companion matrix of a polynomial. If p is a multivariate polynomial, `CompanionMatrix(A,`p`,`x`)` returns the companion matrix of p as a polynomial in the variable x.

`ConditionNumber`

The condition number $\text{cond}(A)$ of a square matrix is relative to a matrix norm:

$$\text{cond}(A) = ||A||\,||A^{-1}||.$$

`ConditionNumber(A,`p`)` computes a condition number relative to a specified p-norm, where p is a nonnegative number, or `infinity`, `Frobenius`, or `Euclidean`.

`CreatePermutation`

Creates a permutation matrix or vector for a NAG pivot vector. See `?CreatePermutation` for more information.

`CrossProduct`

`CrossProduct(`$\vec{v}_1$`, `$\vec{v}_2$`)` computes the cross product $\vec{v}_1 \times \vec{v}_2$ of two column vectors $\vec{v}_1$, $\vec{v}_2$.

`DeleteColumn`

Deletes a column or list of columns from a matrix.

`DeleteRow`

Deletes a row or list of rows from a matrix.

`Dimension`

Returns the number of rows and columns of a matrix.

Equal

`Equal(A,B)` returns `true` if the two matrices (or vectors) A and B are equal.

ForwardSubstitute

Solves a linear system whose given augmented matrix is in lower row echelon form.

FrobeniusForm

Returns the rational canonical form of a square matrix. As an example, we compute the rational canonical form of the matrix

$$A = \begin{bmatrix} 0 & -1 & 2 \\ 3 & -4 & 6 \\ 2 & -2 & 3 \end{bmatrix}.$$

```
> with(LinearAlgebra):
> A := Matrix(3,3,[[0,-1,2],[3,-4,6],[2,-2,3]]);
> FrobeniusForm(A);
> (F,Q) := FrobeniusForm(A,output=['F','Q']);
> MatrixInverse(Q).A.Q;
```

The matrix F should be the rational canonical form of A, and the matrix Q should satisfy $Q^{-1}AQ = F$.

GenerateEquations

Generates a system of equations from a coefficient matrix of a given augmented matrix. As an example consider the matrix

$$A = \begin{bmatrix} 1 & 2 & 3 & 4 \\ 2 & -3 & 4 & 5 \\ 3 & -7 & 8 & 9 \end{bmatrix}.$$

The syntax of `GenerateEquations` has the form

`GenerateEquations(`A`,` *list_of_variables*`)`

Here the number of variables in the list either equals the number of columns of A or is one less. When it equals the number of columns, a homogeneous linear system is generated. Try

```
> with(LinearAlgebra):
> A:=Matrix(3,4,[[1,2,3,4],[2,-3,4,5],[3,-7,8,9]]);
> GenerateEquations(A,[x,y,z,w]);
```

When the number of variables is one less than the number of columns of A, A is interpreted as an augmented matrix so that the last column corresponds to the right side of the linear system. Now try

```
> GenerateEquations(A,[x,y,z]);
```

GetResultDataType

Gives a compatible data type of two input data types for matrices or vectors. See ?GetResultDataType for more information.

GetResultShape

Gives the shape of the resulting data type of two input data types for matrices or vectors relative to some specified operation. See ?GetResultShape for more information.

HermiteForm

Returns the Hermite normal form (reduced row echelon form) of a matrix whose entries are polynomials in a single variable x over the field $\mathbb{Q}$ or a field of rational functions. Try the following example:

```
> with(LinearAlgebra):
> A:=Matrix(3,3,[[5-x,5-2*x,2-x],
    [-x^2+x-4,-2*x^2+2*x-1,-x^2+x],
    [-x^3-5,-2*x^3-10,-x^3-5]]);
> (H,U) := HermiteForm(A,x,output=['H','U']);
> HH:= map(expand,U.A);
> Equal(H,HH);
```

The matrix H is the Hermite normal form of A, and $H = U\,A$.

HermitianTranspose

Returns the Hermitian transpose of a matrix. The Hermitian transpose of a matrix M is sometimes denoted by M^H and defined by $M^H = (\overline{M})^T$. Try

```
> with(LinearAlgebra):
> A:=Matrix(2,[[2,1-I],[1+I,1]]);
> U:=1/sqrt(3)*Matrix(2,[[1-I,-1],[1,1+I]]);
> HermitianTranspose(U).U;
> HermitianTranspose(U).A.U;
```

HessenbergForm

Computes the Hessenberg form of a square matrix. Computation is done within the floating point domain so that results are not exact. A matrix is in Hessenberg form if it is upper triangular except for the first subdiagonal. The function

`HessenbergForm(A)` computes a Hessenberg matrix H and a unitary matrix Q so that $Q^H AQ = H$. Try

```
> with(LinearAlgebra):
> A:=Matrix(3,[[2,1+I,I],[1-I,1,3],[-I,3,1]]);
> (H,Q) := HessenbergForm(A, output=['H','Q']);
> map(fnormal[6],H);
> HH := map(simplify[zero],%);
> Q.H.HermitianTranspose(Q);
> map(fnormal[6],%);
> map(simplify[zero],%);
```

<u>`IntersectionBasis`</u>

Computes a basis for the intersection of given subspaces of $\mathbb{R}^n$. Each subspace is a given by a list of spanning vectors. In the example below

$$W_1 = \text{Span}(\vec{v}_1, \vec{v}_2, \vec{v}_3),$$
$$W_2 = \text{Span}(\vec{v}_4, \vec{v}_5).$$

You will compute a basis for the intersection $W = W_1 \cap W_2$.

```
> with(LinearAlgebra):
> V1:=<2|3|5|-1>;
> V2:=<3|9|6|-1>;
> V3:=<6|32|10|-1>;
> V4:=<8|21|17|-3>;
> V5:=<19|52|26|-4>;
> IntersectionBasis([ [V1,V2,V3], [V4,V5] ]);
```

<u>`IsOrthogonal`</u>

Determines whether a given matrix A is orthogonal (i.e., $AA^T = I$). The call `IsOrthogonal(A, M)` determines whether A is orthogonal with respect to the innerproduct

$$\langle \vec{x}, \vec{y} \rangle = \vec{x}^T M \vec{y}.$$

<u>`IsSimilar`</u>

Determines whether two given matrices A, B are similar (i.e., whether there is an invertible matrix Q such that $QA = BQ$). In the case of matrices with floating point entries, there is a `tolerance` option which can be set when comparing eigenvalues numerically. See `?IsSimilar` for more details.

<u>`IsUnitary`</u>

Determines whether a given matrix A is unitary (i.e., $AA^H = I$). The call `IsUnitary(A, M)` determines whether A is unitary with respect to the complex innerproduct

$$\langle \vec{u}, \vec{v} \rangle = \vec{u}^H M \vec{v}.$$

Map
Map2

The call Map(f,M) applies the function f to each entry of the matrix M and assigns M to the result. Try

```
> with(LinearAlgebra):
> M := Matrix([[1,2],[3,4]]);
> Map(x->1/x,M);
> M;
```

Observe how the entries of M have changed. The function Map2 is analogous to map2.

MatrixAdd

MatrixAdd(A, B, c_1, c_2) computes the linear combination $c_1A + c_2B$, where A and B are both matrices with the same dimensions. If used with the inplace option, the first argument is overwritten. Try

```
> with(LinearAlgebra):
> A:=<<1|2|3>,<4|5|6>,<7|8|9>>;
> B:=<<a|b|c>,<d|e|f>,<g|h|i>>;
> MatrixAdd(A,B,1,-3,inplace);
> A;
```

Observe how A is replaced by $A - 3B$.

MatrixMatrixMultiply

Computes the product of two matrices. Syntax is analogous to MatrixAdd.

MatrixScalarMultiply

MatrixScalarMultiply(A,c) computes cA if c is a scalar and A is a matrix.

MatrixVectorMultiply

Computes the product of a matrix and a column vector.

MinimalPolynomial

Computes the minimal polynomial of a square matrix A. As an example, we compute the minimal polynomial of

$$A = \begin{bmatrix} 0 & -1 & 2 \\ 3 & -4 & 6 \\ 2 & -2 & 3 \end{bmatrix}.$$

```
> with(LinearAlgebra):
> A := Matrix(3,3,[[0,-1,2],[3,-4,6],[2,-2,3]]);
```

$$A := \begin{bmatrix} 0 & -1 & 2 \\ 3 & -4 & 6 \\ 2 & -2 & 3 \end{bmatrix}$$

```
> mpoly := MinimalPolynomial(A,x);
```

$$mpoly := -1 + x^2$$

```
> charpoly := CharacteristicPolynomial(A,x);
```

$$charpoly := x^2 + x^3 - 1 - x$$

```
> normal(charpoly/mpoly);
```

$$x + 1$$

Observe that the characteristic polynomial divides the minimal polynomial. We check that A satisifies its minimal polynomial.

```
> P := unapply( mpoly, x );
```

$$P := x \mapsto -1 + x^2$$

```
> P(A);
```

$$\begin{bmatrix} 0 & 0 & 0 \\ 0 & 0 & 0 \\ 0 & 0 & 0 \end{bmatrix}$$

<u>`Norm`</u>

`Norm(`A`,p)` computes a matrix norm if A is a matrix, and a vector norm if A is a vector. See sections 9.11.2 and 9.11.3.

<u>`Normalize`</u>

`Normalize(`$\vec{v}$`, `p`)` normalizes a vector $\vec{v}$ relative to the specified norm (i.e., it returns the vector $\vec{v}/||\vec{v}||_p$). Here p corresponds to a vector norm, so p is either a nonnegative number, `infinity`, `Euclidean`, or `Frobenius`. If p is not specified, the infinity-norm is assumed. Try

```
> with(LinearAlgebra):
> V := <1 | 2 | 3 | 4>;
> W := Normalize(V);
> U := Normalize(V,2);
> DotProduct(U,U);
```

<u>`OuterProductMatrix`</u>

If $\vec{u}$ and $\vec{v}$ are column vectors, then `OuterProductMatrix(`$\vec{u}$`,`$\vec{v}$`)` returns the matrix $\vec{u}\,\vec{v}^T$.

<u>`Permanent`</u>

Computes the permanent of a square matrix.

Pivot

The call Pivot(A,i,j) pivots the matrix A about the nonzero ijth entry of A (i.e., multiples of the ith row are added to the other rows to obtain zeros in all other entries in the jth column). Try

```
> with(LinearAlgebra):
> A := Matrix(4,[[1,2,3,4],[0,-2,5,7],[0,3,5,6],
    [0,11,-9,3]]);
> Pivot(A,2,2);
```

To change entries only in the third and fourth rows, try

```
> Pivot(A,2,2,[3,4]);
```

RowDimension

Returns the number of rows in a matrix.

ScalarMatrix

Returns a scalar multiple of the identity matrix. ScalarMatrix(λ,n) returns λI, where I is the $n \times n$ identity matrix. Try

```
> with(LinearAlgebra):
> ScalarMatrix(lambda,3);
> ScalarMatrix(lambda,3,4);
```

ScalarMultiply

ScalarMultiply is the same function as MatrixScalarMultiply.

ScalarVector

Let $\vec{e}_j$ denote the jth column of the $n \times n$ identity matrix I. Then for a scalar c, ScalarVector(c,j,n) returns the vector $c\,\vec{e}_j$. Try

```
> with(LinearAlgebra):
> ScalarVector(x,3,4);
```

SchurForm

Computes the Schur form of a square matrix. Computation is done within the floating point domain so that results are not exact. A matrix is in Schur form if it is upper triangular with eigenvalues along the diagonal. The function SchurForm(A) computes an upper triangular matrix T and a unitary matrix Z so that $Z^H AZ = T$. Try

```
> with(LinearAlgebra):
> A:=Matrix(3,[[2,1+I,I],[1-I,1,3],[1,0,1]]);
> (T,Z) := SchurForm(A, output=['T','Z']);
> map(fnormal[6],T);
```

```
> TT := map(simplify[zero],%);
> Z.T.HermitianTranspose(Z);
> map(fnormal[6],%);
> map(simplify[zero],%);
```

SingularValues

Computes the singular values of a matrix. As an example we compute the singular values of a random 4×3 matrix:

```
> with(LinearAlgebra):
> A := RandomMatrix(4,3,outputoptions=[datatype=float]);
```

$$A := \begin{bmatrix} -34.0 & -56.0 & 62.0 \\ -62.0 & -8.0 & -79.0 \\ -90.0 & -50.0 & -71.0 \\ -21.0 & 30.0 & 28.0 \end{bmatrix}$$

```
> S := SingularValues(A, output='list');
```

$$S := [158.0588783049174442, 95.5064017072014764, 44.1238962687336666, 0.0]$$

```
> map(evalf[7],%);
```

$$[158.0589, 95.50640, 44.12390, 0.0]$$

We found that singular values of our matrix A are

$$\sigma_1 \approx 158.0589 \geq \sigma_2 \approx 95.50640 \geq \sigma_3 \approx 44.12390 \geq 0.$$

If A is an $m \times n$ matrix, then there are orthogonal matrices U and V such that

$$A = U\Sigma V^T,$$

where Σ is an $m \times n$ matrix of zeros except for singular values along the diagonal. We can compute U, V using the `SingularValues` function. Try

```
> Sig := DiagonalMatrix( S[1..3], 4, 3 );
> U, Vt := SingularValues(A, output=['U', 'Vt']);
> U.Sig.Vt;
```

The matrix `Vt` corresponds to V^T, and `Sig` is Σ. Did you get $A = U\Sigma V^T$?

SmithForm

Computes the Smith normal form of a matrix A whose entries are polynomials in a single variable. The Smith normal form is a diagonal matrix obtained by doing elementary row and column operations. `SmithForm(`A`)` returns the Smith normal of a matrix A.

```
> with(LinearAlgebra):
> A:=Matrix(3,3,[[5-x,5-2*x,2-x],
    [-x^2+x-4,-2*x^2+2*x-1,-x^2+x],
    [-x^3-5,-2*x^3-10,-x^3-5]]);
> S := SmithForm(A);
```

We can find invertible matrices U, V (corresponding to the row and column operations) such that UAV is the Smith normal form.

```
> (U,V) := SmithForm(A,x,output=['U','V']);
> U.A.V;
> map(simplify,%);
```

Did `U.A.V` simplify to S, the Smith normal form of A?

<u>`SubMatrix`</u>

Returns a submatrix of a matrix. Let r be a list of row numbers and c be a list of column numbers, then `SubMatrix(`A`,`r`,`c`)` returns the submatrix with entries $A[i, j]$, where i is from r, and j is from c. Try

```
> with(LinearAlgebra):
> A:=Matrix(6,[seq([seq(a[i,j],j=1..6)],i=1..6)]);
> SubMatrix(A,[1,3,5],[1..3,6]);
```

<u>`SubVector`</u>

Returns a subvector of a vector. Let L be a list of component places and $\vec{v}$ a vector. Then `SubVector(`$\vec{v}$`, `L`)` returns the vector with components $\vec{v}[j]$, where j is in L.

```
> with(LinearAlgebra):
> V:=Vector([2,4,6,8,10]);
> SubVector(V,[1,4,5]);
```

<u>`SumBasis`</u>

Computes a basis for the sum of given subspaces of $\mathbb{R}^n$. Each subspace is given by a list of spanning vectors. In the example below

$$W_1 = \text{Span}(\vec{v}_1, \vec{v}_2, \vec{v}_3),$$
$$W_2 = \text{Span}(\vec{v}_4, \vec{v}_5).$$

You will compute a basis for the intersection $W = W_1 + W_2$.

```
> with(LinearAlgebra):
> V1:=<2|3|5|-1>;
> V2:=<3|9|6|-1>;
> V3:=<6|32|10|-1>;
```

```
> V4:=<8|21|17|-3>;
> V5:=<19|52|26|-4>;
> B1:=SumBasis([ [V1,V2,V3], [V4,V5] ]);
> B2:=Basis([V1,V2,V3,V4,V5]);
```

What do you notice about $B1$ and $B2$?

SylvesterMatrix

Returns the Sylvester matrix of two polynomials. The Sylvester matrix is used in the computation of the resultant. In fact, the determinant of the Sylvester matrix is the resultant of the two polynomials. Try

```
> with(LinearAlgebra):
> p:=x -> x^2-2:
> q:=x -> x^2-3:
> SylvesterMatrix(p(t),q(x-t),t);
> Determinant(%);
> solve(%=0,x);
```

What do you notice about these roots?

TridiagonalForm

Computes the tridiagonal form matrix of a real symmetric or complex Hermitian matrix. Computation is done within the floating point domain so that results are not exact. A tridiagonal matrix is a square matrix of zeros, except on the main diagonal and on the subdiagonal above and below the main one. Let's compute the tridiagonal form of a random symmetric 3×3 matrix.

```
> with(LinearAlgebra):
> A := RandomMatrix(3,3,outputoptions=[datatype=float,
   shape=symmetric]);
> TridiagonalForm(A);
> TridiagonalForm(A, output=NAG);
> (T,Q) := TridiagonalForm(A, output=['T','Q']);
> Q.T.Transpose(Q);
> map(fnormal[6],%);
> map(simplify[zero],%);
```

In the computation above, QTQ^T should simplify to A. When A is complex Hermitian, Q will be a unitary matrix and QTQ^H should simplify to A.

UnitVector

`UnitVector(`j`,`n`)` gives the jth column vector of the $n \times n$ identity matrix I. Try

```
> with(LinearAlgebra):
> UnitVector(3,4);
```

VectorAdd

`VectorAdd` is analogous to `MatrixAdd`. It computes the linear combination of two vectors. Try

```
> with(LinearAlgebra):
> U := <a | b | c>;
> V := <i | j | k>;
> VectorAdd(U,V,3,4);
```

VectorAngle

Computes the angle θ between two vectors $\vec{u}$, $\vec{v}$, using the formula

$$\cos\theta = \frac{\vec{u}\cdot\vec{v}}{||\vec{u}||\,||\vec{v}||}.$$

Try

```
> with(LinearAlgebra):
> U := <1 | 2 | 3>;
> V := <4 | 1 | -2>;
> W := <1 | 1 | 1>;
> VectorAngle(U,2*U);
> VectorAngle(U,V);
> VectorAngle(U,W);
```

VectorMatrixMultiply

If A is a matrix and $\vec{v}$ is a row vector, `VectorMatrixMultiply(`$\vec{v}$`, `A`)` computes the product $\vec{v}\,A$.

VectorScalarMultiply

`VectorScalarMultiply(`$\vec{v}$`, `c`)` computes $c\vec{v}$ if c is a scalar and $\vec{v}$ is a vector.

Zip

If f is a function of two variables and A, B are two vectors or matrices of the same size and shape, then `Zip(`f`,`A`,`B`)` is the vector (or matrix) obtained by applying f component-wise to A and B. Try

```
> with(LinearAlgebra):
> A:=<<a | b>, <c | d>>;
> B:=<<x | y>, <z | w>>;
> Zip(f,A,B);
> Zip(`+`,A,B);
> Zip(`*`,A,B);
```

9.16 The *linalg* package

For the bulk of this chapter we have concentrated on the *LinearAlgebra* package, which is a great package for doing numerical matrix computations. For abstract or exact computations it is advisable to use the *linalg* package. In this section we give an overview of the *linalg* package. To see all the functions in the *linalg* package try

```
> with(linalg);
```

You will notice many functions in common with the *LinearAlgebra* package.

9.16.1 Matrices and vectors

In the *linalg* package, matrices and vectors are defined as in the *LinearAlgebra* package, except that `matrix( )` and `vector( )` are used instead of `Matrix( )` and `Vector( )`.

```
> with(linalg):
> v:=vector([1,2,3]);
```

$$V := [1, 2, 3]$$

```
> A := matrix(2,3,[a,b,c,d,e,f]);
```

$$A := \begin{bmatrix} a & b & c \\ d & e & f \end{bmatrix}$$

```
> A := matrix(2,3,[[a,b,c],[d,e,f]]);
```

$$A := \begin{bmatrix} a & b & c \\ d & e & f \end{bmatrix}$$

```
> v;
```

$$v$$

```
> A;
```

$$A$$

```
> print(v);
```

$$[1, 2, 3]$$

```
> print(A);
```

$$\begin{bmatrix} a & b & c \\ d & e & f \end{bmatrix}$$

We used the `vector` and `matrix` functions in the *linalg* package to define the three-dimensional vector v and the 2×3 matrix A. Notice that typing `v` or `A` did not cause the vector or matrix to be displayed. We displayed them using the `print command`. Also, try

```
> op(A);
> eval(A);
> evalm(A);
```

9.16.2 Conversion between *linalg* and *LinearAlgebra*

Try

```
> with(linalg):
> with(LinearAlgebra):
> A := matrix(3,3,(i,j)->(i+j));
```

$$A := \begin{bmatrix} 2 & 3 & 4 \\ 3 & 4 & 5 \\ 4 & 5 & 6 \end{bmatrix}$$

```
> Determinant(A);
```

```
Error, LinearAlgebra:-Determinant expects its 1st
argument, A, to be of type Matrix, but received A
> det(A);
```

$$0$$

The function `Determinant` is in the *LinearAlgebra* package and expects a `Matrix`, not a `matrix`. `det` is the determinant function in the *linalg* package. It easy to convert a `matrix` to a `Matrix`.

```
> B := convert(A, Matrix);
```

$$B := \begin{bmatrix} 2 & 3 & 4 \\ 3 & 4 & 5 \\ 4 & 5 & 6 \end{bmatrix}$$

```
> Determinant(B);
```

$$0$$

It is easy to convert a `Matrix` to a `matrix`.

```
> C := convert(B, matrix);
```

$$C := \begin{bmatrix} 2 & 3 & 4 \\ 3 & 4 & 5 \\ 4 & 5 & 6 \end{bmatrix}$$

```
> det(C);
```

$$0$$

Symbolic or abstract computations are performed better using the *linalg* package. Let's perform a symbolic computation in the *LinearAlgebra* package.

```
> with(LinearAlgebra):
> A := Matrix(3,3,(i,j)->x^(i+j));
```

$$A := \begin{bmatrix} x^2 & x^3 & x^4 \\ x^3 & x^4 & x^5 \\ x^4 & x^5 & x^6 \end{bmatrix}$$

```
> B := A - y*IdentityMatrix(3);
```

$$B := -y \begin{bmatrix} 1 & 0 & 0 \\ 0 & 1 & 0 \\ 0 & 0 & 1 \end{bmatrix} + \begin{bmatrix} x^2 & x^3 & x^4 \\ x^3 & x^4 & x^5 \\ x^4 & x^5 & x^6 \end{bmatrix}$$

```
> simplify(B);
```

$$y \begin{bmatrix} 0 & 0 & 0 \\ 0 & 0 & 0 \\ 0 & 0 & 0 \end{bmatrix} + \begin{bmatrix} x^2 & x^3 & x^4 \\ x^3 & x^4 & x^5 \\ x^4 & x^5 & x^6 \end{bmatrix}$$

```
> C := A - 5*IdentityMatrix(3);
```

$$C := \begin{bmatrix} x^2 - 5 & x^3 & x^4 \\ x^3 & x^4 - 5 & x^5 \\ x^4 & x^5 & x^6 - 5 \end{bmatrix}$$

Observe that the command `A - y*IdentityMatrix(3)` did not return a simplified matrix. It did, however, return a simplified matrix when `y` was given the numeric value 5. Note also that `simplify(B)` not only failed to simplify B, but it gave an incorrect result.

Let's try the same calculation using *linalg*.

```
> with(linalg):
> A := matrix(3,3,(i,j)->x^(i+j));
```

$$A := \begin{bmatrix} x^2 & x^3 & x^4 \\ x^3 & x^4 & x^5 \\ x^4 & x^5 & x^6 \end{bmatrix}$$

```
> I3 := diag(1,1,1);
```

$$I3 := \begin{bmatrix} 1 & 0 & 0 \\ 0 & 1 & 0 \\ 0 & 0 & 1 \end{bmatrix}$$

```
> B := A - y*I3;
```

$$B := A - yI3$$

```
> evalm(B);
```

$$\begin{bmatrix} x^2 - y & x^3 & x^4 \\ x^3 & x^4 - y & x^5 \\ x^4 & x^5 & x^6 - y \end{bmatrix}$$

The call `evalm(B)` gave a single matrix.

9.16.3 Matrix operations in *linalg*

Matrix Operation	Mathematical Notation	MAPLE Notation
Addition	$A + B$	`A + B`
Subtraction	$A - B$	`A - B`
Scalar multiplication	$c\,A$	`c*A`
Matrix multiplication	$A\,B$	`A &* B` or `multiply(A,B)`
Matrix power	A^n	`A^n`
Inverse	A^{-1}	`A^(-1)` or `1/A` or `inverse(A)`
Transpose	A^T	`transpose(A)`
Trace	$\operatorname{tr} A$	`trace(A)`

Look at the following example:

```
> with(linalg):
> A := matrix(2,2,[1,2,3,4]):
> B := matrix(2,2,[-2,3,-5,1]):
> A+B;
```

$$A + B$$

```
> evalm(%);
```

$$\begin{bmatrix} -1 & 5 \\ -2 & 5 \end{bmatrix}$$

Notice that we had to use the function `evalm` to display the matrix $A+B$. Now try the following:

```
> with(linalg):
> A:=matrix(2,3,[1,2,3,4,5,6]);
> B:=matrix(3,2,[2,4,-7,3,5,1]);
> C:=matrix(2,2,[1,-2,-3,4]);
> A&*B;
> evalm(%);
```

```
> multiply(A,B);
> evalm(A&*B-2*C);
```

Check your results with pencil and paper. You should have found that

$$AB - 2C = \begin{bmatrix} 1 & 17 \\ 9 & 29 \end{bmatrix}$$

9.16.4 The functions in the *linalg* package

In this section we have seen a few of the *linalg* functions. Below we list all the functions in the package.

`addcol`	linear combination of matrix columns
`addrow`	linear combination of matrix rows
`adjoint`	adjoint of a matrix
`angle`	angle between two vectors
`augment`	augmented matrix
`backsub`	back substitution
`band`	band matrix
`basis`	basis for a span of vectors
`bezout`	Bezout matrix of two polynomials
`BlockDiagonal`	see `diag`
`blockmatrix`	block matrix
`charmat`	characteristic matrix
`charpoly`	characteristic polynomial of a matrix
`cholesky`	Cholesky decomposition
`col`	extract columns from a matrix
`coldim`	number of columns in a matrix
`colspace`	basis for a column space
`colspan`	spanning vectors of a column space
`companion`	companion matrix for a polynomial
`cond`	standard condition number
`copyinto`	copies a matrix into another
`crossprod`	cross-product of two vectors
`curl`	curl of a vector field
`definite`	test for positive or negative definite
`delcols`	delete columns of a matrix
`delrows`	delete rows of a matrix
`det`	determinant
`diag`	block diagonal matrix
`diverge`	divergence of a vector field
`dotprod`	dot-product of two vectors
`eigenvals`	eigenvalues of a matrix
`eigenvectors`	bases for eigenspaces
`entermatrix`	interactive matrix entry
`equal`	determine whether two matrices are equal

`exponential`	matrix exponential
`extend`	enlarge a matrix
`ffgausselim`	fraction-free Gaussian elimination
`fibonacci`	Fibonacci matrix
`forwardsub`	forward substitution
`frobenius`	see `ratform`
`gausselim`	Gaussian elimination
`gaussjord`	Gauss-Jordan elimination
`geneqns`	generate system of equations
`genmatrix`	generate augmented matrix
`grad`	gradient of a function
`GramSchmidt`	Gram-Schmidt orthogonalization process
`hadamard`	an upper bound for determinant
`hermite`	Hermite normal form of matrix with polynomial entries
`hessian`	Hessian matrix
`hilbert`	Hilbert matrix
`htranspose`	Hermitian transpose
`ihermite`	integer only Hermite normal form
`indexfunc`	indexing function of an array
`innerprod`	innerproduct $\mathbf{u}^T A\,\mathbf{v}$
`intbasis`	basis for intersection of subspaces
`inverse`	inverse of a matrix
`ismith`	integer-only Smith normal form
`issimilar`	determine if two matrices are similar
`iszero`	determine whether a matrix is the zero matrix
`jacobian`	Jacobian matrix of a vector function
`jordan`	Jordan form
`JordanBlock`	Jordan block matrix
`kernel`	basis for the nullspace of a matrix
`leastsqrs`	least squares problem
`linsolve`	solve a linear system
`LUdecomp`	LU-decomposition
`matadd`	computes a matrix sum
`minpoly`	minimal polynomial of a matrix
`mulcol`	multiply a column by an expression
`mulrow`	multiply a row by an expression
`multiply`	product of two matrices
`norm`	norm of a matrix or vector
`normalize`	normalize a vector
`nullspace`	see `kernel`
`orthog`	determine whether a matrix is orthogonal
`permanent`	permanent of a matrix
`pivot`	pivot about a matrix entry
`potential`	potential function of a vector field

`QRdecomp`	QR-decomposition of a matrix
`randmatrix`	random matrix generator
`randvector`	random vector generator
`rank`	rank of a matrix
`ratform`	rational canonical form
`row`	extract rows from a matrix
`rowdim`	number of rows in a matrix
`rowspace`	basis for a rowspace
`rref`	see `gaussjord`
`scalarmul`	multiply a matrix by an expression
`singularvals`	singular values of a matrix
`smith`	Smith normal form
`stack`	stacks two matrices vertically
`submatrix`	extract a submatrix
`subvector`	extract a vector from a matrix
`sumbasis`	basis for sum of subspaces
`swapcol`	swap two columns in a matrix
`swaprow`	swap two rows in a matrix
`sylvester`	Sylvester matrix of two polynomials
`toeplitz`	Toeplitz matrix
`trace`	trace of a matrix
`vandermonde`	Vandermonde matrix
`vecpotent`	vector potential of a vector field
`vectdim`	number of components in a vector
`wronskian`	Wronskian matrix

10. Multivariable and Vector Calculus

10.1 Vectors

In Chapter 9, we saw how to define and manipulate vectors using the *LinearAlgebra* and *linalg* packages. We will use the *linalg* package in this chapter because it contains more functions for handling vectors.

10.1.1 Vector operations

Let's define two vectors

$$\begin{aligned}\vec{u} &= (1, -4, 5),\\ \vec{v} &= (2, 3, 7).\end{aligned}$$

We are able to add and subtract vectors and perform scalar multiplication:

```
> u := vector([1,-4,5]);
```

$$u := [1, -4, 5]$$

```
> v := vector([2,3,7]);
```

$$v := [2, 3, 7]$$

```
> u + v;
```

$$u + v$$

```
> evalm(u + v);
```

$$[3, -1, 12]$$

```
> evalm(u - v);
```

$$[-1, -7, -2]$$

```
> evalm(5*u - 3*u);
```

$$[2, -8, 10]$$

Remember, we must use the `evalm` function when doing vector operations. We found

$$\begin{aligned}\vec{u} + \vec{v} &= (3, -1, 12),\\ \vec{u} - \vec{v} &= (-1, -7, -2),\\ 5\vec{u} - 3\vec{v} &= (2, -8, 10).\end{aligned}$$

10.1.2 Length, dot product, and cross product

We use the `norm` function from the *linalg* package to compute the length of a vector.

```
> with(linalg):
> u := vector([x,y,z]);
```

$$u := [x, y, z]$$

```
> v := vector([2,-5,6]);
```

$$v := [2, -5, 6]$$

```
> norm(u);
```

$$\max(|x|, |y|, |z|)$$

```
> norm(u,2);
```

$$\sqrt{(|x|)^2 + (|y|)^2 + (|z|)^2}$$

```
> norm(v,2);
```

$$\sqrt{65}$$

In *linalg* the default norm is the infinity norm. To obtain the length of a vector $\vec{u}$, we use the command `norm(`$\vec{u}$`, 2)`. For $\vec{v} = (2, -5, 6)$, we found the length $||\vec{v}|| = \sqrt{65}$.

To find the dot product of two vectors, we use the `dotprod` function from the *linalg* package.

```
> with(linalg):
> U := vector([u[1],u[2],v[2]]);
```

$$U := [u_1, u_2, v_2]$$

```
> V := vector([v[1],v[2],v[2]]);
```

$$V := [v_1, v_2, v_2]$$

```
> dotprod(U,V);
```

$$u_1\,\overline{v_1} + u_2\,\overline{v_2} + u_3\,\overline{v_3}$$

```
> a := vector([1,2,3]);
```

$$[1, 2, 3]$$

```
> b := vector([-3,5,7]);
```

$$[-3, 5, 7]$$

```
> dotprod(a,b);
```

$$28$$

Notice how MAPLE defined the dot product in terms of the conjugate to cover complex vectors. If the vector is real, this corresponds to the usual dot product. For $\vec{a} = (1, 2, 3)$, $\vec{b} = (-3, 5, 7)$ we found that

$$\vec{a} \cdot \vec{b} = 28.$$

To find the angle between two vectors $\vec{u}$, $\vec{v}$, use the function `angle(`$\vec{u}$`, `$\vec{v}$`)` in the *linalg* package. Find the angle θ between $\vec{u} = (2, 1, 2)$ and $\vec{v} = (1, 1, 0)$:

```
> with(linalg):
> u := vector([2,1,2]);
> v := vector([1,1,0]);
> angle(u,v);
> simplify(%);
```

Did you get $\theta = \frac{\pi}{4}$?

To find the cross product of two vectors, we use the `crossprod` function. We find the cross product of $\vec{u} = (1, 2, 3)$ and $\vec{v} = (5, -2, 1)$.

```
> with(linalg):
> u := vector([1,2,3]);
```

$$u := [1, 2, 3]$$

```
> v := vector([5,-2,1]);
```

$$v := [5, -2, 1]$$

```
> w := crossprod(u,v);
```

$$[8, 14, -12]$$

We found that

$$\vec{w} = \vec{u} \times \vec{v} = (8, 14, -12).$$

Now try

```
> dotprod(u,w);
> dotprod(v,w);
```

What did you find? What does this imply about the three vectors $\vec{u}$, $\vec{v}$, and $\vec{w}$?

10.1.3 Plotting vectors

To plot vectors, we use the `arrow` function in the *plottools* package. More details of the *plottools* package can be found in Section 14.1. When plotting a two-dimensional vector, the syntax of the `arrow` function has the form

`arrow( [`a`, `b`], `$\vec{v}$`, `wb`, `wh`, `hh`)`

This plots an arrow (vector) in the direction $\vec{v}$ with initial point (a, b). Here wb is the width of the body of the arrow, wh is the width of the head of the arrow, and hh is ratio of the head to the body of the arrow.

To plot the vector $\vec{v} = [2,3]$ try

```
> with(plottools):
> with(plots):
> v := vector([2,3]);
```

$$v := [2,3]$$

```
> vec := arrow([0,0],v,.1,.2,.2,color=red):
> display(vec):
```

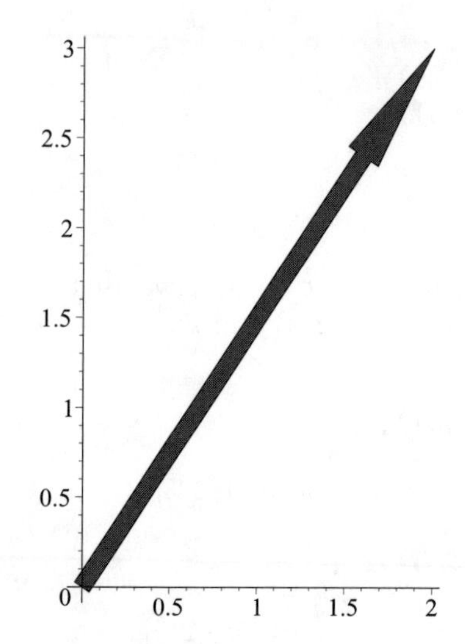

Figure 10.1 Plot of a two-dimensional vector.

Here our initial point was `[0,0]`, so that the vector is drawn from the origin.

When plotting a three-dimensional vector, the syntax of the `arrow` function has the form

arrow($[a,\ b,\ c]$, $\vec{v}$, wb, wh, hh)
arrow($[a,\ b,\ c]$, $\vec{v}$, $\vec{m}$, wb, wh, hh)

This plots an arrow (vector) in the direction $\vec{v}$ with initial point (a,b). As before, wb is the width of the body of the arrow, wh is the width of the head of the arrow, and hh is ratio of the head to the body of the arrow. The vector $\vec{m}$ specifies the plane containing the vector. This plane passes through the point (a,b,c) and has normal vector $\vec{n} = \vec{v} \times (\vec{v} \times \vec{m})$. The vector $\vec{m}$ will be a normal vector for this plane if it is orthogonal to $\vec{v}$. Try the following

```
> with(plots):
> with(plottools):
> with(linalg):
> u := normalize(vector([1,2,3]));
> v := normalize(vector([5,-2,1]));
> w := crossprod(u,v);
> uvec:=arrow([0,0,0],u,w,.1,0.2,.1,color=red):
```

```
> vvec:=arrow([0,0,0],v,w,.1,.2,.1,color=blue):
> wvec:=arrow([0,0,0],w,u,.1,.2,.1,color=green):
> utext:=textplot3d([u[1],u[2],u[3]," u "],color=black):
> vtext:=textplot3d([v[1],v[2],v[3]," v "],align=LEFT,
    color=black):
> wtext:=textplot3d([w[1],w[2],w[3]," u x v "],align=LEFT,
    color=black):
> c := sphere([0,0,0], 0.1,color=black):
> display(uvec,utext,vvec,vtext,wvec,wtext,c,
    scaling=constrained,axes=boxed,orientation=[25,60]);
```

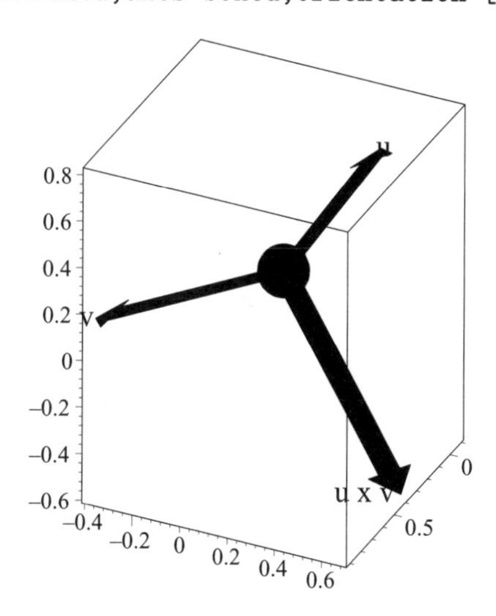

Figure 10.2 Illustrating vectors in space.

10.2 Lines and planes

Besides *linalg*, another useful package here is *geom3d*. The *geom3d* package handles computations for geometry in three-dimensional Euclidean space. We will need only a few functions from this package. An overview of this package can be found in Section 17.7.9.

10.2.1 Lines

We find the vector equation of the line ℓ passing through the points $P\ (1,2,3)$ and $Q\ (4,-7,2)$. First we define the points P and Q:

```
> with(geom3d):
> point(P,1,2,3);
```

$$P$$

```
> point(Q,4,-7,2);
```

$$Q$$

Next, we define the line ℓ that passes through P and Q:

```
> line(l,[P,Q]);
```

$$l$$

Now we use `Equation` function to find the equation of the line:

```
> Equation(l,t);
```

$$[1 + 3t, 2 - 9t, 3 - t]$$

Here t is the parameter used in the vector equation. Thus the vector equation of the line ℓ is

$$\vec{r} = (1 + 3t)\vec{i} + (2 - 9t)\vec{j} + (3 - t)\vec{k}.$$

Alternatively, we could have found the vector $\overrightarrow{PQ}$, which is the direction of the required line. Try

```
> P := vector([1,2,3]);
> Q := vector([4,-7,2]);
> PQ := evalm(Q - P);
```

Did you obtain $\overrightarrow{PQ} = (3, -9, -1)$?

In our next example, we find the distance d between the point Q $(1, 2, 3)$ and the line given parametrically by $x = 2t$, $y = 1 - 3t$, $z = 2 + 5t$. The distance d is given by

$$d = \frac{||\vec{L} \times \overrightarrow{PQ}||}{||\vec{L}||},$$

where $\vec{L}$ is the direction vector of the line and P is any point on the line. We take P $(0, 1, 2)$ by putting $t = 0$. Here $\vec{L} = (2, -3, 5)$. Try

```
> with(linalg):
> Q := vector([1,2,3]);
> P := vector([0,1,2]);
> L := vector([2,-3,5]);
> PQ := Q - P;
> LPQ := crossprod(L,PQ);
> dLPQ := norm(LPQ,2);
> dL := norm(L,2);
> dist := dLPQ/dL;
```

Did you obtain

$$d = \frac{7}{\sqrt{19}}?$$

10.2.2 Planes

The equation of a plane takes the form

$$ax + by + cz = d.$$

In the previous section we used the `Equation` function in the *geom3d* package to find the equation of a line. We can find the equation of a plane in a similar fashion. Let's find the equation of the plane $\mathcal{P}$ passing through the three points P_0 $(1,1,1)$, P_1 $(2,1,3)$, P_2 $(3,2,1)$. First we define the three points:

```
> with(geom3d):
> point(P0,1,1,1):
> point(P1,2,1,3):
> point(P2,3,2,1):
```

If x, y, z denote the variables of our coordinate system, the plane $\mathcal{P}$ through the three points P_0, P_1, P_2 is given by `plane(`$\mathcal{P}$`, [`P_0`, `P_1`, `P_2`], [`x`, `y`, `z`])`. We define the plane $\mathcal{P}$ and find its equation:

```
> plane(P,[P0,P1,P2],[x,y,z]):
> Equation(P);
```

$$-3 - 2\,x + 4\,y + z = 0$$

We find that the equation of the plane is

$$-2x + 4y + z = 3.$$

Let $\vec{N}$ be the normal vector of a plane $\mathcal{P}$. The distance d between a point Q and the plane $\mathcal{P}$ is given by

$$d = \frac{\left|\vec{N} \cdot \overrightarrow{PQ}\right|}{||\vec{N}||},$$

where P is any point on the plane $\mathcal{P}$. We find the distance d between the point Q $(1,2,3)$ and the plane with equation

$$-2x + 4y + z = 3.$$

Here $\vec{N} = (-2,4,1)$, and we take P $(0,0,3)$ which is clearly a point on the plane. Try

```
> with(linalg):
> P := vector([0,0,3]);
> Q := vector([1,2,3]);
> N := vector([-2,4,1]);
> PQ := Q - P;
> NPQ := dotprod(N,PQ);
> dist := abs(NPQ)/norm(N,2);
```

Did you obtain

$$d = \frac{2\sqrt{21}}{7}?$$

10.3 Vector-valued functions

We can represent a vector-valued function of t,

$$\vec{F}(t) = f_1(t)\vec{\imath} + f_2(t)\vec{\jmath} + f_3(t)\vec{k},$$

as a `vector` whose components are functions of t. Let

$$\vec{F}(t) = \vec{\imath} + t\,\vec{\jmath} + \sqrt{t}\,\vec{k},$$
$$\vec{G}(t) = \sin t\,\vec{\imath} + \cos t\,\vec{\jmath} + t\,\vec{k}.$$

We use MAPLE to find the cross product $(\vec{F} \times \vec{G})(t)$:

```
> with(linalg):
> F := vector([1, t, sqrt(t)]);
```

$$F := [1, t, \sqrt{t}]$$

```
> G := vector([sin(t), cos(t), t]);
```

$$G := [\sin(t), \cos(t), t]$$

```
> crossprod(F,G);
```

$$[t^2 - \sqrt{t}\cos(t), \sqrt{t}\sin(t) - t, \cos(t) - t\sin(t)]$$

We found that

$$(\vec{F} \times \vec{G})(t) = (t^2 - \sqrt{t}\cos(t))\,\vec{\imath} + (\sqrt{t}\sin(t) - t)\,\vec{\jmath} + (\cos(t) - t\sin(t))\,\vec{k}.$$

Now try finding $(\vec{F} + \vec{G})(t)$ and $(\vec{F} \cdot \vec{G})(t)$:

```
> F + G;
> dotprod(F,G);
```

10.3.1 Differentiation and integration of vector functions

Probably the best way to compute the derivative of a vector-valued function in MAPLE is to `map diff` with respect to t onto the vector:

```
> F := vector([f[1](t),f[2](t),f[3](t)]);
```

$$[f_1(t), f_2(t), f_3(t)]$$

```
> map(diff,F,t);
```

$$[\frac{d}{dt}f_1(t), \frac{d}{dt}f_2(t), \frac{d}{dt}f_3(t)]$$

Suppose the position vector of an object at time t is given by

$$\vec{r}(t) = 2\cos t\,\vec{\imath} + 3\sin t\,\vec{\jmath} + t\,\vec{k}.$$

We use MAPLE to find the velocity $\vec{v}(t)$ and acceleration $\vec{a}(t)$:

```
> r := vector([2*cos(t), 3*sin(t), t]);
```

$$r := [2\cos(t), 3\sin(t), t]$$

```
> v := map(diff,r,t);
```

$$v := [-2\sin(t), 3\cos(t), 1]$$

```
> a := map(diff,v,t);
```

$$a := [-2\cos(t), -3\sin(t), 0]$$

Indefinite and definite integration of a vector-valued function of t can be done in similar fashion by replacing `diff` with `int`. Let

$$\vec{r}(t) = t^2\,\vec{\imath} + \ln(1+t)\,\vec{\jmath} + \sqrt{1-t}\,\vec{k}.$$

Use MAPLE to find

$$\int \vec{r}(t)\,dt,$$

and

$$\int_0^1 \vec{r}(t)\,dt:$$

```
> r := vector([t^2, ln(1+t), sqrt(1-t)]);
> map(int,r,t);
> map(int,r,t=0..1);
```

Did you obtain

$$\int_0^1 \vec{r}(t)\,dt = \frac{1}{3}\vec{\imath} + (2\ln 2 - 1)\,\vec{\jmath} + \frac{2}{3}\vec{k}?$$

10.3.2 Space curves

In Section 6.2.3 we saw how to plot space curves using the `spacecurve` function in the *plots* package. Try plotting the helix parameterized by

$$\vec{r}(t) = \cos t\,\vec{\imath} + 3\sin t\,\vec{\jmath} + t\,\vec{k}, \quad 0 \le t \le 4\pi.$$

```
> with(plots):
> spacecurve([cos(t),3*sin(t),t],t=0..4*Pi,color=black,
    thickness=3, numpoints=200, axes=boxed,
    orientation=[30,65]);
```

Now consider a point moving through space whose direction vector $\vec{r}(t)$ is given by

$$\vec{r}(t) = \cos t\,\vec{\imath} + \sin t\,\vec{\jmath} + t\,\vec{k}, \quad 0 \le t \le 4\pi.$$

We can visualize this moving point using the **animate3d** function. To help plot this moving point, we define three functions:

```
> fx:=(rho,phi,theta)->rho*sin(phi)*cos(theta);
```

$$fx := (\rho, \phi, \theta) \mapsto \rho \sin(\phi) \cos(\theta)$$

```
> fy:=(rho,phi,theta)->rho*sin(phi)*sin(theta);
```

$$fy := (\rho, \phi, \theta) \mapsto \rho \sin(\phi) \sin(\theta)$$

```
> fz:=(rho,phi,theta)->rho*cos(phi);
```

$$fz := (\rho, \phi, \theta) \mapsto \rho \cos(\phi)$$

The reader should recognize these three functions as giving the (x, y, z) coordinates of a point with the given spherical coordinates (ρ, ϕ, θ). To plot a sphere of radius 1 try

```
> plot3d([fx(1,phi,theta),fy(1,phi,theta),fz(1,phi,theta)],
   phi=-Pi..Pi,theta=0..2*Pi);
```

We now produce an animation of a sphere (radius 1/10) moving along the helix:

```
> with(plots):
> S := spacecurve([cos(t),sin(t),t],t=0..4*Pi,color=black,
   numpoints=200,axes=boxed):
> A:= animate3d([cos(t)+fx(1/10,phi,theta),sin(t)+
   fy(1/10,phi,theta), t+ fz(1/10,phi,theta)],phi=-Pi..Pi,
   theta=0..2*Pi,t=0..4*Pi, frames=32,
   scaling=constrained):
> display(S,A,scaling=constrained,orientation=[60,30]);
```

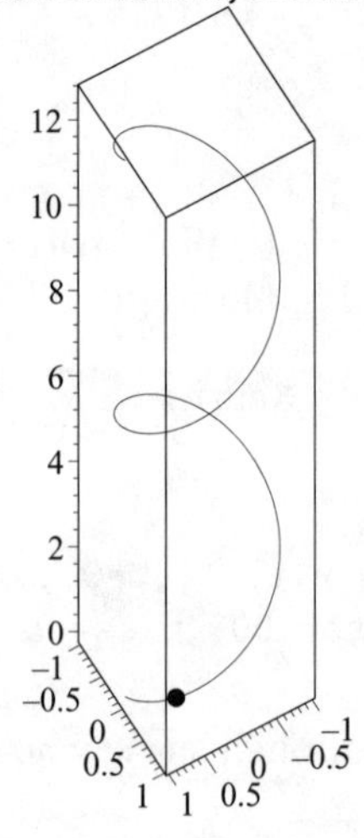

Figure 10.3 Animation of a point on a helix.

Click on the display, press [▶], and watch the point go up the helix. Now let's compute the length of our helix. In general, the length L of a curve parameterized by $\vec{r}(t)$, $a = let \le b$, is given by

$$L = \int_a^b ||\vec{r}\,'(t)||\, dt.$$

```
> with(linalg):
> r := vector([cos(t),sin(t),t]);
```

$$[\cos(t), \sin(t), t]$$

```
> dr := map(diff,r,t);
```

$$[-\sin(t), \cos(t), 1]$$

```
> nr := simplify(norm(dr,2));
```

$$\sqrt{1 + (|\sin(t)|)^2 + (|\cos(t)|)^2}$$

```
> L := int(nr,t=0..4*Pi);
```

$$4\,\pi\,\sqrt{2}$$

10.3.2 Tangents and normals to curves

First we define some simple MAPLE functions that will prove useful when simplifying and manipulating vector-valued functions. Often a vector-valued function is written in the form

$$\vec{r}(t) = f_1(t)\,\vec{i} + f_2(t)\,\vec{j} + f_3(t)\,\vec{k},$$

where $\vec{i}$, $\vec{j}$, $\vec{k}$ are the usual standard basis vectors for $\mathbb{R}^3$. We could represent the three vectors $\vec{i}$, $\vec{j}$, $\vec{k}$ by unknowns `i`, `j`, `k`. Here is an example.

```
> r := t -> 6*t*i + 3* sqrt(2)*t^2*j + 2*t^3*k;
```

$$r := t \mapsto 6\,t i + 3\,\sqrt{2} t^2 j + 2\,t^3 k$$

See how we defined the function

$$\vec{r}(t) = 6t\,\vec{i} + 3\sqrt{2}t^2\vec{j} + 2t^3\,\vec{k}.$$

We define four functions. Enter the following MAPLE functions into a text file *vecfuncs* and save it.

```
r2vec := r -> vector([coeff(r,i,1),coeff(r,j,1),
         coeff(r,k,1)]):
normv := v -> radsimp(sqrt(factor(v[1]^2+v[2]^2+
         v[3]^2))):
vec2r := v -> v[1]*i + v[2]*j + v[3]*k:
rsimp := rt -> collect(numer(rt),[i,j,k])/denom(rt):
```

So you should have a text file called *vecfuncs* containing these four functions. Alternatively, you can type these functions directly into a MAPLE worksheet.

The function `r2vec(r)` converts a vector-valued function

$$\vec{r}(t) = f_1(t)\,\vec{\imath} + f_2(t)\,\vec{\jmath} + f_3(t)\,\vec{k},$$

into the form

$$[f_1(t), f_2(t), f_3(t)].$$

Here is an example.

```
> read vecfuncs:
> r := t -> 6*t*i + 3* sqrt(2)*t^2*j + 2*t^3*k:
> vec := r2vec(r(t));
```

$$vec := [6\,t, 3\,\sqrt{2}t^2, 2\,t^3]$$

The function `vec2r(v)` does the opposite of `r2vec`.

```
> vec2r(vec);
```

$$6\,ti + 3\,\sqrt{2}t^2 j + 2\,t^3 k$$

The function `normv(v)` computes the usual norm of a vector $\vec{v} = v_1\,\vec{\imath} + v_2\,\vec{\jmath} + v_3\,\vec{k}$:

$$||\vec{v}|| = \sqrt{v_1^2 + v_2^2 + v_3^2}.$$

We compute $||\vec{r}(t)||$.

```
> normv(vec);
```

$$\sqrt{2}t\sqrt{18 + 9\,t^2 + 2\,t^4}$$

We found that

$$||\vec{r}(t)|| = \sqrt{2}t\sqrt{18 + 9\,t^2 + 2\,t^4}.$$

You should verify this calculation by hand. We could have used the `norm` function in the *linalg* package. Try

```
> with(linalg):
> norm(2,vec);
```

and you will see why we chose to use `normv` instead. We will use the `rsimp` function to simplify a vector-valued function. For our function $\vec{r}(t)$ we will compute the unit tangent vector $\vec{T}$, which is defined by

$$\vec{T}(t) = \frac{\vec{r}\,'(t)}{||\vec{r}\,'(t)||}.$$

First we compute the derivative $\vec{r}\,'(t)$.

```
> rp := diff(r(t),t);
```

$$rp := 6\,i + 6\,\sqrt{2}tj + 6\,t^2k$$

We found that

$$\vec{r}\,'(t) = 6\,\vec{i} + 6\sqrt{2}t\,\vec{j} + 6t^2\,\vec{k}.$$

Now we compute the norm $||\vec{r}\,'(t)||$, but first we use `r2vec` to convert $\vec{r}$ to component form.

```
> rpv := r2vec(rp);
```

$$rpv := [6, 6\,\sqrt{2}t, 6\,t^2]$$

```
> n := normv(rpv);
```

$$n := 6 + 6\,t^2$$

We found that

$$||\vec{r}\,'(t)|| = 6(1 + t^2).$$

Now we are ready to compute the unit tangent vector $\vec{T}(t)$.

```
> read vecfuncs:
> T := rp/n;
```

$$T := \frac{6\,i + 6\,\sqrt{2}tj + 6\,t^2k}{6 + 6\,t^2}$$

```
> T := rsimp(T);
```

$$T := \frac{i + \sqrt{2}tj + t^2k}{1 + t^2}$$

Notice how we used our function `rsimp` to simplify $\vec{T}(t)$ to find that

$$\vec{T}(t) = \frac{\vec{i} + \sqrt{2}t\,\vec{j} + t^2\,\vec{k}}{1 + t^2}.$$

To check our answer we make sure that $||\vec{T}(t)|| = 1$.

```
> normv(r2vec(T));
```

$$1$$

Now we are ready to compute the principal normal vector

$$\vec{N}(t) = \frac{\vec{T}\,'(t)}{||\vec{T}\,'(t)||}.$$

First we compute the derivative $\vec{T}\,'(t)$.

```
> nt := diff(T,t);
```

$$nt := \frac{\sqrt{2}j + 2\,tk}{1+t^2} - 2\,\frac{\left(i + \sqrt{2}tj + t^2k\right)t}{\left(1+t^2\right)^2}$$

```
> nt := rsimp(nt);
```

$$nt := \frac{-2\,ti + \left(\sqrt{2} - \sqrt{2}t^2\right)j + 2\,tk}{\left(1+t^2\right)^2}$$

```
> nn := normv(r2vec(nt));
```

$$nn := \frac{\sqrt{2}}{1+t^2}$$

```
> N := nt/nn;
```

$$N := 1/2\,\frac{\left(-2\,ti + \left(\sqrt{2} - \sqrt{2}t^2\right)j + 2\,tk\right)\sqrt{2}}{1+t^2}$$

So according to MAPLE we have

$$\vec{N}(t) = 1/2\,\frac{\left(-2t\,\vec{i} + \left(\sqrt{2} - \sqrt{2}t^2\right)\vec{j} + 2t\,\vec{k}\right)\sqrt{2}}{1+t^2}.$$

To check our results we make sure that $||\vec{N}(t)||$ and that $\vec{T} \cdot \vec{N} = 0$.

```
> normv(r2vec(N));
```

$$1$$

```
> with(linalg):
> dotprod(r2vec(T),r2vec(N),orthogonal);
```

$$-\frac{\sqrt{2}t}{\left(1+t^2\right)^2} + \frac{t\left(\sqrt{2} - \sqrt{2}t^2\right)}{\left(1+t^2\right)^2} + \frac{t^3\sqrt{2}}{\left(1+t^2\right)^2}$$

```
> normal(%);
```

$$0$$

We used the `dotprod` function (with the `orthogonal` option) in the *linalg* package to compute the dot product $\vec{T} \cdot \vec{N}$. Observe that after using `normal`, we see that the dot product simplifies to 0 as expected.

10.3.3 Curvature

The curvature κ of a curve parameterized by a vector-valued function $\vec{r}(t)$ is given by

$$\kappa(t) = \frac{||\vec{T}\,'(t)||}{||\vec{r}\,'(t)||}.$$

In the previous section we considered

$$\vec{r}(t) = 6t\,\vec{\imath} + 3\sqrt{2}t^2\vec{\jmath} + 2t^3\,\vec{k}.$$

In the previous section we computed `nn := ` $||\vec{T}\,'(t)||$ and `n := ` $||\vec{r}\,'(t)||$. Continue the calculation and compute the curvature.

```
> kappa := normal(nn/n);
```

$$\kappa := 1/6\,\frac{\sqrt{2}}{(1+t^2)^2}$$

For the example above, verify the formula

$$\kappa = \frac{||\vec{v}\times\vec{a}||}{||\vec{v}||^3},$$

where, as usual, $\vec{v}(t) = \vec{r}\,'(t)$, and $\vec{a}(t) = \vec{v}\,'(t)$.

```
> read vecfuncs:
> r := t -> 6*t*i + 3* sqrt(2)*t^2*j + 2*t^3*k:
> v := r2vec(diff(r(t),t));
> a := r2vec(diff(r(t),t,t));
> with(linalg):
> crossprod(v,a);
> normv(%)/normv(v)^3;
> normal(%);
```

Remember, the file *vecfuncs* was created in the previous section and contains the `r2vec` and `normv` functions among other things.

10.4 The gradient and directional derivatives

For a real-valued function $f(x,y,z)$, the gradient of f is defined by

$$\operatorname{grad} f(x,y,z) = \frac{\partial f}{\partial x}\vec{\imath} + \frac{\partial f}{\partial y}\vec{\jmath} + \frac{\partial f}{\partial z}\vec{k}.$$

In MAPLE it is computed using the `grad` function in the *linalg* package. We compute the gradient of $f(x,y,z) = x^3 + \sin(x + yz^2)$.

```
> with(linalg):
> f := x^3 + sin(x +y*z^2);
```

$$f := x^3 + \sin(x + yz^2)$$

```
> grad(f, [x,y,z]);
```

$$[3\,x^2 + \cos(x+yz^2), \cos(x+yz^2)z^2, 2\,\cos(x+yz^2)yz]$$

We found that

$$\text{grad}\, f = (3\,x^2 + \cos(x + yz^2))\vec{\imath} + \cos(x + yz^2)z^2\vec{\jmath} + 2\,\cos(x + yz^2)yz\vec{k}.$$

If $\vec{u}$ is a unit vector, then the directional derivative of f at (x_0, y_0, z_0) is given by

$$D_{\vec{u}}f(x_0, y_0, z_0) = \text{grad}\, f(x_0, y_0, z_0) \cdot \vec{u}.$$

As an example we compute the directional derivative of $f(x, y, z) = xy + 3yz^3$ at $(1, -1, 2)$ in the direction of $\vec{v} = \vec{\imath} + 2\vec{\jmath} + 3\vec{k}$.

```
> with(linalg):
> f := x*y+ 3*y*z^3;
```

$$f := xy + 3\,yz^3$$

```
> v := vector([1,2,3]);
```

$$v := [1, 2, 3]$$

```
> u := normalize(v);
```

$$u := [1/14\,\sqrt{14}, \frac{1}{7}\,\sqrt{14}, 3/14\,\sqrt{14}]$$

```
> grad(f,[x,y,z]);
```

$$[y, x + 3\,z^3, 9\,yz^2]$$

```
> g := subs(x=1,y=-1,z=2,%);
```

$$g := [-1, 25, -36]$$

```
> dotprod(g,u);
```

$$-\frac{59}{14}\,\sqrt{14}$$

We found that

$$D_{\vec{u}}f(1, -1, 2) = -59\sqrt{14}.$$

10.5 Extrema

10.5.1 Local extrema and saddle points

In this section we consider the problem of determining the nature of critical points of a function $f(x, y)$. We determine the critical points of the function $f(x, y) = x^3 - 3yx + y^3$.

```
> f := (x,y) -> x^3 - 3*y*x + y^3;
```

$$f := (x, y) \mapsto x^3 - 3\,yx + y^3$$

```
> criteqs := {diff(f(x,y),x)=0, diff(f(x,y),y)=0};
```

$$criteqs := \{3\,x^2 - 3\,y = 0, -3\,x + 3\,y^2 = 0\}$$

```
> solve(criteqs, {x,y});
```

$$\{y = 0, x = 0\}\,\{y = 1, x = 1\}\,\{y = RootOf(\_Z^2 + \_Z + 1, label = \_L2),$$
$$x = -1 - RootOf(\_Z^2 + \_Z + 1, label = \_L2)\}$$

We solved the equations

$$\frac{\partial f}{\partial x} = \frac{\partial f}{\partial y} = 0,$$

to find the critical points $(x, y) = (0, 0), (1, 1)$. We use the second partials test to determine the nature of the critical points (x_0, y_0). This involves the discriminant of f at (x_0, y_0) given by

$$D(x, y) = \begin{vmatrix} f_{xx}(x_0, y_0) & f_{xy}(x_0, y_0) \\ f_{xy}(x_0, y_0) & f_{yy}(x_0, y_0) \end{vmatrix}.$$

This matrix of second-order partials is called the Hessian. We use the `hessian` function in the *linalg* package. We are now ready to determine the nature of the critical points of our function $f(x, y)$ given above.

```
> f := (x,y) -> x^3 - 3*y*x + y^3;
```

$$f := (x, y) \mapsto x^3 - 3\,y x + y^3$$

```
> with(linalg):
> h := hessian(f(x,y), [x,y]);
```

$$h := \begin{bmatrix} 6\,x & -3 \\ -3 & 6\,y \end{bmatrix}$$

```
> det(h);
```

$$36\,y x - 9$$

```
> des := unapply(%,x,y);
```

$$des := (x, y) \mapsto 36\,y x - 9$$

```
> des(0,0);
```

$$-9$$

```
> des(1,1);
```

$$27$$

We find $D(0,0) = -9 < 0$, so that $(0,0)$ is a saddle point of f. We can confirm this by plotting the function $f(x,y)$ near the point $(0,0)$.

```
> f := (x,y) -> x^3 - 3*y*x + y^3;
> plot3d(f(x,y),x=-0.1..0.1,y=-0.1..0.1,axes=boxed,
    style=patch,orientation=[20,70]);
```

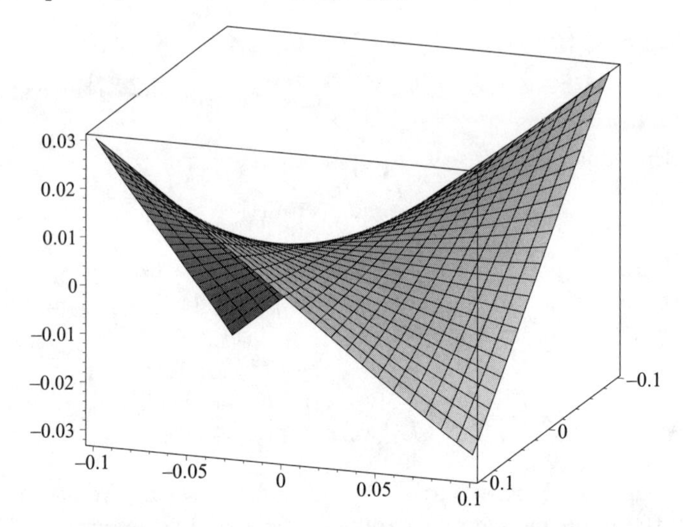

Figure 10.4 A saddle point.

Also, $D(1,1) = 27 > 0$ and $f_{xx}(1,1) = 6 > 0$, so that f has a local minimum at $(1,1)$. Plot $f(x,y)$ near $(1,1)$:

```
> plot3d(f(x,y),x=0.9..1.1,y=0.9..1.1,axes=boxed,
    style=patch,orientation=[20,70]);
```

Did your plot confirm that f has a local minimum near $(1,1)$?

10.5.2 Lagrange multipliers

Lagrange multipliers are used to calculate extreme values of a function subject to a constraint. In particular, if the function $f(x,y,z)$, subject to the constraint $g(x,y,z) = c$, has an extreme value at (x_0, y_0, z_0), then

$$\operatorname{grad} f(x_0, y_0, z_0) = \lambda \operatorname{grad} g(x_0, y_0, z_0),$$

for some scalar λ. This number λ is called the Lagrange multiplier.

We find the minimum distance between the surface $xy + xz = 4$ and the origin $(0,0,0)$ using Lagrange multipliers. We let

$$f(x,y,z) = x^2 + y^2 + z^2.$$

We want the minimum value of the function $f(x,y,z)$ subject to the constraint $g(x,y,z) = xy + xz = 4$.

```
> with(linalg):
> f := x^2 +y^2 + z^2;
```

$$f := x^2 + y^2 + z^2$$

```
> g := x*y + x*z;
```

$$g := xy + xz$$

```
> grad(f,[x,y,z])-lambda*grad(g,[x,y,z]);
```

$$[2\,x, 2\,y, 2\,z] - \lambda\,[y + z, x, x]$$

```
> evalm(%);
```

$$[2\,x - \lambda\,(y + z), 2\,y - \lambda\,x, 2\,z - \lambda\,x]$$

```
> map(V->V=0,%):
> convert(%,set):
> EQNS := % union {g=4};
```

$$EQNS := \{2\,x - \lambda\,(y + z) = 0, 2\,y - \lambda\,x = 0, 2\,z - \lambda\,x = 0, xy + xz = 4\}$$

The set EQNS contains the equations we must solve. They correspond to the equations

$$\begin{aligned} \operatorname{grad} f(x, y, z) &= \lambda \operatorname{grad} g(x, y, z), \\ g(x, y, z) &= 4. \end{aligned}$$

We set

```
> _EnvExplicit := true;
```

$$true$$

so that the solutions will be more explicit.

```
> SOL := solve(EQNS,x,y,z,lambda);
```

$$SOL := \left\{\left\{\lambda = \sqrt{2}, x = 2^{\frac{3}{4}}, z = 2^{1/4}, y = 2^{1/4}\right\},\right.$$

$$\left\{\lambda = -\sqrt{2}, x = -2\,\frac{1}{\sqrt{-\sqrt{2}}}, z = -\sqrt{-\sqrt{2}}, y = -\sqrt{-\sqrt{2}}\right\},$$

$$\left\{z = \sqrt{-\sqrt{2}}, y = \sqrt{-\sqrt{2}}, \lambda = -\sqrt{2}, x = 2\,\frac{1}{\sqrt{-\sqrt{2}}}\right\},$$

$$\left.\left\{\lambda = \sqrt{2}, z = -2^{1/4}, y = -2^{1/4}, x = -2^{\frac{3}{4}}\right\}\right\}$$

The only real solutions occur when

$$(x, y, z) = \pm(2^{\frac{3}{4}}, 2^{\frac{1}{4}}, 2^{\frac{1}{4}}).$$

```
> subs(SOL[1],sqrt(f));
```

$$\sqrt{4}\ 2^{1/4}$$

```
> radsimp(%);
```

$$2\ 2^{1/4}$$

The minimum distance is $2^{\frac{5}{4}}$.

10.6 Multiple integrals

10.6.1 Double integrals

As an example, we compute the double integral

$$\int_0^1 \int_{\sqrt{y}}^1 \sin(x^3)\, dx\, dy = \iint_R \sin(x^3)\, dx\, dy,$$

where R is the region given by

$$0 \le y \le 1, \quad \sqrt{y} \le x \le 1.$$

Try

```
> I1 := Int(Int(sin(x^3),x=sqrt(y)..1),y=0..1);
```

$$I1 := \int_0^1 \int_{\sqrt{y}}^1 \sin(x^3) dx\, dy$$

```
> value(%);
```

Did you get a terrible mess? It is clear, we should reverse the order of integration:

$$\iint_R \sin(x^3)\, dx\, dy = \int_0^1 \int_0^{x^2} \sin(x^3)\, dy\, dx.$$

```
> I2:=Int(Int(sin(x^3),y=0..x^2),x=0..1);
```

$$I2 := \int_0^1 \int_0^{x^2} \sin(x^3) dy\, dx$$

```
> value(%);
```

$$-\frac{1}{3}\cos(1) + \frac{1}{3}$$

We found that

$$\iint_R \sin(x^3)\, dx\, dy = \frac{1}{3}(1 - \cos 1).$$

Check your answer.

```
> evalf(I1);
> evalf(I2);
```

10.6.2 Triple integrals

As an example, we compute the triple integral

$$\iiint\limits_S z^2\sqrt{x^2+y^2}\,dV = \int_{-1}^{1}\int_{-\sqrt{1-x^2}}^{\sqrt{1-x^2}}\int_{-\sqrt{1-x^2-y^2}}^{\sqrt{1-x^2-y^2}} z^2\sqrt{x^2+y^2}dz\,dy\,dx,$$

where S is the unit ball centered at the origin. Try

```
> I1:=Int(z^2*sqrt(x^2+y^2),z=-sqrt(1-x^2-y^2)..
   sqrt(1-x^2-y^2)):
> I2:=Int(I1,y=-sqrt(1-x^2)..sqrt(1-x^2)):
> I3:=Int(I2,x=-1..1);
```

$$\int_{-1}^{1}\int_{-\sqrt{1-x^2}}^{\sqrt{1-x^2}}\int_{-\sqrt{1-x^2-y^2}}^{\sqrt{1-x^2-y^2}} z^2\sqrt{x^2+y^2}dz\,dy\,dx$$

```
> value(%);
```

Did you get another mess? The inner integral I2 can be written as an elliptic integral, which leaves I3 unevaluated. We try converting to cylindrical coordinates.

```
> f:=z^2*r;
```

$$z^2r$$

```
> I1:=Int(f*r,z=-sqrt(1-r^2)..sqrt(1-r^2)):
> I2:=Int(I1,r=0..1):
> I3:=Int(I2,theta=0..2*Pi);
```

$$\int_0^{2\pi}\int_0^1\int_{-\sqrt{1-r^2}}^{\sqrt{1-r^2}} z^2r^2dz\,dr\,d\theta$$

```
> value(I3);
```

$$1/24\,\pi^2$$

We found that

$$\iiint\limits_S z^2\sqrt{x^2+y^2}\,dV = \int_0^{2\pi}\int_0^1\int_{-\sqrt{1-r^2}}^{\sqrt{1-r^2}} z^2r(r\,dz\,dr\,d\theta) = \frac{\pi^2}{24}.$$

Let's try doing this integral in spherical coordinates:

$$
\begin{aligned}
r &= \rho \sin \phi, \\
x &= r \cos \theta = \rho \sin \phi \cos \theta, \\
y &= r \sin \theta = \rho \sin \phi \sin \theta, \\
z &= \rho \cos \phi, \\
dV &= \rho^2 \sin \phi \, d\rho \, d\phi \, d\theta.
\end{aligned}
$$

```
> f :=z^2*r;
```

$$f := z^2 r$$

```
> f := subs(r=rho*sin(phi),z=rho*cos(phi),f);
```

$$f := \rho^3 \left(\cos(\phi)\right)^2 \sin(\phi)$$

```
> d := rho^2*sin(phi):
> I1:=Int(f*d,phi=0..Pi):
> I2:=Int(I1,rho=0..1):
> I3:=Int(I2,theta=0..2*Pi);
```

$$I3 := \int_0^{2\pi} \int_0^1 \int_0^\pi \rho^5 \left(\cos(\phi)\right)^2 \left(\sin(\phi)\right)^2 d\phi \, d\rho \, d\theta$$

```
> value(%);
```

$$1/24 \, \pi^2$$

Again, we found that

$$\iiint_S z^2 \sqrt{x^2+y^2} \, dV = \frac{\pi^2}{24}.$$

10.6.3 The Jacobian

The Jacobian of a transformation is used to calculate a change of variables in multiple integrals. We use the `jacobian` function in the *linalg* package. As an example, we show that in spherical coordinates

$$dV = \rho^2 \sin \phi \, d\rho \, d\phi \, d\theta,$$

which was a formula we used in the previous section.

```
> r := rho*sin(phi):
> x := r*cos(theta):
> y := r*sin(theta):
> z := rho*cos(phi):
> f := vector([x,y,z]);
```

$$f := [\rho \sin(\phi) \cos(\theta), \rho \sin(\phi) \sin(\theta), \rho \cos(\phi)]$$

```
> with(linalg):
> jacobian(f,[rho,phi,theta]);
```

$$\begin{bmatrix} \sin(\phi)\cos(\theta) & \rho\cos(\phi)\cos(\theta) & -\rho\sin(\phi)\sin(\theta) \\ \sin(\phi)\sin(\theta) & \rho\cos(\phi)\sin(\theta) & \rho\sin(\phi)\cos(\theta) \\ \cos(\phi) & -\rho\sin(\phi) & 0 \end{bmatrix}$$

```
> det(%);
```

$$(\sin(\phi))^3(\cos(\theta))^2\rho^2 + (\sin(\phi))^3(\sin(\theta))^2\rho^2$$
$$+\rho^2(\cos(\phi))^2(\cos(\theta))^2\sin(\phi) + \rho^2\sin(\phi)(\sin(\theta))^2(\cos(\phi))^2$$

```
> simplify(%);
```

$$\sin(\phi)\rho^2$$

We found that

$$\frac{\partial(x,y,z)}{\partial(\rho,\phi,\theta)} = \rho^2\sin\phi,$$

which implies that

$$dV = \rho^2\sin\phi\,d\rho\,d\phi\,d\theta.$$

10.7 Vector field

10.7.1 Plotting a vector field

To plot a two-dimensional vector field, we use the `fieldplot` function in the *plots* package. See Section 6.1.10 for an example. To plot a three-dimensional vector field, we use the `fieldplot3d` function in the *plots* package. See Section 6.2.8 for an example.

10.7.2 Divergence and curl

To compute divergence we use the `diverge` function in the *linalg* package, and to compute the curl we use the `curl` function. We compute the divergence and curl of the vector field

$$\vec{F} = xy\,\vec{i} + y^2z^2\,\vec{j} + xyz\vec{k}.$$

```
> F := vector([x*y,y^2*z^2,x*y*z]);
```

$$F := [xy, y^2z^2, xyz]$$

```
> with(linalg):
> diverge(F,[x,y,z]);
```

$$y + 2\,yz^2 + xy$$

```
> curl(F, [x,y,z]);
```

$$[xz - 2\,y^2 z, -yz, -x]$$

We found that

$$\operatorname{div} \vec{F} = y + 2\,yz^2 + xy,$$
$$\operatorname{curl} \vec{F} = (xz - 2\,y^2 z)\,\vec{\imath} - yz\,\vec{\jmath} - x\,\vec{k}.$$

10.7.3 Potential functions

The function $\Phi(x, y, z)$ is a *potential function* of the vector field $\vec{F}(x, y, z)$ if

$$\operatorname{grad} \Phi = \vec{F}.$$

The function `potential` in the *linalg* package determines whether a given vector field has a potential function and calculates one if it exists. We compute a potential function for the vector field

$$\vec{F} = (yz + z^3 - 4\,y^2)\vec{\imath} + (xz - 8\,yx)\vec{\jmath} + (yx + 3\,xz^2)\,\vec{k}.$$

```
> with(linalg):
> F := vector([y*z+z^3-4*y^2, x*z-8*y*x, y*x+3*x*z^2]);
```

$$F := [yz + z^3 - 4\,y^2, xz - 8\,yx, yx + 3\,xz^2]$$

```
> potential(F,[x,y,z],'Phi');
```

$$true$$

```
> Phi;
```

$$\left(yz + z^3 - 4\,y^2\right) x$$

The call `potential(F,[x,y,z],'Phi')` assigns the potential function the name `Phi` if it exists. We found that $\vec{F}$ does have a potential function

$$\Phi = x\left(yz + z^3 - 4\,y^2\right).$$

Check your answer:

```
> grad(Phi,[x,y,z])-F;
> evalm(%);
> simplify(%);
```

Were you able to verify that $\operatorname{grad} \Phi = \vec{F}$?

A vector field $\vec{G}$ is a vector potential of a vector field $\vec{F}$ if $\vec{F} = \operatorname{curl} \vec{G}$. The MAPLE function `vecpotent` in the *linalg* package determines whether a given vector field has a vector potential and calculates one if it exists. We find a vector potential for the vector field

$$\vec{B} = (x - y)\,\vec{\imath} - y\,\vec{\jmath} - x\vec{k}.$$

```
> with(linalg):
> B := vector([x-y,-y,-x]);
```

$$B := [x - y, -y, -x]$$

```
> vecpotent(B,[x,y,z],'G');
```

$$true$$

```
> evalm(G);
```

$$[-yz + xy, -xz + yz, 0]$$

We found that $\vec{G} = (xy - yz)\,\vec{\imath} + (yz - xz)\,\vec{\jmath}$ is a vector potential for $\vec{B}$. Now check the answer:

```
> curl(G,[x,y,z]);
```

10.8 Line integrals

Let $\vec{F}(x, y, z)$ be a vector field and $\mathcal{C}$ a smooth, oriented curve parameterized by $\vec{r}(t)$. The line integral

$$\int_{\mathcal{C}} \vec{F} \cdot d\vec{r} = \int_a^b F(\vec{\vec{r}}(t)) \cdot \vec{r}\,'(t)\,dt.$$

We calculate the work done by the force field

$$\vec{F}(x, y) = 2xy\,\vec{\imath} + (x - y)\,\vec{\jmath},$$

in moving a particle once around the unit circle $\mathcal{C}$ $(x^2 + y^2 = 1)$ counterclockwise. The circle is parameterized by

$$\vec{r}(t) = \cos t\,\vec{\imath} + \sin t\,\vec{k},$$

where $0 \le t \le 2\pi$. The work done is given by the line integral

$$W = \int_{\mathcal{C}} \vec{F} \cdot d\vec{r}.$$

```
> with(linalg):
> F := (x,y) -> vector([2*x*y,x-y]);
```

$$F := (x, y) \mapsto [2\,xy, x - y]$$

```
> r := t -> vector([cos(t),sin(t)]);
```

$$r := t \mapsto [\cos(t), \sin(t)]$$

```
> dr := map(diff,r(t),t);
```

$$dr := [-\sin(t), \cos(t)]$$

```
> Int(dotprod(F(r(t)[1],r(t)[2]),dr,orthogonal),t=0..2*Pi);
```

$$\int_0^{2\pi} -2\cos(t)(\sin(t))^2 + (\cos(t) - \sin(t))\cos(t)dt$$

```
> value(%);
```

$$\pi$$

We found that

$$W = \int_C \vec{F} \cdot d\vec{r} = \pi.$$

10.9 Green's theorem

Green's theorem states that

$$\int_C M(x,y)\,dx + N(x,y)\,dy = \iint\limits_R \left(\frac{\partial N}{\partial x} - \frac{\partial N}{\partial y}\right) dA,$$

where C is a piecewise, smooth curve that encloses a simply connected region R in the plane. In the previous section, we calculated the work done by a force field using a line integral. We calculate this line integral using Green's theorem. Here,

$$\int_C \vec{F} \cdot d\vec{r} = \int_C M(x,y)\,dx + N(x,y)\,dy,$$

where $M(x,y) = 2xy$, $N(x,y) = x - y$. We apply Green's theorem with R being the unit circle given in polar coordinates by the inequalities

$$0 \le r \le 1, \quad 0 \le \theta \le 2\pi.$$

Thus we write the double integral in polar coordinates and use $dA = r\,dr\,d\theta$:

```
> M := 2*x*y;
```

$$M := 2\,xy$$

```
> N := x - y;
```

$$N := x - y$$

```
> diff(N,x)-diff(M,y);
```

$$1 - 2\,x$$

```
> subs(x=cos(theta),y=sin(theta),%);
```

$$1 - 2\cos(\theta)$$

```
> Int(Int(%*r,r=0..1),theta=0..2*Pi);
```

$$\int_0^{2\pi}\int_0^1 (1 - 2\cos(\theta))\, r dr\, d\theta$$

```
> value(%);
```

$$\pi$$

We found that

$$W = \iint_R (1 - 2x)\, dA = \pi,$$

confirming our earlier result.

10.10 Surface integrals

Let $R \subset \mathbb{R}^2$ and suppose $\vec{r} : R \longrightarrow \mathbb{R}^3$ gives a parameterization of a surface Σ. In other words, a point (x, y, z) on the surface is given by three functions:

$$x = r_1(u, v),$$
$$y = r_2(u, v),$$
$$z = r_3(u, v).$$

Let $G : \Sigma \longrightarrow \mathbb{R}$. The surface integral of G over Σ is given by

$$\iint_\Sigma G(x, y, z)\, dS = \iint_R G(\vec{r}(u, v))\, |\mathbf{T}_u \times \mathbf{T}_v|\, dA,$$

where $\mathbf{T}_u = \frac{\partial x}{\partial u}\vec{i} + \frac{\partial y}{\partial u}\vec{j} + \frac{\partial z}{\partial u}\vec{k}$, and $\mathbf{T}_v = \frac{\partial x}{\partial v}\vec{i} + \frac{\partial y}{\partial v}\vec{j} + \frac{\partial z}{\partial v}\vec{k}$.

We compute the surface area of a generic torus parameterized by

$$x = (R + r\cos u)\cos v,$$
$$y = (R + r\cos u)\sin v,$$
$$z = r\sin u,$$

where $0 \le u, v \le 2\pi$. This is the torus obtained by rotating a circle centered at $(R, 0, 0)$ radius r (in the xz-plane) about the z-axis. Try plotting an example with $R = 3$, $r = 1$:

```
> x:=(3+cos(u))*cos(v);
> y:=(3+cos(u))*sin(v);
> z:=sin(u);
```

```
> plot3d([x,y,z],u=0..2*Pi,v=0..2*Pi,scaling=constrained,
    orientation=[45,65]);
```

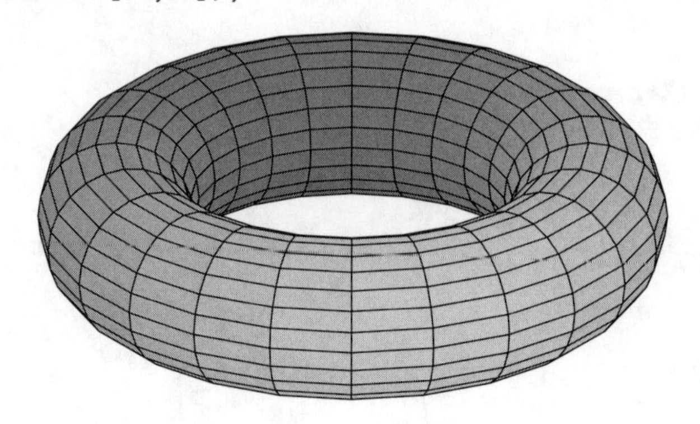

Figure 10.5 A torus.

We now compute the surface area of our generic torus $\mathcal{T}$:

```
> with(linalg):
> x:=(R+r*cos(u))*cos(v);
```

$$x := (R + r\cos(u))\cos(v)$$

```
> y:=(R+r*cos(u))*sin(v);
```

$$y := (R + r\cos(u))\sin(v)$$

```
> z:=r*sin(u);
```

$$z := r\sin(u)$$

```
> rv:=vector([x,y,z]);
```

$$rv := [(R + r\cos(u))\cos(v), (R + r\cos(u))\sin(v), r\sin(u)]$$

```
> Tu := map(diff,rv,u);
```

$$Tu := [-r\sin(u)\cos(v), -r\sin(u)\sin(v), r\cos(u)]$$

```
> Tv := map(diff,rv,v);
```

$$Tv := [-(R + r\cos(u))\sin(v), (R + r\cos(u))\cos(v), 0]$$

```
> cp := crossprod(Tu,Tv);
```

$$cp := [-r\cos(u)(R + r\cos(u))\cos(v), -r\cos(u)(R + r\cos(u))\sin(v),$$
$$-r\sin(u)(\cos(v))^2(R + r\cos(u)) - r\sin(u)(\sin(v))^2(R + r\cos(u))]$$

```
> simplify(dotprod(cp,cp,orthogonal)):
> n := radsimp(sqrt(factor(%)));
```

$$n := r\,(R + r\cos(u))$$

```
> int(int(n,u=0..2*Pi),v=0..2*Pi);
```

$$4\,R\pi^2 r$$

We find that

$$\begin{aligned}\text{Surface Area} &= \iint_T 1\,dS = \int_0^{2\pi}\int_0^{2\pi} |\mathbf{T}_u \times \mathbf{T}_v|\,du\,dv \\ &= \int_0^{2\pi}\int_0^{2\pi} r\,(R + r\cos(u))\,du\,dv = 4\pi^2\,R\,r.\end{aligned}$$

10.10.1 Flux of a vector field

As above, let $R \subset \mathbb{R}^2$ and suppose $\vec{r} : R \longrightarrow \mathbb{R}^3$ gives a parameterization of an oriented surface Σ, with unit normal $\vec{n}(u,v)$ at the point $\vec{r}(u,v)$. Let $\vec{F} : \Sigma \longrightarrow \mathbb{R}^3$ be a vector field on Σ. The flux of $\vec{F}$ over Σ is given by

$$\iint_\Sigma \vec{F}\cdot d\vec{S} = \iint_\Sigma \vec{F}\cdot\vec{n}\,dS = \iint_R \vec{F}\cdot(\mathbf{T}_u \times \mathbf{T}_v)\,dA,$$

where $\mathbf{T}_u$, $\mathbf{T}_v$ are defined as above.

As an example, we find the flux of the vector field

$$\vec{F}(x,y,z) = y^4\,\vec{\imath} + z^4\,\vec{\jmath} + x^4\,\vec{k},$$

over the unit hemisphere $x^2 + y^2 + z^2 = 1$, $z \geq 0$. We use spherical coordinates to parameterize the unit sphere:

$$\vec{r}(\phi,\theta) = \sin\phi\cos\theta\,\vec{\imath} + \sin\phi\sin\theta\,\vec{\imath} + \cos\phi\,\vec{k},$$

where $0 \leq \theta \leq 2\pi$, and $0 \leq \phi \leq \pi/2$. First we compute $\mathbf{T}_\phi \times \mathbf{T}_\theta$:

```
> with(linalg):
> x:=sin(phi)*cos(theta);
```

$$x := \sin(\phi)\cos(\theta)$$

```
> y:=sin(phi)*sin(theta);
```

$$y := \sin(\phi)\sin(\theta)$$

```
> z:=cos(phi);
```

$$z := \cos(\phi)$$

```
> rv:=vector([x,y,z]);
```

$$rv := [\sin(\phi)\cos(\theta), \sin(\phi)\sin(\theta), \cos(\phi)]$$

```
> Tphi := map(diff,rv,phi);
```

$$Tphi := [\cos(\phi)\cos(\theta), \cos(\phi)\sin(\theta), -\sin(\phi)]$$

```
> Ttheta := map(diff,rv,theta);
```

$$Ttheta := [-\sin(\phi)\sin(\theta), \sin(\phi)\cos(\theta), 0]$$

```
> cp := simplify(crossprod(Tphi,Ttheta));
```

$$cp := [-\left(-1 + (\cos(\phi))^2\right)\cos(\theta), -\left(-1 + (\cos(\phi))^2\right)\sin(\theta),$$
$$\cos(\phi)\sin(\phi)]$$

We found

$$\mathbf{T}_\phi \times \mathbf{T}_\theta = \sin^2\phi\,\cos\theta\,\vec{\imath} + \sin^2\phi\,\sin\theta\,\vec{\jmath} + \cos\phi\,\sin\phi\,\vec{k}.$$

We are now ready to compute the flux:

```
> F := (x,y,z) -> vector([y^4,z^4,x^4]);
```

$$F := (x, y, z) \mapsto [y^4, z^4, x^4]$$

```
> dp := dotprod(F(x,y,z),cp,orthogonal);
```

$$dp := -(\sin(\phi))^4(\sin(\theta))^4\left(-1 + (\cos(\phi))^2\right)\cos(\theta) -$$
$$(\cos(\phi))^4\left(-1 + (\cos(\phi))^2\right)\sin(\theta) + (\sin(\phi))^5(\cos(\theta))^4\cos(\phi)$$

```
> int(int(dp,phi=0..Pi/2),theta=0..2*Pi);
```

$$1/8\,\pi$$

We found

$$\text{Flux} = \int_0^{2\pi}\int_0^{\pi/2} \vec{F}\cdot(\mathbf{T}_\phi \times \mathbf{T}_\theta)\,d\phi\,d\theta = \frac{\pi}{8}.$$

10.10.2 Stoke's theorem

Stoke's theorem states that

$$\int_C \vec{F}\cdot d\vec{r} = \iint_{\Sigma} \operatorname{curl}\vec{F}\cdot d\vec{S},$$

where Σ is an oriented surface bounded by a simple closed curve $\mathcal{C}$ with positive orientation, and $\vec{F}$ is a vector field.

We apply Stoke's theorem to the problem considered in the previous section. Luckily, the vector field

$$\vec{F}(x,y,z) = y^4\,\vec{i} + z^4\,\vec{j} + x^4\,\vec{k},$$

has a vector potential.

```
> with(linalg):
> F := (x,y,z) -> vector([y^4,z^4,x^4]);
```

$$F := (x,y,z) \mapsto [y^4, z^4, x^4]$$

```
> vecpotent(F(x,y,z),[x,y,z],G);
```

$$true$$

```
> evalm(G);
```

$$[\frac{1}{5}\,z^5 - x^4 y, -y^4 z, 0]$$

We found that

$$\operatorname{curl}\vec{G} = \vec{F},$$

where

$$\vec{G} = \left(\frac{1}{5}\,z^5 - x^4 y\right)\vec{i} - y^4 z\,\vec{j}.$$

We check our answer.

```
> curl(G,[x,y,z]);
```

$$[y^4, z^4, x^4]$$

Let Σ be the upper unit hemisphere $x^2+y^2+z^2=1$, $z \geq 0$. Then by Stoke's theorem, we have

$$\iint_{\Sigma} \vec{F}\cdot d\vec{S} = \iint_{\Sigma} \operatorname{curl}\vec{G}\cdot d\vec{S} = \int_C \vec{G}\cdot d\vec{r},$$

where $\mathcal{C}$ is the positive unit circle $x^2+y^2=1$ in the xy-plane. We use this result to compute the flux integral:

```
> r:=t->vector([cos(t),sin(t),0]);
```

$$r := t \mapsto [\cos(t), \sin(t), 0]$$

```
> dr := map(diff,r(t),t);
```

$$dr := [-\sin(t), \cos(t), 0]$$

```
> rsubs := J -> subs({x=cos(t),y=sin(t),z=0},J);
```

$$rsubs := J \mapsto subs(\{x = cos(t), y = sin(t), z = 0\}, J);$$

```
> GG := map(rsubs,G);
```

$$GG := [-(\cos(t))^4 \sin(t), 0, 0]$$

```
> Gdr := dotprod(GG,dr,orthogonal);
```

$$Gdr := (\cos(t))^4 (\sin(t))^2$$

```
> int(Gdr,t=0..2*Pi);
```

$$1/8\,\pi$$

We found

$$\text{Flux} = \iint_{\Sigma} \vec{F} \cdot d\vec{S} = \int_C \vec{G} \cdot d\vec{r}, = \int_0^{2\pi} \cos^4 t \sin^2 t \, dt = \frac{\pi}{8},$$

which confirms the result obtained in the previous section.

10.10.3 The divergence theorem

The divergence theorem states that

$$\iint_{\Sigma} \vec{F} \cdot d\vec{S} = \iint_{\Sigma} \vec{F} \cdot \vec{n}\, dS = \iiint_{D} \operatorname{div} \vec{F}\, dV,$$

where D is a simple solid region whose boundary surface Σ is oriented by the normal $\vec{n}$ directed outward from D, and $\vec{F}$ is a vector field.

We verify the divergence theorem for the vector field

$$\vec{F}(x, y, z) = x^3 \,\vec{\imath} + y^3 \,\vec{\jmath} + z^3 \,\vec{k},$$

over the solid unit sphere D, $x^2 + y^2 + z^2 \le 1$. In spherical coordinates this is given by $0 \le \theta \le 2\pi$, $0 \le \phi \le \pi$, $0 \le \rho \le 1$. Clearly,

$$\operatorname{div} \vec{F} = 3(x^2 + y^2 + z^2) = 3\rho^2.$$

We compute the divergence integral.

```
> divF := 3*rho^2;
```

$$divF := 3\,\rho^2$$

```
> dV := rho^2*sin(phi);
```

$$dV := \rho^2 \sin(\phi)$$

```
> int(int(int(divF*dV,theta=0..2*Pi),phi=0..Pi),rho=0..1);
```

$$\frac{12}{5}\,\pi$$

We found that

$$\iiint_D \operatorname{div} \vec{F}\, dV = \frac{12\pi}{5}.$$

Let Σ be the unit sphere $x^2 + y^2 + z^2 = 1$. The surface integral

$$\iint_\Sigma \vec{F} \cdot d\vec{S} = \iint_R \vec{F} \cdot (\mathbf{T}_\phi \times \mathbf{T}_\theta)\, dA,$$

where $\mathbf{T}_\phi$, $\mathbf{T}_\theta$ are given in Section 10.10.1 and R is given by $0 \le \phi \le \pi$, $0 \le \theta \le 2\pi$. In Section 10.10.1 we found that

$$\mathbf{T}_\phi \times \mathbf{T}_\theta = \sin^2\phi\, \cos\theta\, \vec{i} + \sin^2\phi\, \sin\theta\, \vec{j} + \cos\phi\, \sin\phi\, \vec{k}.$$

We are now ready to compute the surface integral:

```
> with(linalg):
> x:=sin(phi)*cos(theta):
> y:=sin(phi)*sin(theta):
> z:=cos(phi):
> rv:=vector([x,y,z]):
> Tphi := map(diff,rv,phi):
> Ttheta := map(diff,rv,theta):
> cp := simplify(crossprod(Tphi,Ttheta)):
> F := (x,y,z) -> vector([x^3,y^3,z^3]):
> dp := dotprod(F(x,y,z),cp,orthogonal):
> int(int(dp,phi=0..Pi),theta=0..2*Pi);
```

$$\frac{12}{5}\,\pi$$

We found that

$$\iint_\Sigma \vec{F} \cdot d\vec{S} = \frac{12\pi}{5},$$

and thus

$$\iiint_D \operatorname{div} \vec{F}\, dV = \frac{12\pi}{5} = \iint_\Sigma \vec{F} \cdot d\vec{S},$$

confirming the divergence theorem for our example.

11. COMPLEX ANALYSIS

11.1 Arithmetic of complex numbers

In MAPLE the complex number i (also known as j to electrical engineers) is represented by the symbol I.

```
> I^2;
```

$$-1$$

Observe that I^2 returned -1 as expected. Arithmetic of complex numbers is easy for MAPLE:

```
> z1 := 2 + 3*I;
```

$$z1 := 2 + 3\,I$$

```
> z2 := 4 - I;
```

$$z2 := 4 - I$$

```
> z1 + z2;
```

$$6 + 2\,I$$

```
> z1 - z2;
```

$$-2 + 4\,I$$

```
> z1 * z2;
```

$$11 + 10\,I$$

```
> z1/z2;
```

$$\frac{5}{17} + \frac{14}{17}\,I$$

```
> abs(z1);
```

$$\sqrt{13}$$

```
> Re(z1);
```

$$2$$

```
> Im(z1);
```

$$3$$

```
> conjugate(z1);
```

$$2 - 3\,I$$

For $z_1 = 2 + 3i$ and $z_2 = 4 - i$, MAPLE easily found that

$$\begin{aligned} z_1 + z_2 &= 6 + 2i, \\ z_1 - z_2 &= -2 + 4i, \\ z_1\, z_2 &= 11 + 10i, \\ \frac{z_1}{z_2} &= \frac{5}{17} + \frac{14}{17}i, \\ |z_1| &= \sqrt{13}, \\ \Re z_1 &= 2, \\ \Im z_2 &= 3, \\ \overline{z_1} &= 2 - 3i. \end{aligned}$$

Notice that all complex arithmetic was performed automatically. In MAPLE the functions `Re` and `Im` give the real and imaginary parts, respectively. Naturally, `conjugate` is the conjugate function.

Now try:

```
> z := x + I*y;
```

$$z := x + Iy$$

```
> Re(z);
```

$$\mathit{Re}(x + Iy)$$

```
> Im(z);
```

$$\mathit{Im}(x + Iy)$$

Notice this time that `Re(z)` did not return x for the real part of $z = x + iy$. The problem is that we have not told MAPLE that x and y are real. Try again:

```
> assume(x,real);
> assume(y,real);
> z := x + y*I;
```

$$z := x\tilde{} + y\tilde{}\,I$$

```
> Re(z);
```

$$x\tilde{}$$

```
> Im(z);
```

$$y\tilde{}$$

Alternatively, we can use the `evalc` function. This function attempts to split a complex number into its real and imaginary parts. First we restart:

```
> restart;
> x,y;
```

$$x, y$$

Now try

```
> z := x + I*y;
```

$$x + Iy$$

```
> evalc(Re(z));
```

$$x$$

```
> evalc(Im(z));
```

$$y$$

```
> evalc(abs(z));
```

$$\sqrt{x^2 + y^2}$$

```
> evalc(conjugate(z));
```

$$x - Iy$$

Note that the `evalc` function assumed that x and y were real.

11.2 Polar form

To convert the complex number z to polar form we use the command `convert(z,polar)`. Let's find the polar form of $z = \sqrt{3} + i$:

```
> z := sqrt(3) + I;
```

$$z := \sqrt{3} + I$$

```
> convert(z,polar);
```

$$\mathit{polar}(2, \frac{1}{6}\pi)$$

This means that

$$z = \sqrt{3} + i = 2\,e^{\pi i/6} = 2\,(\cos\pi/6 + i\sin\pi/6).$$

In general, `polar(r,`θ`)` corresponds to the complex number $r\,e^{i\theta}$. We use `evalc` to convert polar form to Cartesian form:

```
> w := polar(sqrt(2),Pi/4);
```

$$w := \mathit{polar}(\sqrt{2}, \frac{1}{4}\pi)$$

```
> evalc(w);
```

$$1 + I$$

We found

$$\sqrt{2}\,e^{\pi i/4} = 1 + i.$$

In MAPLE the principal value of the argument is given by the `argument` function.

```
> argument(1-I);
```

$$-\frac{1}{4}\pi$$

```
> argument(polar(4,5*Pi/7));
```

$$5/7\,\pi$$

```
> argument(polar(4,12*Pi/7));
```

$$-\frac{2}{7}\,\pi$$

We found that

$$\begin{aligned}\operatorname{Arg}(1-i) &= -\frac{\pi}{4},\\ \operatorname{Arg}(4\,e^{5\pi/7}) &= \frac{5\pi}{7},\\ \operatorname{Arg}(4\,e^{12\pi/7}) &= -\frac{2\pi}{7}.\end{aligned}$$

A related function is `arctan`. For real numbers a, b, `arctan(b,a)` returns $\operatorname{Arg}(a+bi)$.

```
> argument(sqrt(3) - I);
```

$$-\frac{1}{6}\,\pi$$

```
> arctan(-1,sqrt(3));
```

$$-\frac{1}{6}\,\pi$$

11.3 *n*th roots

As an example, let's find the 4th roots of $-16i$. We use the `solve` function:

```
> solve(z^4=-16*I);
```

$$(-16\,I)^{1/4}, I(-16\,I)^{1/4}, -(-16\,I)^{1/4}, -I(-16\,I)^{1/4}$$

More explicit solutions would be nice. Let's try `simplify`:

```
> map(simplify,[%]);
```

$$[(-16\,I)^{1/4}, I(-16\,I)^{1/4}, -(-16\,I)^{1/4}, -I(-16\,I)^{1/4}]$$

Let's try `evalc` and `simplify`:

```
> solve(z^4=-16*I):
> map(simplify,[%]):
```

```
> map(evalc,%):
> map(simplify,%);
```

$$[2\cos(\frac{1}{8}\pi) - 2I\sin(\frac{1}{8}\pi), 2\sin(\frac{1}{8}\pi) + 2I\cos(\frac{1}{8}\pi),$$
$$-2\cos(\frac{1}{8}\pi) + 2I\sin(\frac{1}{8}\pi), -2\sin(\frac{1}{8}\pi) - 2I\cos(\frac{1}{8}\pi)]$$

That's better. Looks like we should have used polar form. Try

```
> p := convert(-16*I,polar);
> solve(z^4 = p);
> map(simplify,[%]);
```

In any case, the 4th roots of $-16i$ are

$$z = \pm 2\,(\cos\tfrac{\pi}{8} \pm i\sin\tfrac{\pi}{8}).$$

11.4 The Cauchy-Riemann equations and harmonic functions

Let $z = x + iy$, and suppose

$$f(z) = u(x,y) + i\,v(x,y),$$

is analytic on some domain. Then the Cauchy-Riemann equations,

$$\frac{\partial u}{\partial x} = \frac{\partial v}{\partial y}, \qquad \frac{\partial v}{\partial x} = -\frac{\partial u}{\partial y},$$

hold on this domain. The converse holds, assuming all the partial derivatives are continuous. As an example, we show, using MAPLE, that $f(z) = z^7$ satisfies the Cauchy-Riemann equations.

```
> z := x + I*y;
```

$$x + iy$$

```
> u := evalc(Re(z^7));
```

$$x^7 - 21\,x^5y^2 + 35\,x^3y^4 - 7\,xy^6$$

```
> v := evalc(Im(z^7));
```

$$7\,x^6y - 35\,x^4y^3 + 21\,x^2y^5 - y^7$$

We see that

$$f(z) = z^7 = u + i\,v,$$

where

$$u = x^7 - 21\,x^5y^2 + 35\,x^3y^4 - 7\,xy^6,$$
$$v = 7\,x^6y - 35\,x^4y^3 + 21\,x^2y^5 - y^7.$$

```
> diff(u,x)-diff(v,y);
```

$$0$$

```
> diff(v,x)+diff(u,y);
```

$$0$$

We see that the function $f(z) = z^7$ satisfies the Cauchy-Riemann equations and is thus analytic because the partial derivatives are clearly continuous.

A real-valued function $u(x, y)$ is harmonic if it satisfies Laplace's equation

$$\frac{\partial^2}{\partial x^2}u + \frac{\partial^2}{\partial y^2}u = 0,$$

and $u(x, y)$ has continuous first- and second-order partial derivatives. The left side of the equation above is the Laplacian of $u(x, y)$. This can be computed using the `laplacian` function in the *linalg* package. We use MAPLE to show that

$$u(x, y) = \cos^3 x \cosh^3 y - 3 \cos x \cosh y \sin^2 x \sinh^2 y,$$

is harmonic.

```
> with(linalg):
> u:=cos(x)^3*cosh(y)^3-3*cos(x)*cosh(y)*sin(x)^2*sinh(y)^2;
```

$$(\cos(x))^3 (\cosh(y))^3 - 3 \cos(x) \cosh(y) (\sin(x))^2 (\sinh(y))^2$$

```
> laplacian(u,[x,y]);
```

$$0$$

We see that $u(x, y)$ satisfies Laplace's equation and is harmonic because the first- and second-order partial derivatives are clearly continuous.

If $u(x, y)$ is harmonic on a simply connected domain, then there is a harmonic function $v(x, y)$ such that

$$f(z) = u(x, y) + i\,v(x, y)$$

is analytic. The function $v(x, y)$ is called the *harmonic conjugate* of $u(x, y)$. We use MAPLE to find a harmonic conjugate of our function $u(x, y)$ above.

We want

$$\frac{\partial u}{\partial x} = \frac{\partial v}{\partial y},$$

so that

$$v = \int \frac{\partial u}{\partial x}\,dy.$$

```
> u:=cos(x)^3*cosh(y)^3-3*cos(x)*cosh(y)*sin(x)^2
  *sinh(y)^2;
```

$$(\cos(x))^3 (\cosh(y))^3 - 3 \cos(x) \cosh(y) (\sin(x))^2 (\sinh(y))^2$$

```
> v:=simplify(int(diff(u,x),y)+ K(x));
```

$$-4\ (\cos(x))^2 \sin(x) \sinh(y) (\cosh(y))^2 + (\cos(x))^2 \sin(x) \sinh(y) \\ + \sin(x) \sinh(y) (\cosh(y))^2 - \sin(x) \sinh(y) + K(x)$$

We also require that

$$\frac{\partial v}{\partial x} = -\frac{\partial u}{\partial y}.$$

```
> simplify(diff(v,x)+diff(u,y));
```

$$\frac{d}{dx} K(x)$$

We can take $K(x) = 0$ and

$$v(x, y) = -\sin x \sinh y \left(4 \cos^2 x \cosh^2 y - \cos^2 x - \cosh^2 y + 1\right),$$

is a harmonic conjugate of $u(x, y)$. Do you recognize the analytic function

$$f(z) = u(x, y) + i\, v(x, y)?$$

11.5 Elementary functions

MAPLE knows the complex exponential function

$$e^z = e^x (\cos y + i\ \sin y),$$

where $z = x + i\, y$.

```
> z := x + I*y;
```

$$z := x + I\, y$$

```
> evalc(exp(z));
```

$$e^x \cos(y) + ie^x \sin(y)$$

```
> exp(3/2*ln(2) + Pi/4*I);
```

$$e^{\frac{3}{2}\ \ln(2) + \frac{1}{4}\ i\pi}$$

```
> evalc(%);
```

$$2 + 2\, i$$

We found that for $z = \frac{3}{2}\ \ln 2 + \frac{\pi}{4}\, i$, $e^z = 2(1 + i)$.

MAPLE knows the complex trigonometric functions.

```
> z := x + I*y;
```

$$z := x + I\, y$$

```
> evalc(sin(z));
```

$$\sin(x)\cosh(y) + i\cos(x)\sinh(y)$$

```
> evalc(cos(z));
```

$$\cos(x)\cosh(y) + -i\sin(x)\sinh(y)$$

```
> cos(ln(2)*I);
```

$$5/4$$

We found that $\cos(i \ln 2) = \frac{5}{4}$. Try the following:

```
> z := x + I*y;
> evalc(tan(z));
> evalc(cot(z));
> evalc(sec(z));
> evalc(csc(z));
> evalc(cosh(z));
> evalc(sinh(z));
```

MAPLE knows the principal value of the complex logarithm

$$\operatorname{Log} z = \ln|z| + i \operatorname{Arg} z.$$

```
> z := x + I*y;
```

$$z := x + I\,y$$

```
> log(z);
```

$$\ln(x + I\,y)$$

```
> evalc(%);
```

$$1/2\ \ln(x^2 + y^2) + I\ \arctan(y, x)$$

```
> log(-1);
```

$$I\pi$$

```
> log(I);
```

$$1/2\,I\pi$$

```
> w :=log(exp(2+101*I*Pi/3));
```

$$\ln(e^{2+\frac{101}{3}\,I\pi})$$

```
> evalc(w);
```

$$2 - \frac{1}{3}\,I\pi$$

We found

$$\mathrm{Log}\,(-1) = \pi\, i,$$
$$\mathrm{Log}\,(i) = \frac{\pi i}{2},$$
$$\mathrm{Log}\,(e^{2+101\pi i/3}) = 2 - \frac{1}{3}\, i\pi.$$

Now try

```
> z:=x+I*y;
> evalc(exp(log(z)));
> evalc(log(exp(z)));
```

Did you get the results you expected?

MAPLE is able to compute complex exponents.

```
> z := I^(2*I);
```

$$z := I^{2\,I}$$

```
> evalc(z);
```

$$e^{-\pi}$$

Here MAPLE computed the principal value of i^{2i}.

$$i^{2i} = e^{2i\,\mathrm{Log}\, i} = e^{2i(\pi i/2)} = e^{-\pi}.$$

11.6 Conformal mapping

Let $D \subset \mathbb{C}$. A mapping $f : D \longrightarrow \mathbb{C}$ is conformal if it preserves angles in size and sense. If $f(z)$ is analytic and $f'(z)$ is nonzero on D, then the mapping f is conformal. The `conformal` function in the *plots* package is used to plot the image of rectangular regions under a complex function $f(z)$. The syntax of the `conformal` function has the form

```
conformal(f, z=z1..z2)
conformal(f, z=z1..z2, grid=[m,n])
conformal(f, z=z1..z2, grid=[m,n], numxy=[a,b])
```

Here f is an expression in the variable z. This plots the image of a rectangle with corners at $z = z_1$, z_2. It actually plots the image of horizontal and vertical grid lines in the rectangle. The option `grid=[m,n]` specifies the size of the grid. The option `numxy=[a,b]` specifies the number of points to plot on the image of each grid line.

We plot the image of the rectangle $R = [-1, 1] \times [0, 1]$ under the mapping $w = z^2$. First we plot the rectangle R:

```
> with(plots):
> conformal(z,z=(-1)..(1+I),grid=[11,6],labels=[x,y],
    scaling=constrained);
```

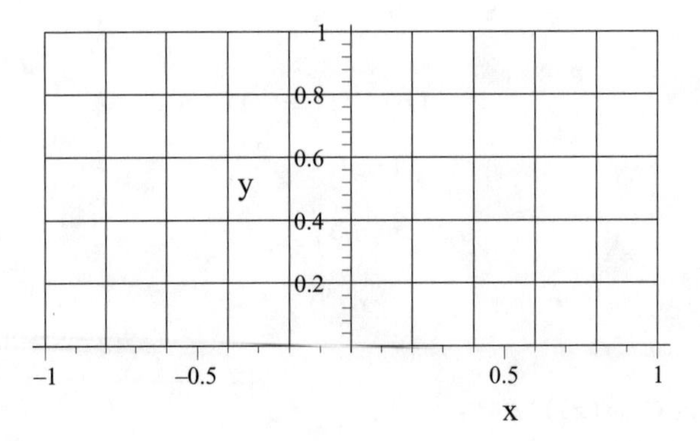

Figure 11.1 Rectangle R in the xy-plane.

Now we plot the image under the mapping $w = z^2$:

```
> with(plots):
> conformal(z^2,z=(-1)..(1+I),grid=[11,6],labels=[u,v],
   scaling=constrained);
```

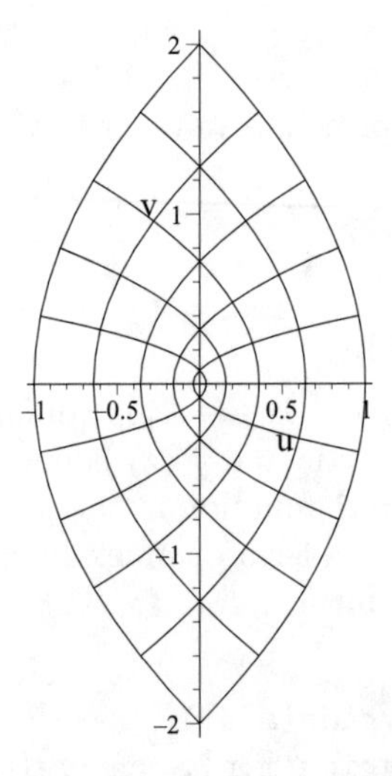

Figure 11.2 Image of R under $w = z^2$ in the uv-plane.

Try plotting the image of a rectangle $[0, 1] \times [0, 1]$ under the mapping $w = e^{2\pi iz}$.

```
> with(plots):
> conformal(exp(2*Pi*I*z),z=0..(1+I));
> conformal(exp(2*Pi*I*z),z=0..(1+I),grid=[20,20],
   numxy=[30,30]);
```

With MAPLE 7, the *plots* package contains a new function `conformal3d`, that projects a conformal map onto the Riemann sphere. Try plotting the image of the rectangle $[-2, 2] \times [0, 2]$, under that map $w = z^2$, onto the Riemann sphere:

```
> with(plots):
> conformal3d(z^2,z=-2..2+2*I,grid=[40,40]);
```

We now consider the problem of plotting the image of a curve $\mathcal{C}$ under a complex mapping $f(z)$. Suppose $\mathcal{C}$ is parameterized by

$$x = x(t), \quad y = y(t),$$

$0 \leq t \leq b$. Then the image $f(\mathcal{C})$ is parameterized by

$$u = \Re(f(x(t) + i\,y(t)), \quad v = \Im(f(x(t) + i\,y(t)), \quad (a \leq t \leq b),$$

in the uv-plane. As an example we consider the Joukowski airfoil. Here

$$f(z) = z + \frac{1}{z}.$$

```
> f := z -> z + 1/z;
```

$$f := z \mapsto z + z^{-1}$$

```
> factor(diff(f(z),z));
```

$$\frac{(z-1)\,(z+1)}{z^2}$$

We see that $f(z)$ is conformal except at $z = \pm 1$. Now we consider any circle centered on the imaginary axis that passes through ± 1. Let $z = b\,i$ be the center. Then the circle is parameterized by

$$x = \sqrt{1+b^2}\cos\theta, \quad y = b + \sqrt{1+b^2}\sin\theta,$$

where $0 \leq \theta \leq 2\pi$. Try plotting the circle with $b = 1$.

```
> b := 1:
> x := sqrt(1+b^2)*cos(t);
> y := b + sqrt(1+b^2)*sin(t);
> plot([x,y,t=0..2*Pi],scaling=constrained);
```

Did you get the correct circle? Now we plot the image of this circle under the mapping $w = f(z) = z + 1/z$.

```
> z := x + I*y;
> u := simplify(evalc(Re(f(z))));
> v := simplify(evalc(Im(f(z))));
> plot([u,v,t=0..2*Pi],scaling=constrained);
```

Did you get a single arc joining the points $z = \pm 2$? For the Joukowski airfoil we must a consider a circle passing through $z = -1$ but whose center is slightly to the right of the imaginary axis. The proc `JoukowskiP(b,` ϵ`)` plots the image of the circle center $b + i\epsilon$ and passing through -1, under the mapping $w = f(z) = z + 1/z$.

```
> JoukowskiP := proc(b,epsilon)
>   local r,x,y,z,u,v,t:
>   r := sqrt( (1+epsilon)^2 + b^2):
>   x := r*cos(t)+epsilon:
>   y := b + r*sin(t):
>   z := x + I*y:
>   u := simplify(evalc(Re(z + 1/z))):
>   v := simplify(evalc(Im(z + 1/z))):
>   return plot([u,v,t=0..2*Pi]):
> end proc:
```

We plot the Joukowski airfoil with $b = \epsilon = 1/10$.

```
> with(plots):
> display(JoukowskiP(1/10,1/10),scaling=constrained,
   thickness=2);
```

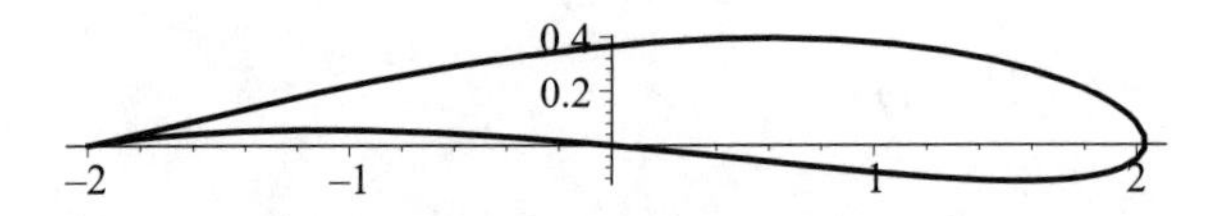

Figure 11.3 The Joukowski airfoil.

Try plotting the airfoil for other values of b and ϵ.

```
> display(JoukowskiP(0,1/10),scaling=constrained);
> display(JoukowskiP(1,1/10),scaling=constrained);
> display(seq(JoukowskiP(k/5,1/10),k=0..5),
   scaling=constrained);
```

11.7 Taylor series and Laurent series

If $f(z)$ is analytic for $|z - z_0| < r$, then $f(z)$ has a Taylor series expansion

$$f(z) = \sum_{n=0}^{\infty} \frac{f^{(n)}(z_0)}{n!}(z - z_0)^n,$$

valid for $|z - z_0| < r$. In MAPLE, the first T terms of the Taylor series of $f(z)$ near $z = z_0$ is computed using the command `taylor(f(z), z=`z_0`, T)`. The function

$$f(z) = \frac{2 - z}{(1 - z)^2},$$

is analytic for $z \neq 1$. We compute the first few terms of the Taylor expansion near $z = 0$.

```
> f := (2-z)/(1-z)^2;
```

$$f := \frac{2-z}{(1-z)^2}$$

```
> taylor(f, z=0, 10);
```

$$2 + 3\,z + 4\,z^2 + 5\,z^3 + 6\,z^4 + 7\,z^5 + 8\,z^6 + 9\,z^7 + 10\,z^8 + 11\,z^9 + O\left(z^{10}\right)$$

It would seem that

$$f(z) = \frac{2-z}{(1-z)^2} = \sum_{n=0}^{\infty}(n+2)\,z^n,$$

for $|z| < 1$. Because our function $f(z)$ is analytic at $z = 2$ we compute a Taylor expansion near $z = 2$.

```
> taylor(f, z=2);
```

$$-(z-2) + 2\,(z-2)^2 - 3\,(z-2)^3 + 4\,(z-2)^4 - 5\,(z-2)^5 + O\left((z-2)^6\right)$$

It would seem that

$$f(z) = \frac{2-z}{(1-z)^2} = \sum_{n=1}^{\infty}(-1)^n n\,(z-2)^n,$$

for $|z-2| < 1$. Our function $f(z)$ is not analytic at $z = 1$. See what happens when we use `taylor` near $z = 1$.

```
> taylor(f, z=1, 10);
Error, does not have a taylor expansion, try series()
```

However, $f(z)$ does have a Laurent series expansion near $z = 1$ in powers of $(z-1)$.

```
> S := series(f, z=1,10);
```

$$S := (z-1)^{-2} - (z-1)^{-1}$$

It would seem that the Laurent series has only two terms,

$$f(z) = \frac{2-z}{(1-z)^2} = (z-1)^{-2} - (z-1)^{-1}.$$

This is easy to check.

```
> S := series(f, z=1,10);
> normal(f - S);
```

Did this simplify to 0? The problem is that S is still a series.

```
> whattype(S);
```

$$\texttt{series}$$

First we must convert S to a polynomial.

```
> P := convert(S, polynom);
```

$$\frac{1}{(z-1)^2} - \frac{1}{(z-1)}$$

```
> normal(f - P);
```

$$0$$

In general, if $f(z)$ is analytic on an annulus, $r_1 < |z - z_0| < r_2$, then $f(z)$ has a Laurent series expansion of the form

$$f(z) = \sum_{n=-\infty}^{\infty} a_n (z - z_0)^n,$$

valid for $r_1 < |z - z_0| < r_2$. We use the **series** function to compute Laurent series.

The function

$$g(z) = \frac{-2z + 3}{z(z-1)(z-2)},$$

is analytic for $z \neq 1, 2$. We compute Laurent series expansions for $g(z)$ in powers of z. There are three different Laurent series because there are three possible annuli centered at $z = 0$: $0 < |z| < 1$, $1 < |z| < 2$, and $|z| > 2$.

$\underline{0 < |z| < 1}$

```
> g := (-2*z + 3)/z/(z-1)/(z-2);
```

$$g := \frac{-2\,z + 3}{z\,(z-1)\,(z-2)}$$

```
> series(g, z=0, 6);
```

$$\frac{3}{2}z^{-1} + \frac{5}{4} + \frac{9}{8}z + \frac{17}{16}z^2 + \frac{33}{32}z^3 + \frac{65}{64}z^4 + \frac{129}{128}z^5 + O\left(z^6\right)$$

It would seem that

$$g(z) = \frac{-2z + 3}{z(z-1)(z-2)} = \sum_{n=-1}^{\infty} \left(1 + \frac{1}{2^{n+2}}\right) z^n,$$

for $0 < |z| < 1$.

$1 < |z| < 2$

To compute the Laurent expansion on this region, we first need the partial fraction expansion of $g(z)$.

```
> PF := convert(g, parfrac, z);
```

$$PF := \frac{3}{2}\frac{1}{z} - \frac{1}{z-1} - \frac{1}{2}\frac{1}{z-2}$$

We assign names to each term of the partial fraction expansion.

```
> g1 := op(1,PF);
```

$$g1 := \frac{3}{2}\frac{1}{z}$$

```
> g2 := op(2,PF);
```

$$g2 := -\frac{1}{z-1}$$

```
> g3 := op(3,PF);
```

$$g3 := -\frac{1}{2}\frac{1}{z-2}$$

Thus

$$g(z) = g_1(z) + g_2(z) + g_3(z),$$

where $g_1(z) = 3/(2z)$, $g_1(z) = -1/(z-1)$, and $g_3(z) = -1/(2(z-2))$. The function $g_2(z)$ is analytic for $z \neq 1$, so we need the Laurent series expansion valid for $|z| > 1$. The function $g_2(1/z)$ is analytic for $0 < |z| < 1$, so we compute its Laurent expansion near $z = 0$.

```
> gg2 := subs(z=1/z,g2);
```

$$gg2 := -\frac{1}{z^{-1}-1}$$

```
> series(gg2, z=0);
```

$$-z - z^2 - z^3 - z^4 - z^5 + O\left(z^6\right)$$

So it seems that

$$g_2(z) = -\sum_{n=1}^{\infty} \frac{1}{z^n},$$

for $|z| > 1$. The function $g_3(z)$ is analytic for $|z| < 2$, so we compute the series expansion near $z = 0$.

```
> series(g3, z=0);
```

$$\frac{1}{4} + \frac{1}{8}z + \frac{1}{16}z^2 + \frac{1}{32}z^3 + \frac{1}{64}z^4 + \frac{1}{128}z^5 + O\left(z^6\right)$$

Thus it seems that

$$g_3(z) = \sum_{n=0} \frac{1}{2^{n+2}} z^n,$$

for $|z| < 2$. We have the Laurent expansion

$$g(z) = \frac{-2z+3}{z(z-1)(z-2)} = \sum_{n=-\infty}^{-2} z^n + \sum_{n=-1}^{\infty} \frac{1}{2^{n+2}} z^n,$$

for $1 < |z| < 2$.

$|z| > 2$

The remaining region to consider is given by $|z| > 2$. We leave this as an exercise for the reader. The extra computation involves computing the Laurent series of $g_3(z)$ valid for $|z| > 2$, because we already have the Laurent series of $g_2(z)$ for $|z| > 1$.

11.8 Contour integrals

Let $\mathcal{C}$ be a piecewise, smooth curve with parameterization $z(t) = x(t) + i\,y(t)$, $a \le t \le b$. Let $f : \mathcal{C} \longrightarrow \mathbb{C}$ be a continuous function. The contour integral of $f(z)$ over the contour $\mathcal{C}$ is given by

$$\int_{\mathcal{C}} f(z)\,dz = \int_a^b f(z(t))\,z'(t)\,dt,$$

where

$$z'(t) = x'(t) + i\,y'(t).$$

Let $\mathcal{C}$ be the parabolic contour $y = x^2$, $0 \le x \le 1$. $\mathcal{C}$ is parameterized by $z(t) = t + i\,t^2$, $0 \le t \le 1$. We compute the contour integral of $f(z) = \Re(z^2)$ over $\mathcal{C}$:

```
> f := z -> Re(z^2);
```

$$f := z \mapsto Re(z^2)$$

```
> Z := t-> t + I*t^2;
```

$$Z := t \mapsto t + it^2$$

```
> dZ := diff(Z(t),t);
```

$$1 + 2\,it$$

```
> Int(f(Z(t))*dZ,t=0..1)=
> int(evalc(f(Z(t))*dZ),t=0..1);
```

$$\int_0^1 Re\left(\left(t + It^2\right)^2\right)(1 + 2\,It)\,dt = \frac{2}{15} + \frac{1}{6}\,I$$

We found

$$\int_{\mathcal{C}} \Re(z^2)\,dz = \frac{2}{15} + \frac{1}{6}\,i.$$

11.9 Residues and poles

A complex function $f(z)$ has a singularity at $z = z_0$ if $f(z)$ is not analytic at z_0. In MAPLE the `singular` function will return points at which a function or expression is not defined. Let

$$f(z) = \frac{(e^{z^2} - 1)}{z^4(z-1)^3}.$$

Clearly, f has singularities at $z = 0,\ 1$.

```
> f := z -> (exp(z^2) - 1)/z^4/(z-1)^3;
```

$$f := z \mapsto \frac{e^{z^2} - 1}{z^4\,(z-1)^3}$$

```
> singular(f(z));
```

$$\{z = 0\}\,\{z = 1\}\,\{z = -\infty\}\,\{z = \infty\}$$

To determine the nature of these singularities, we compute a Laurent expansion near each singularity.

```
> series(f(z),z=0);
```

$$-z^{-2} - 3\,z^{-1} - \frac{13}{2} - \frac{23}{2}z + O\left(z^2\right)$$

We see that f has a double pole at $z = 0$.

```
> f := z -> (exp(z^2) - 1)/z^4/(z-1)^3:
> series(f(z),z=1);
```

$$\left(e^1 - 1\right)(z-1)^{-3} + \left(-2\,e^1 + 4\right)(z-1)^{-2} + \left(5\,e^1 - 10\right)(z-1)^{-1} - \frac{26}{3}\,e^1 + 20 + \left(\frac{89}{6}\,e^1 - 35\right)(z-1) + \left(-\frac{341}{15}\,e^1 + 56\right)(z-1)^2 + O\left((z-1)^3\right)$$

We see that f has a pole of order 3 at $z = 1$.

Suppose $f(z)$ has an isolated singularity at $z = z_0$. The **residue** of $f(z)$ at $z = z_0$ is the coefficient of $(z - z_0)^{-1}$ in the Laurent series expansion of $f(z)$ near $z = z_0$. This is given in MAPLE by the command `residue(f(z), z=`z_0`)`. We compute the residue of our function $f(z)$ above at each singularity.

```
> residue(f(z),z=0);
```

$$-3$$

```
> residue(f(z),z=1);
```

$$5\,e^1 - 10$$

We found

$$\operatorname*{Res}_{z=0} f(z) = -3,$$
$$\operatorname*{Res}_{z=1} f(z) = 5e - 10.$$

This agrees with the Laurent series computations of $f(z)$ near $z = 0$ and $z = 1$ done earlier.

As an application we compute the contour integral

$$\int_{\mathcal{C}} f(z)\,dz = \int_{\mathcal{C}} \frac{(e^{z^2} - 1)}{z^4(z-1)^3}\,dz,$$

where $\mathcal{C}$ is the simple counterclockwise circle $|z| = 2$. The function $f(z)$ is analytic on the contour, and inside the countour except for singularities at $z = 0, 1$. By Cauchy's residue theorem

$$\int_{\mathcal{C}} f(z)\,dz = 2\pi i \left(\operatorname*{Res}_{z=0} f(z) + \operatorname*{Res}_{z=1} f(z) \right).$$

```
> CI := 2*Pi*I*(residue(f(z),z=0)+residue(f(z),z=1));
```

$$CI := 2\,I\pi\,\left(-13 + 5\,e^1\right)$$

```
> evalf(CI);
```

$$3.715933220\,I$$

We found

$$\int_{\mathcal{C}} f(z)\,dz = 2\,i\pi\,(5\,e - 13) \approx 3.715933220\,i.$$

We check this result by computing the integral using brute force. The contour $\mathcal{C}$ is parameterized by $z(t) = 2e^{it}$, $0 \le t \le 2\pi$.

```
> Z := t -> 2*exp(I*t);
```

$$Z := t \mapsto 2\,e^{(I\,t)}$$

```
> dZ := diff(Z(t),t);
```

$$dZ := 2\,I e^{(I\,t)}$$

```
> CI2 := int(evalc(f(Z(t))*dZ),t=0..2*Pi);
> evalf(CI2);
```

$$-.88\,10^{-11} + 3.715933233\,I$$

MAPLE was unable to evaluate this integral, so we found an approximation using `evalf`. This approximation agrees with the result found using the residue theorem.

12. Opening, Saving, and Exporting Worksheets

In Section 1.8 we saw how to save our work in a MAPLE worksheet. In Section 12.1 we will learn how to open an existing worksheet. In Section 12.2 we will learn how to save a worksheet as an *mws* file and as a text file of different types. In Section 12.3 we will see how to open a MAPLE text file, and in Section 12.4 we consider MAPLE's export features.

12.1 Opening an existing worksheet

To learn how to open an existing worksheet, we need a worksheet that was created elsewhere. Quite often one acquires new worksheets from the Web. A rich source of MAPLE worksheets is the *Maple Application Center* page with url:
`http://www.mapleapps.com`
See Appendix A for more information. Use your favorite browser to download the following *mws* (MAPLE worksheet) file:
`http://www.math.ufl.edu/~frank/maple-book/mwsfiles/ch11_1.mws`
You can open a url by clicking on . Save this file on your computer as *ch11_1.mws.*

Now start MAPLE if you haven't done so already and click on . An **Open** window should appear. In the File name box, type the name of the file you just downloaded. You may have to search for it first. Then press `Open`. The new worksheet should open. See Figure 12.1 below.

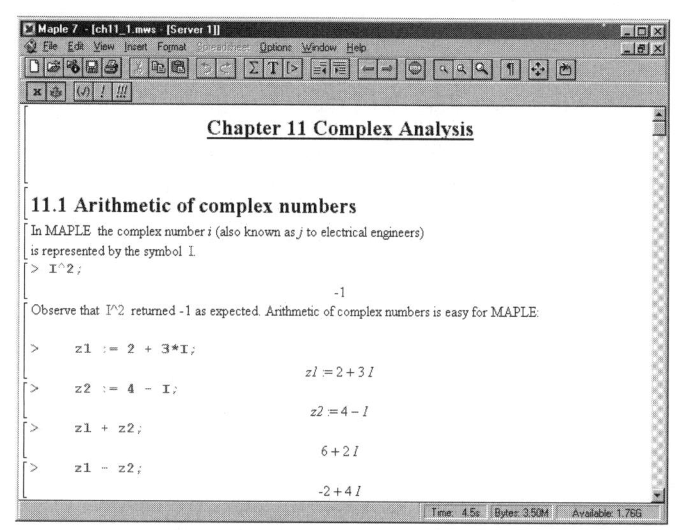

Maple 7 - [ch11_1.mws - [Server 1]]
File Edit View Insert Format Spreadsheet Options Window Help
Chapter 11 Complex Analysis
11.1 Arithmetic of complex numbers
In MAPLE the complex number i (also known as j to electrical engineers)
is represented by the symbol I.
> I^2;
-1
Observe that I^2 returned -1 as expected. Arithmetic of complex numbers is easy for MAPLE:
> z1 := 2 + 3*I;
z1 := 2 + 3 I
> z2 := 4 - I;
z2 := 4 - I
> z1 + z2;
6 + 2 I
> z1 - z2;
-2 + 4 I
Time: 4.5s Bytes: 3.50M Available: 1.76G

Figure 12.1 A downloaded worksheet.

The new worksheet is a worksheet version of the first part of Chapter 11 of this book. Notice that we have defined two complex numbers, z1 and z2. Now scroll down to the bottom of the worksheet. You will see

$$\overline{(z_1)} = 2 - 3i$$

\>
\>
Click after a MAPLE prompt, type "**z1;**" and press **Return**.

```
> z1;
```

$$z1$$

Notice that z1 did not return the complex number 2 - 3 I. The problem is that although we have opened a worksheet, the commands in the worksheet have not been executed. To execute the commands in the worksheet, press !!!. This time each command in the worksheet is reexecuted and we get

```
> z1;
```

$$2 + 3I$$

Add the following to the worksheet:

```
> 1/z1 + 1/z2;
```

$$\frac{86}{221} - \frac{38}{221}I$$

12.2 Saving a worksheet

To save our new worksheet, click on File, then Save As.... A **Save As** window should appear. In the File Name box type **ch11_1a**, and press Save. Our worksheet has been saved as the file *ch11_1a.mws*. This can be opened in a later MAPLE session by pressing .

Our worksheet can also be saved in different text formats. Again we click on File, and then Save As.... In the **Save As** window, click on ▼ in the Save as type box. We see all the possible types:

```
Maple Worksheet
Maple Text
HTML Source
Rich Text Format
Text
LaTeX Source
```

Now select Maple text, type ch11_1a in the File name box, and press Save. This time our worksheet was saved as the text file *ch11_1a.txt*. Open a text editor and take a look at this file.

```
# Chapter 11 Complex Analysis
#
# 11.1 Arithmetic of complex numbers
# In MAPLE the complex number i (also known as j to electrical
# engineers)
# is represented by the symbol I.
> I^2;
                                -1
# Observe that I^2 returned -1 as expected.  Arithmetic of complex
# numbers is easy for MAPLE:
#
> z1 := 2 + 3*I;
                           z1 := 2 + 3 I
```

Notice that everything in the worksheet has been translated into easy-to-read text. MAPLE command lines begin with the MAPLE prompt >. Output of commands is converted to text and centered. Lines of text in the worksheet all begin with comment symbol #.

Now let's save again but this time select `HTML Source`. This time when you press `Save` an **HTML Options** window should open. When saving a worksheet in HTML there are options for how MAPLE output is saved. The Image Location box is used to specify a subdirectory where images will be saved. Click next to GIF, deselect the Use Frames option, and press `OK`. There should be a new file *ch11_1a.html*. Use your favorite Web browser to open it. What you will see looks like a MAPLE worksheet but it is just a Web document. Each piece of MAPLE output has been saved as a *gif* file in the *images* subdirectory. Any animated plots will be saved as animated *gifs*. Anyway, it is handy for posting your work on the Web so that others can view it. Instead of GIF you can select one of the MathML options. Hopefully sometime soon Web browsers will understand MathML.

To save your worksheet as a LaTeX file, select `LaTeX Source`. You will get a TeX file named *ch11_1a.tex*. Most output will be saved as LaTeX commands. Any graphics or plots will be saved as EPS files.

12.3 Opening a MAPLE text file

Opening a MAPLE text file is similar to opening a worksheet. We will open the file *ch11_1a.txt*, which was created in the previous section. Click on . An **Open** window should appear. In the Files of type box type select `Maple text`, type `ch11_1a.txt` in the File name box, and then press `Open`. A **Text Format Choice** window will open. Under Text Format select `Maple Text`, and press `OK`. Lines that begin with # appear as text in the worksheet. Lines that begin with > appear as MAPLE command lines. MAPLE output is not included in the worksheet.

We cannot read this text file using read. Look what happens:

```
> read "ch11_1a.txt";
```

```
on line 7, syntax error, > unexpected
```

The problem is that lines that begin with the MAPLE prompt > will cause an error. If you are keen to use the read command, you must strip out the prompts. On a UNIX machine you could do something like

```
grep '^>' ch11_1a.txt | sed 's/^>//' > ch11_1b.txt
```

This bit of UNIX code selects MAPLE command lines, strips out the MAPLE prompts, and saves the result in the file *ch11_1b.txt*.

```
> read "ch11_1b.txt";
```

$$-1$$

This time reading was successful. MAPLE commands are executed. Output appears in the worksheet, but the MAPLE commands do not. For the regular use of read, see Section 7.10.3.

All of the MAPLE commands used in this book are available on the Web as MAPLE text files. The following url contains links to these files:
http://www.math.ufl.edu/~frank/maple-book/mtxtfiles/index.html
Use your favorite browser to go to this page and click on the link ch11-maple.txt. This will give you the MAPLE text file containing all the MAPLE commands used in Chapter 11. Save it as *ch11-maple.txt* and open it in a MAPLE session.

12.4 Exporting worksheets and LaTeX

In Section 12.2 we saw how a MAPLE worksheet can be saved as a *.mws* file and opened in a later session. We also saw how it can be saved as different types of text files. An alternative method is to use the Export As submenu. Click on File, click on Export As, slide to the right, and a submenu should appear:

HTML...
HTML with MathML...
LaTeX...
Maple Text...
Plain Text...
RTF...

These are basically the same choices we got in Section 12.2 when we used Save As.... In fact, when you make a selection in the Export As submenu, a **Save As** window will appear, and we proceed as before. Try selecting LaTeX... to save the worksheet as a LaTeX file.

Selecting RTF... converts a worksheet to an *rtf* (rich text format) file. This can be used later in a Microsoft® Word document.

We can export a MAPLE worksheet as MathML by selecting `HTML with MathML...`. When you make this selection, a **Save As** window will appear. In the Save as type box select `HTML Source`. Press `Save` and an **HTML Options** window will open. Try selecting MathML 2.0 with WebEQ. Press `OK` and the worksheet will be saved as an *html* file with embedded MathML using WebEQ. This time MAPLE output is converted to MathML. This avoids saving output as *gif* files which can use a lot of memory. Even if your browser does not support MathML, MAPLE output can be viewed using WebEQ. For more information, see the Web site:
`http://www.maplesoft.com/standards/MathML/info.html`

We can convert MAPLE output into LaTeX directly in the worksheet using the `latex` function. Try

```
> with(linalg):
> A:=matrix(3,3,(i,j)->sin(Pi*i*j/6));
> latex(A);
```

13. Document Preparation

MAPLE has many features for creating documents. It is possible to add MAPLE output to text and create technical documents. There are also facilities for adding headings; changing fonts; inserting expandable subsections, bookmarks, and hyperlinks.

We now demonstrate some of these features with a specific example. Suppose we have the following

Problem. Reduce the weight of a ball bearing with a diameter of 2 cm by 50% by drilling a hole through the center. Determine the diameter of the required drill bit.

This problem can be solved easily in MAPLE by computing a certain integral and solving an equation. Start MAPLE and type in the following.

```
> v:=Int(4*Pi*x*sqrt(1-x^2),x=0..r);
```

$$v := \int_0^r 4\,\pi\,x\sqrt{1-x^2}dx$$

```
> v:=value(v);
```

$$v := -\frac{4}{3}\,\pi\,\left(1-r^2\right)^{\frac{3}{2}} + \frac{4}{3}\,\pi$$

```
> rrs:=solve(v=2*Pi/3,r);
```

$$rrs := \sqrt{1-\frac{1}{2}\,2^{(1/3)}},\ -\sqrt{1-\frac{1}{2}\,2^{(1/3)}}$$

```
> 2*radsimp(%[1]);
```

$$\sqrt{4-2\,2^{(1/3)}}$$

```
> evalf(%);
```

$$1.216617401$$

The desired diameter is

$$2r = \sqrt{2}\,\sqrt{2-2^{1/3}} \approx 1.217\,\text{cm}.$$

You may be wondering what is going on in this problem. We can make a much clearer document by adding text.

13.1 Adding text

First we add some text to our document. Click the cursor on the first line of MAPLE input. Then in the Insert menu, select Execution Group and Before

Cursor. A MAPLE prompt > should appear above the first line of input. Now click on [T] and type

```
Reduce the volume of a ball bearing with a diameter of 2 cm by 50%
by drilling a hole through the center.  Determine the diameter of
the required drill-bit.
```

To create a new paragraph, click on [>] and then [T]. Type

```
First we observe that the ball bearing is the solid obtained by
rotating a circle of radius 1cm about the y-axis.    If we let
r be the radius of the drill-bit then, by the shell method, the
volume of material removed is given by
```

Now we would like to add some in-line math.

13.2 Inserting math into text

In the Insert menu, select Standard Math and a ? should appear. Type

```
Int(4*Pi*x*sqrt(1-x^2),x=0..r)
```

What was MAPLE input should now appear as math in your document. Click on [T] and type

```
.  We compute the integral
```

Let's add a title.

13.3 Adding titles and headings

Click on the first line of the worksheet. In the Insert menu, select Execution Group and Before Cursor. Then click on [T]. In the box [P Normal ▾] select `P Title`. Now type

```
The Ball Bearing Problem
```

The document should now have a title. Press enter and type your name

```
William E. Wilson
```

Your name should now be underneath the title. Press enter again. To make a heading this time, we select `P Heading 2`. Type

```
Statement of the problem
```

To underline this heading, select `Statement of the problem` with the left mouse button, and click on [U].

Now make a heading entitled **Solution** for the next paragraph. Start by clicking on the line "`First we ...`".

Let's move some of the MAPLE computations into a new subsection.

13.4 Creating a subsection

Use the first mouse button to highlight the MAPLE inputs

```
v:=Int(4*Pi*x*sqrt(1-x^2),x=0..r);
```

and

```
v:=value(v);
```

together with their output. Now click on [button]. A little button [-] should appear. Try clicking on it. Pretty neat! Now see if you can add a heading to this subsection using the `P Heading 3` selection.

Next we shall add some more text and math by cutting and pasting.

13.5 Cutting and pasting

First we create a new region. Click on the vertical bar attached to [-] and click on [> and then T. There should now be a new text region below the new subsection. Now type

```
Our computation gave
```

At this point, we would like to add an equation to our document. This time we will use the mouse to cut and paste. First click on [> and type

```
> 'v' =
```

Instead of retyping MAPLE input, we move the cursor to the MAPLE output above and use the mouse to highlight

$$-\frac{4}{3}\left(1-r^2\right)^{3/2}\pi+\frac{4}{3}\pi$$

Use the mouse or hot keys to copy the selection and paste it to the right of the equal sign. The hot keys are system dependent. In Windows, use `Control C` to copy and `Control V` to paste. Observe how the displayed math has been converted to MAPLE input. Now type a semicolon and press enter:

```
>  'v' =-4/3*(1-r^2)^(3/2)*Pi+4/3*Pi;
```

$$v=-\frac{4}{3}\left(1-r^2\right)^{3/2}\pi+\frac{4}{3}\pi$$

Now click the mouse on the MAPLE input line

```
>  'v' =-4/3*(1-r^2)^(3/2)*Pi+4/3*Pi;
```

and hit `Control Delete` and this line should now be erased. Finally, add enough text and equations so that the document is complete. A rendition of how it might appear is given below. This worksheet can be downloaded using the url:
`http://www.math.ufl.edu/~frank/maple-book/mwsfiles/bbprob.mws`

The Ball Bearing Problem

William E. Wilson

Statement of the problem

Reduce the volume of a ball bearing with a diameter of 2 cm by 50% by drilling a hole through the center. Determine the diameter of the required drill-bit.

Solution

First we observe that the ball bearing is the solid obtained by rotating a circle of radius 1cm about the y-axis. If we let r be the radius of the drill-bit then, by the shell method, the volume v of material removed is given by $\int_0^r 4\,\pi\,x\sqrt{1-x^2}dx$. We compute the integral.

Computation

```
> v:=Int(4*Pi*x*sqrt(1-x^2),x=0..r);
```

$$v := \int_0^r 4\,\pi\,x\sqrt{1-x^2}dx$$

```
> v:=value(v);
```

$$v := -\frac{4}{3}\left(1-r^2\right)^{3/2}\pi + \frac{4}{3}\pi$$

Our computation gave

$$v = -\frac{4}{3}\left(1-r^2\right)^{3/2}\pi + \frac{4}{3}\pi$$

We solve the equation

$$-\frac{4}{3}\left(1-r^2\right)^{3/2}\pi + \frac{4}{3}\pi = \frac{2}{3}\pi$$

Computation

to find that the required diameter is

$$2r = \sqrt{2}\sqrt{2-2^{1/3}}$$

which is approximately 1.217 cm.

13.6 Bookmarks and hypertext

A *bookmark* is a name that marks a location in a worksheet. Selecting this name will move the cursor to the specified location. To create a bookmark at the last equation in our document, click the cursor on the equation. Then, in the View menu, select `Bookmarks` and then `Edit Bookmark...`. An **Add or Modify Bookmark** window should appear. In the Bookmark Text box, type a word, say, `ANSWER` and click on `OK`. Although the worksheet appears no different, it now has a single bookmark. We can access this bookmark by selecting Bookmarks in the View menu. Now `ANSWER` should appear in the submenu:

Edit Bookmark...
ANSWER

Select `ANSWER` and the cursor will move to the specified location. Try moving the cursor to a different place in the worksheet and select `ANSWER` again.

Now we will use our bookmark to create a hyperlink in our worksheet. A hyperlink is a link from one location in the worksheet to a different location in the worksheet or to a different worksheet altogether. The presence of a hyperlink is indicated by green underlined text. Clicking on this text will move the cursor to the new location. In our worksheet we will attach a hyperlink from the word *diameter* in the statement of the problem to our bookmark `ANSWER`.

Move the cursor to the word diameter near the top of the worksheet and in the Insert menu select HyperLink... . A **HyperLink Properties** window should appear:

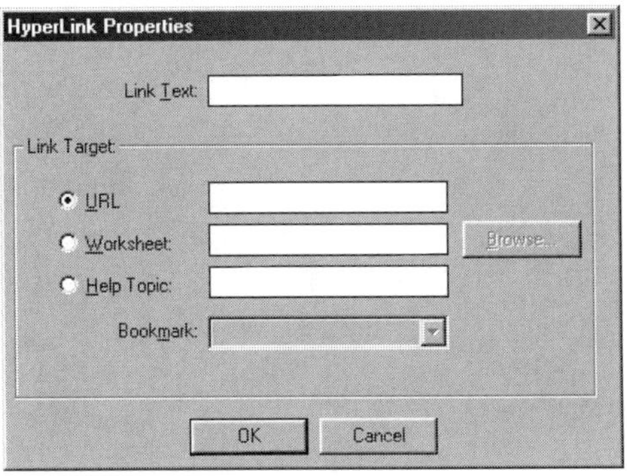

Figure 13.1 Hyperlink Properties window.

In the Link Text box, type `diameter`, and click to the left of Worksheet. Then click on ▼ near the Book Mark box and select `ANSWER` (or type `ANSWER` in the box). Finally, click on `OK`. The worksheet should now contain a green diameter. You will need to delete the old "diameter." Try clicking on diameter. The cursor should move to the last equation in the worksheet where we placed the bookmark `ANSWER`. This worksheet (with the hyperlink and bookmark) can be downloaded using the url:

`http://www.math.ufl.edu/~frank/maple-book/mwsfiles/bbprob2.mws`

Try adding a hyperlink to a different worksheet. First create a new worksheet, say, *shell.mws*, which contains a description of the shell method. Then attach a hyperlink to the phrase "shell method" in the original worksheet.

14. More Graphics

In Chapter 6 we studied MAPLE's basic plot functions `plot`, and `plot3d`, as well as the *plots* package. There are a few more packages used for plotting and creating graphics. They are *DEtools*, *plottools*, *geometry*, *geom3d*, and *statplots*. See Section 8.7.1 for the plotting functions in the *DEtools* package. *statplots* is part of the *stats* package that we will examine in Chapter 16. In this chapter we will concentrate mainly on the *plottools*, *geometry*, and *geom3d* packages.

14.1 The *plottools* package

To see the functions in the *plottools* package type

```
> with(plottools);
```

14.1.1 Two-dimensional plot objects

In this section we examine the functions in the *plotools* package for generating two-dimensional plot objects. Each function produces a `PLOT` data structure that can be rendered using the `display` function in the *plots* package.

arc

The function `arc([a,b], r,` θ_1`..`θ_2`)` gives the arc of a circle centered at (a, b), radius r, and angle θ satisfying $\theta_1 \leq \theta \leq \theta_2$. We plot a one-quarter circle centered at the origin, radius 1, and in the second and third quadrants. See Figure 14.1.

```
> with(plottools):
> qc := arc([0,0],1,Pi/2..Pi):
> plots[display](qc, scaling=constrained);
```

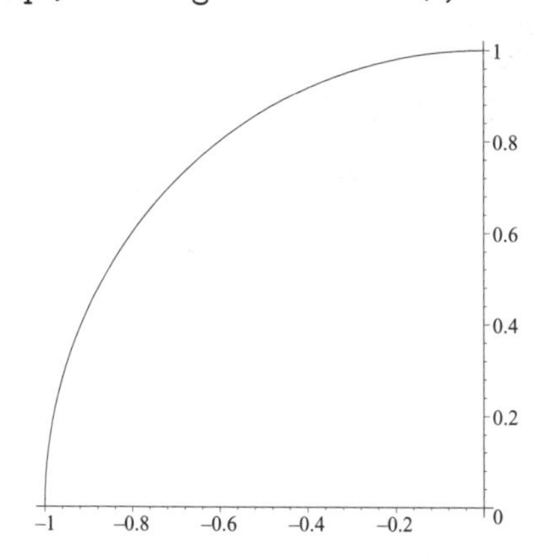

Figure 14.1 Arc of a circle.

arrow

The `arrow` function is used to plot vectors. See Section 10.1.3 for some examples.

circle

The function `circle([a,b],r)` gives the circle centered at (a, b), with radius r. Try

```
> with(plottools):
> circle([1,0],1):
> plots[display](%, scaling=constrained);
```

curve

The function `curve(` $[[x_1, y_1], [x_2, y_2], \ldots, [x_n, y_n]]$`)` gives a sequence of straight line segments joining the points $(x_1, y_1), \ldots, (x_n, y_n)$.

```
> with(plottools):
> pts := [[0, 0], [.93, .80], [1.2, .95], [1.6, 1.],
    [.31, .30], [.62, .60]] :
> curve(pts):
> plots[display](%);
```

disk

The function `disk([a,b],r)` gives a disk centered at (a, b), with radius r. We plot a sequence of disks with varying center, radii, and shade of red.

```
> with(plottools):
> disk_seq := seq(disk([cos(t*Pi/10),sin(t*Pi/10)],t/30,
    color=COLOR(RGB,t/5,0,0)),t=1..5):
> plots[display](disk_seq, scaling=constrained);
```

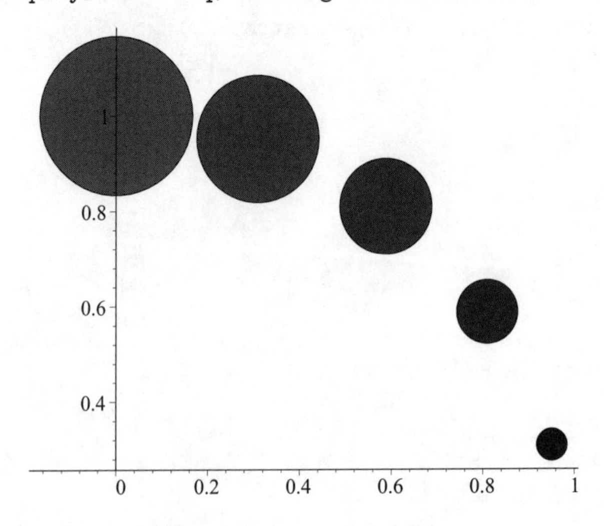

Figure 14.2 A sequence of disks.

hyperbola

The function `hyperbola( [`x_0`, `y_0`], a, b, `x_1`..`x_2`)` gives the hyperbola with equation

$$\frac{(x-x_0)^2}{a^2} - \frac{(y-y_0)^2}{b^2} = 1,$$

where $x_1 \le x \le x_2$.

```
> with(plottools):
> hyperbola([2,1], 1, 1, -3..3):
> plots[display](%);
```

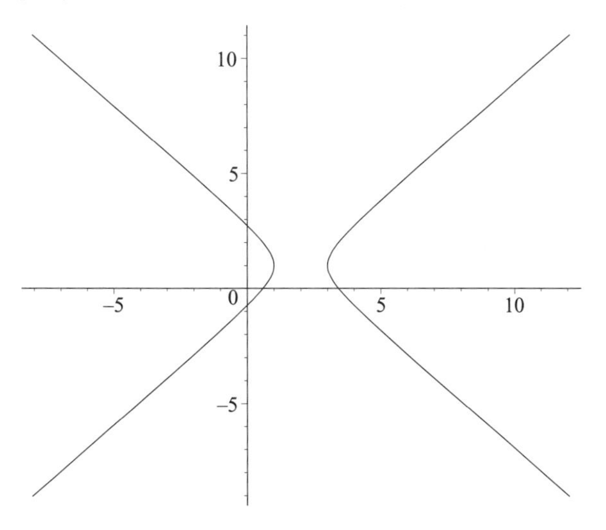

Figure 14.3 Hyperbola.

ellipse

The function `ellipse( [`x_0`, `y_0`], a, b)` gives the ellipse with equation

$$\frac{(x-x_0)^2}{a^2} + \frac{(y-y_0)^2}{b^2} = 1.$$

Try plotting a pink oval:

```
> with(plottools):
> ellipse([2,1], 1, 2, filled=true, color=pink):
> plots[display](%, scaling=constrained);
```

ellipticArc

is the elliptic version of the circular `arc` function. The function `ellipticArc( [`x_0`, `y_0`], a, b, `θ_1`..`θ_2`)` gives the elliptic arc

$$x = x_0 + a\cos\theta, \quad y = y_0 + b\sin\theta,$$

where $\theta_1 \leq \theta \leq \theta_2$. Try

```
> with(plottools):
> ellipticArc([0,1], 2, 1, 0..Pi/3,filled=true,
    color=turquoise):
> plots[display](%,scaling=constrained);
```

line

The function line([a,b], [c,d]) gives the line segment joining the points (a, b) and (c, d). Try

```
> with(plottools):
> line([3,1],[1,4]):
> plots[display](%);
```

pieslice

The function pieslice([a,b], r, $\theta_1..\theta_2$) gives the sector of a circle centered at (a, b), radius r, and where $\theta_1 \leq \theta \leq \theta_2$.

```
> with(plottools):
> pie := pieslice([0,0],1,Pi/6..Pi/3,color=khaki):
> plots[display](%,scaling=constrained);
```

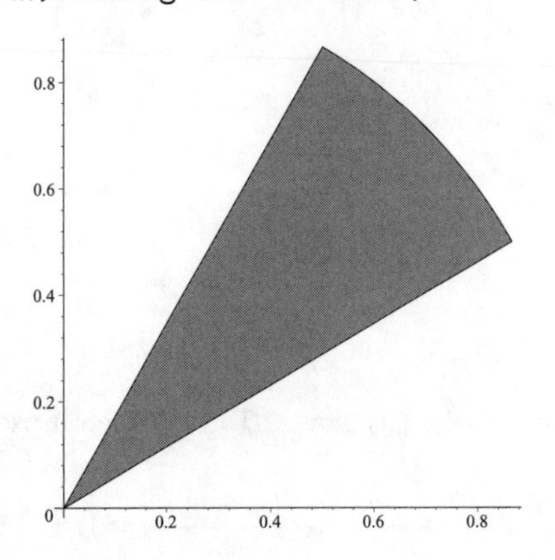

Figure 14.4 The sector of a circle.

point

The function point([x,y]) gives the point (x, y). Try

```
> with(plottools):
> p:=point([2,3]):
> plots[display](p);
```

polygon

The function `polygon( [[`x_1`,`y_1`], [`x_2`,`y_2`], ... ,[`x_n`,`y_n`]])` creates a polygon by joining the points (x_1, y_1), ... , (x_n, y_n).

```
> with(plottools):
> polyg := polygon([[0,0],[1,2],[1,1]]);
> plots[display](polyg);
```

Let's plot a regular pentagon:

```
> with(plottools):
> pt := t -> [cos(t),sin(t)]:
> pts := [seq(pt(2*Pi*k/5+Pi/10),k=0..4)]:
> pentagon := polygon(pts, color=coral):
> plots[display](pentagon,axes=none,scaling=constrained);
```

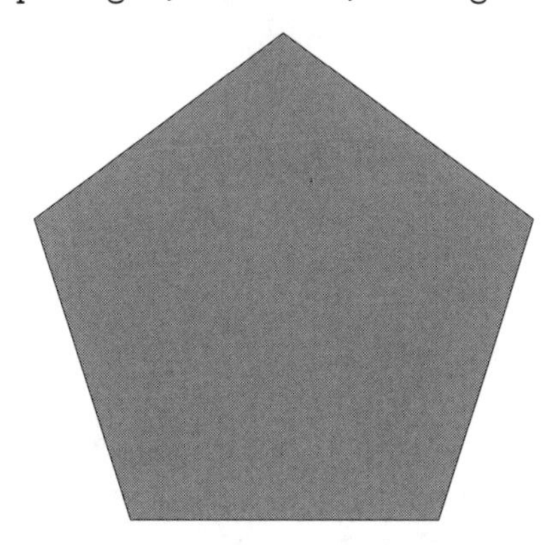

Figure 14.5 A pentagon.

rectangle

The function `rectangle( [a,b], [c,d])` gives a rectangle whose sides are parallel to the coordinate axes and has vertices at the specified points. Try

```
> with(plottools):
> polyg := rectangle([0,0],[1,2]);
> plots[display](polyg, color=blue, scaling=constrained);
```

14.1.2 Three-dimensional plot objects

cone

The `cone` function produces a right circular cone. The syntax has the form `cone([`a`,`b`,`c`], `r`, `h`)`, where (a, b, c) is the vertex, r is the radius of the circular top, and h is the height. The parameters r and h are optional, and have a default value 1.

```
> with(plottools):
> cone1 := cone([0,0,0],1,2):
```

```
> plots[display](cone1, color=wheat, scaling=constrained,
    orientation=[50,75]);
```

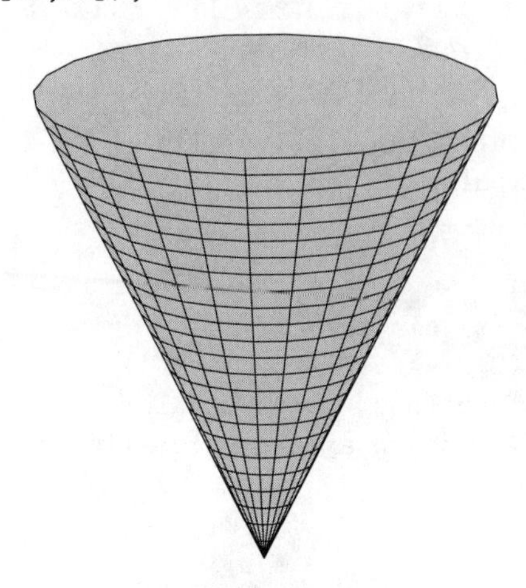

Figure 14.6 A cone.

<u>cuboid</u>

The function `cuboid(`$[x_1, y_1, z_1]$`, `$[x_2, y_2, z_2]$`)` produces a cube where (x_1, y_1, z_1), (x_2, y_2, z_2) are opposite vertices.

```
> with(plottools):
> cube := cuboid([0,0,0],[1,1,1],color=black):
> plots[display](cube,scaling=constrained,axes=boxed,
    style=wireframe,thickness=3,orientation=[40,75]);
```

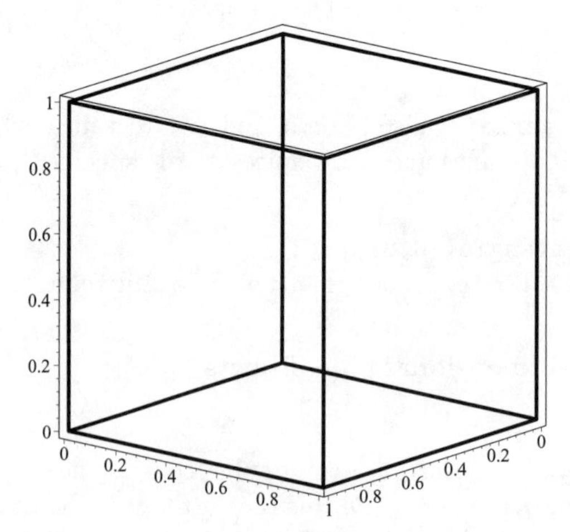

Figure 14.7 A cube.

<u>cylinder</u>

The `cylinder` function produces a right circular cylinder. The syntax has the form `cylinder([a,b,c], r, h)`, where (a, b, c) is the center of the base, r is the radius, and h is the height.

```
> with(plottools):
> c := cylinder([0,0,0], 3,6):
> plots[display](c, scaling=constrained,orientation=[50,75]);
```

Let's plot a sequence of cylinders:

```
> cyl := t -> cylinder([cos(t),sin(t),0],0.2,t):
> cylseq := seq(cyl(2*Pi*k/10),k=1..10):
> plots[display](cylseq);
```

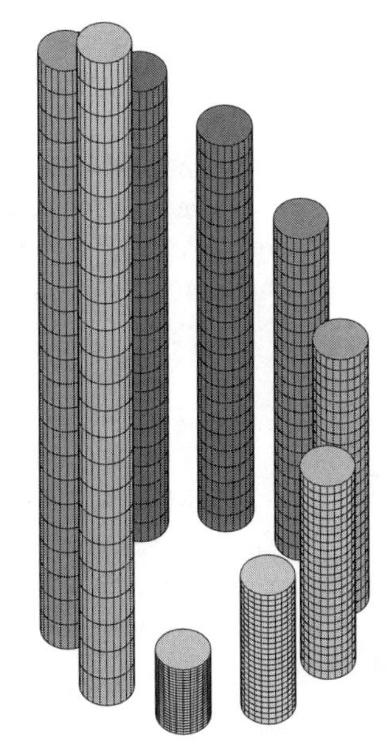

Figure 14.8 A sequence of cylinders.

<u>dodecahedron</u>

The function `dodecahedron(`$[a,b,c]$`,` s`)` gives a dodecahedron centered at (a, b, c) and scale factor s. In section 6.2.8 we saw how to plot a dodecahedron using the `polyhedraplot` function in the *plots* package, with the `polytype=dodecahedron` option. The `docdecahedron` function is really this `polyhedraplot` function in disguise. Try

```
> with(plottools):
> dd :=dodecahedron([0,0,0],1):
> plots[display](dd,scaling=constrained);
```

hemisphere

The function `hemisphere([a,b,c], r)` produces the bottom half of a sphere centered at (a, b, c) with radius r.

```
> with(plottools):
> hs := hemisphere([0,0,0],12):
> plots[display](hs, scaling=constrained, axes=boxed,
    orientation=[40,70]);
```

hexahedron

A hexahedron is really a cube. The function `hexahedron([a,b,c], s)` produces a cube centered at (a, b, c) with scale factor s. Try

```
> with(plottools):
> hex1 := hexahedron([0,0,0],1,color=green):
> plots[display](hex1,scaling=constrained,axes=boxed);
> cub1 := cuboid([1,1,-1],[-1,-1,1],color=red):
> plots[display](cub1,scaling=constrained,axes=boxed);
```

`hex1` and `cub1` should produce the same polyhedron except for color.

icosahedron

The function `icosahedron([a,b,c], s)` gives an icosahedron centered at (a, b, c) and scale factor s. In section 6.2.8 we saw how to plot a icosahedron using the `polyhedraplot` function in the *plots* package, with the `polytype=icosahedron` option. The `icosahedron` function is really this `polyhedraplot` function in disguise. Try

```
> with(plottools):
> ic :=icosahedron([0,0,0],1):
> plots[display](ic,scaling=constrained);
```

octahedron

The function `octahedron([a,b,c], s)` gives an octahedron centered at (a, b, c) and scale factor s. This function is really `plots[polyhedraplot]` with the `polytype=octahedron` option.

```
> with(plottools):
> oc := octahedron([0,0,0],1,color=black,thickness=3):
> plots[display](oc,style=wireframe,scaling=constrained,
    orientation=[30,75]);
```

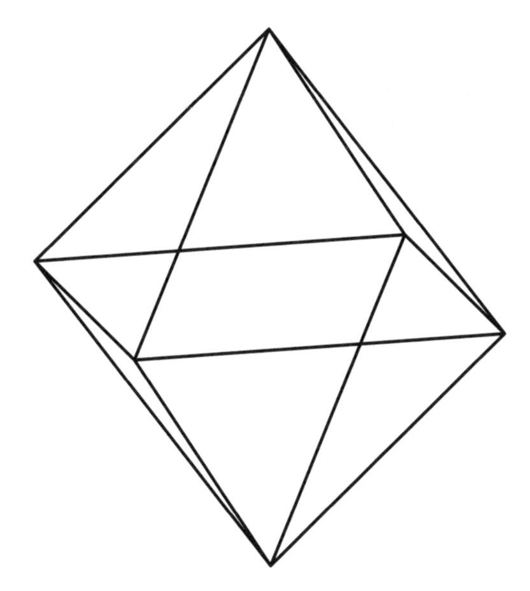

Figure 14.9 An octahedron.

semitorus

A torus can be obtained by rotating a circle radius r about a vertical axis. The function semitorus([a,b,c], θ_1..θ_2, r, R) gives part of torus whose generating circle has radius r, (a,b,c) is the center of the torus, and R is the distance between the center of the generating circle and the axis of rotation. We plot a quarter-torus together with a circle:

```
> with(plottools):
> qtor := semitorus([0,0,0], Pi..3*Pi/2, 1, 4):
> circ := plots[spacecurve]([4*cos(t),4*sin(t),0],t=0..2*Pi,
    color=black,thickness=3):
> plots[display](qtor, circ,scaling=constrained, style=patch,
    axes=boxed,orientation=[50,75]);
```

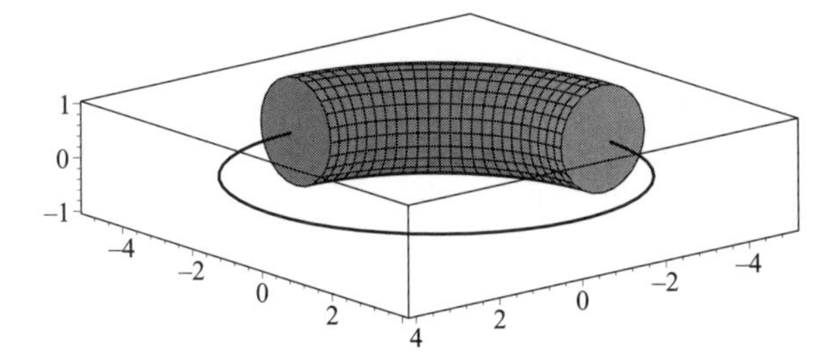

Figure 14.10 A quarter-torus.

<u>sphere</u>

The function `sphere(`$[a,b,c]$`,` r`)` gives a sphere with center (a,b,c) and radius r. We plot a sphere centered at the origin and radius 1:

```
> with(plottools):
> sph := sphere([0,0,0],1):
> plots[display](sph,scaling=constrained,orientation=[30,60]);
```

<u>tetrahedron</u>

The function `tetrahedron(`$[a,b,c]$`,` s`)` gives a tetrahedron centered at (a,b,c) and scale factor s. This function is really `plots[polyhedraplot]` with the `polytype=tetrahedron` option.

```
> with(plottools):
> tet := tetrahedron([0,0,0],1):
> plots[display](tet,scaling=constrained,orientation=[80,80]);
```

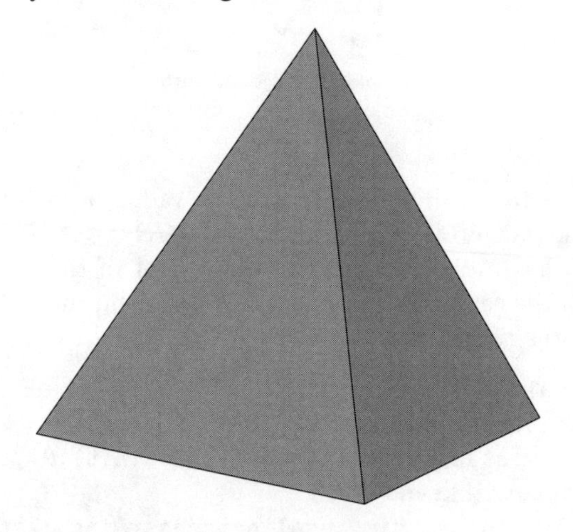

Figure 14.11 A tetrahedron.

<u>torus</u>

The function `torus(`$[a,b,c]$`,` r`,` R`)` gives a torus centered at (a,b,c). It is equivalent to `semitorus([a,b,c], 0..2*Pi,` r`,` R`)`. See the `semitorus` function above. Try

```
> with(plottools):
> tor := torus([0,0,0], 1, 4):
> plots[display](tor, scaling=constrained, style=patch,
    axes=boxed,orientation=[50,75]);
```

14.1.3 Transformation of plots

In this section we consider functions in the *plottools* package that transform two- or three-dimensional plot objects.

cutin

The function `cutin` shrinks each polygonal face of a polyhedron by a specified factor. As an example, we perform this operation on a tetrahedron.

```
> with(plottools):
> tet:=tetrahedron([0,0,0],1):
> plots[display](cutin(tet,2/3),orientation=[70,75]);
```

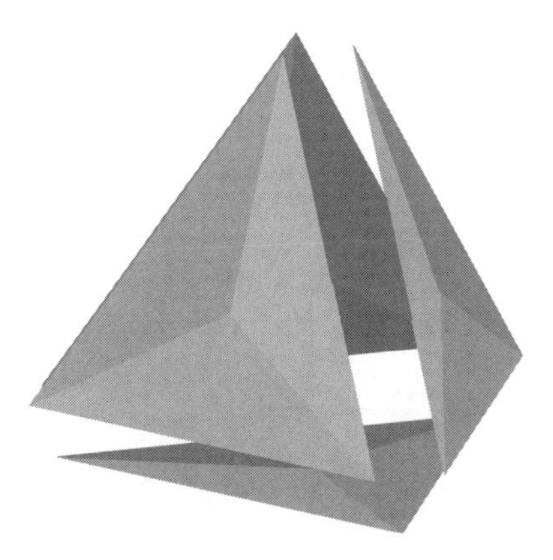

Figure 14.12 A cut tetrahedron.

Figure 14.13 A tetrahedron with holes.

cutout

The function `cutout` cuts a hole in each polygonal face of a polyhedron. The hole is similar in shape to the original face. As an example, we perform this operation on a tetrahedron.

```
> with(plottools):
```

```
> tet:=tetrahedron([0,0,0],1):
> plots[display](cutout(tet,1/3),orientation=[85,75]);
```

The resulting plot is given above in Figure 14.13.

homothety

The homothety function is a special case of the scale function given below, where the scaling for each coordinate is the same. For two-dimensional plot objects p, homothety(p,a) is the same as scale(p,a,a). For three-dimensional plot objects p, homothety(p,a) is the same as scale(p,a,a,a). As an example we use homothety to rescale a dodecahedron so that its vertices coincide with a stellated icosahedron. See below for the stellate function.

```
> with(plottools):
> ico := icosahedron():
> stel_ico:=stellate(ico,2):
> scaled_dodec:=homothety(dodecahedron([0,0,0],1),1.75):
> wf:=plots[display](scaled_dodec,scaling=constrained,
   style=wireframe,color=black,thickness=3):
> plots[display](wf,stel_ico,scaling=constrained);
```

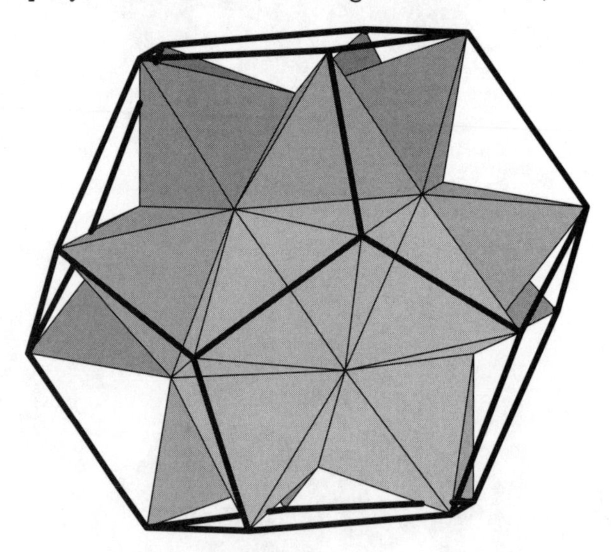

Figure 14.14 A dodecahedron and a stellated icosahedron.

project

This function can project a two-dimensional object onto a line. It can project a three-dimensional object onto a line or a plane. If p is a three-dimensional plot object, then project(p, [[a_1,b_1,c_1],[a_2,b_2,c_2],[a_3,b_3,c_3]]) gives the projection of p onto the plane determined by the points (a_1, b_1, c_1), (a_2, b_2, c_2), (a_3, b_3, c_3). When a vertical cone is projected onto the xy-plane, we get a disk. Try

```
> with(plottools):
> cn :=cone([0,0,0]):
> pcn :=project(cn,[[0,0,0],[0,1,0],[1,0,0]]):
> plots[display](cn,pcn);
```

reflect

This function can reflect a two-dimensional object in a line. It can project a three-dimensional object in a plane. If p is a three-dimensional plot object, then reflect(p, [[a_1,b_1,c_1],[a_2,b_2,c_2],[a_3,b_3,c_3]]) gives the reflection of p in the plane determined by the specified points. We plot a cone and its reflection in the xy-plane.

```
> cone1 :=cone([0,0,0]):
> cone2 :=reflect(cone1,[[0,0,0],[0,1,0],[1,0,0]]):
> plots[display](cone1,cone2,orientation=[45,75],
    scaling=constrained);
```

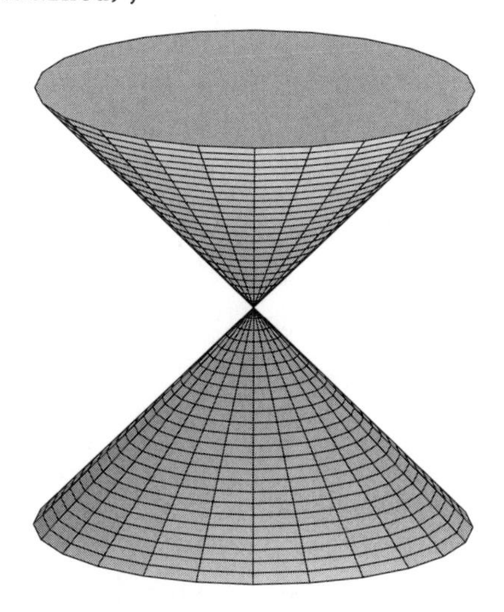

Figure 14.15 Reflecting on a cone.

rotate

The **rotate** function can rotate a two-dimensional plot object about a point, and a three-dimensional plot object about a line. For a two-dimensional plot object p, the syntax takes the form

rotate(p, θ)
rotate(p, θ, [a,b])

Here θ is the angle of rotation, and (a, b) is the center of rotation. The default center of rotation is the origin. Try plotting a hyperbola and its rotation about the origin through an angle of 9°.

```
> with(plottools):
> hyp := hyperbola([0,0],1,1,-2..2):
```

```
> hyprot := rotate(hyp,Pi/20):
> plots[display](hyp,hyprot);
```

Next we plot a sequence of rotated hyperbolas in Figure 14.16.

```
> with(plottools):
> hyp := hyperbola([0,0],1,1,-2..2):
> hypseq := seq(rotate(hyp,Pi*k/20),k=0..10):
> plots[display](hypseq);
```

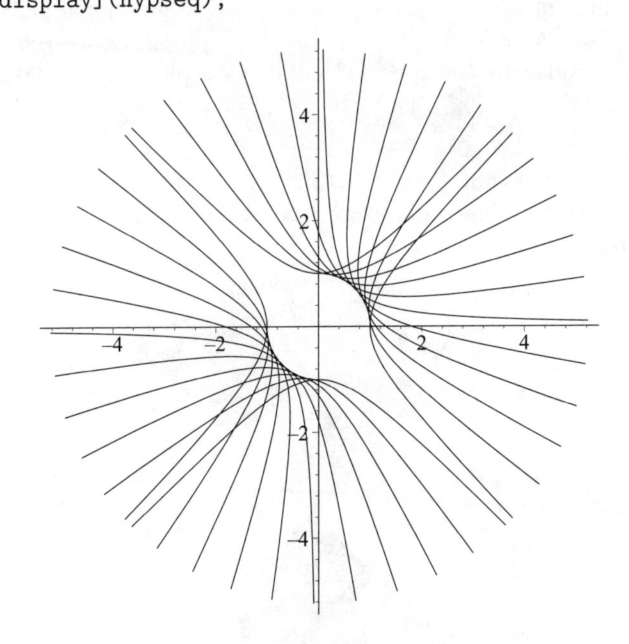

Figure 14.16 A rotating sequence of hyperbolas.

Now try changing the center of rotation:

```
> hypseq2:=seq(rotate(hyp,Pi*k/20,[1,0]),k=0..10):
> plots[display](hypseq2);
```

For a three-dimensional plot object p, the syntax of the `rotate` function has the form:

`rotate(`p`, `α`, `β`, `γ`)`
`rotate(`p`, `α`, [[`a_1`,`b_1`,`c_1`],[[`a_2`,`b_2`,`c_2`]])`

When specified, α, β, γ denote rotation about the x-, y- and z-axis respectively. In the second form, the axis of rotation is the line joining the points (a_1, b_1, c_1) and (a_2, b_2, c_2). We obtain a sequence of cones by rotating a cone in the xz-plane.

```
> with(plottools):
> cone1 := cone([0,0,0],1,3):
> yax := [[0,0,0],[0,1,0]]:
> cone_seq := seq(rotate(cone1,2*Pi*k/10,yax),k=1..10):
```

```
> plots[display](cone_seq,scaling=constrained,
    orientation=[50,60]);
```

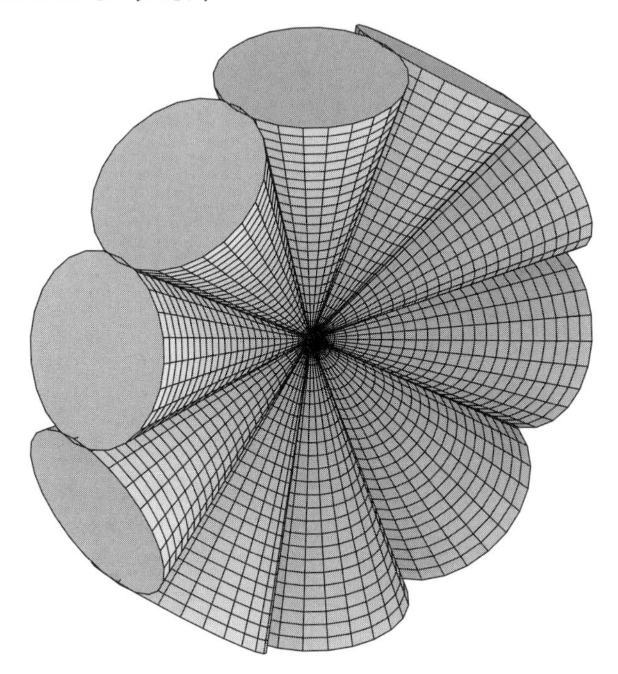

Figure 14.17 A rotating sequence of cones.

scale

The `scale` function is used to rescale a two- or three-dimensional plot object. For a two-dimensional object p, the syntax takes the form

```
scale(p,a,b)
scale(p,a,b,[c,d])
```

In the first form the rescaling corresponds to the transformation $(x,y) \mapsto (ax,by)$; i.e., a rescaling about the point $(0,0)$. In the second form it corresponds to the transformation $(x,y) \mapsto (a(x-c)+c, b(y-d)+d)$; i.e., a rescaling about the point (c,d). For a three-dimensional object p, the syntax is analogous.

```
scale(p,a,b,c)
scale(p,a,b,c,[x,y,z])
```

We next plot a cylinder together with its image under a rotation and rescaling.

```
> with(plottools):
> cyl1:=cylinder([0,0,0],0.4,3):
> yax:=[[0,0,0],[0,1,0]]:
> cyl2:=scale(rotate(cyl1,Pi/5,yax),1/2,1/2,1/2):
```

```
> plots[display](cyl1,cyl2,scaling=constrained,orientation=
   [60,60]);
```

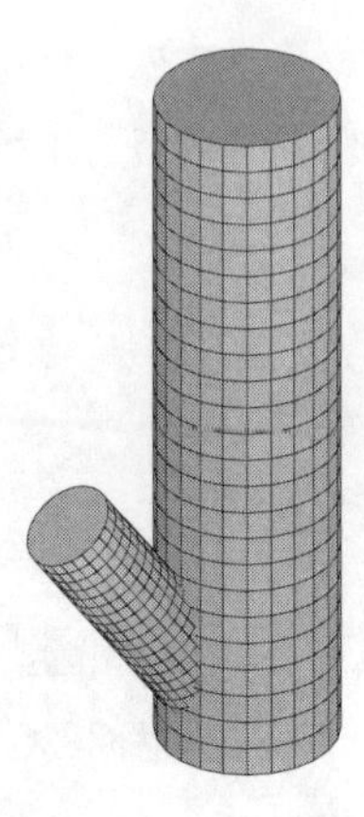

Figure 14.18 Rotated and rescaled cylinder.

stellate

In MAPLE the `stellate` function produces a new polyhedron by adding a pyramid to each polygon face. The syntax has the form

`stellate(`p`, `h`)`

where p is the original polyhedron or three-dimensional plot object and h is the height of the stellate. For $h > 1$, the stellate is directed away from the origin, otherwise it is directed toward the origin. We produce a stellated icosahedron.

```
> with(plottools):
> stel_icosa := stellate(icosahedron([0,0,0],1),4):
> plots[display](stel_icosa,scaling=constrained);
```

Figure 14.19 A stellated icosahedron.

Try making a stellated docdecahedron.

```
> with(plottools):
> plots[display](stellate(dodecahedron([0,0,0],1),3),
    scaling=constrained);
```

<u>`transform`</u>

The **transform** function is used to perform a general transformation on a two- or three-dimensional plot object. We show how it works for three-dimensional plot objects. Two-dimensional plot objects are similar. To apply the transformation

$$F(x,y,z) = (f_1(x,y,z), f_2(x,y,z), f_3(x,y,z))$$

to a three-dimensional plot object p, we define the MAPLE function

`F := transform( (x,y,z) -> [`f_1`(x,y,z), `f_2`(x,y,z), `f_3`(x,y,z)])`
Then the transformed object is obtained by the command **F(p)**. As an example, we plot the image of surface parameterized by

$$x = \sin u, \quad y = \cos v, \quad , z = \cos u + v - 1,$$

where $0 \le u, v \le 2\pi$, under the map

$$F(x,y,z) = (x^2, y^2, z^2).$$

```
> with(plottools):
> p:=plot3d([sin(x),cos(y),cos(x+y-1)],x=0..2*Pi,y=0..2*Pi):
> F:=transform((x,y,z)->[x^2,y^2,z^2]):
> plots[display](F(p));
```

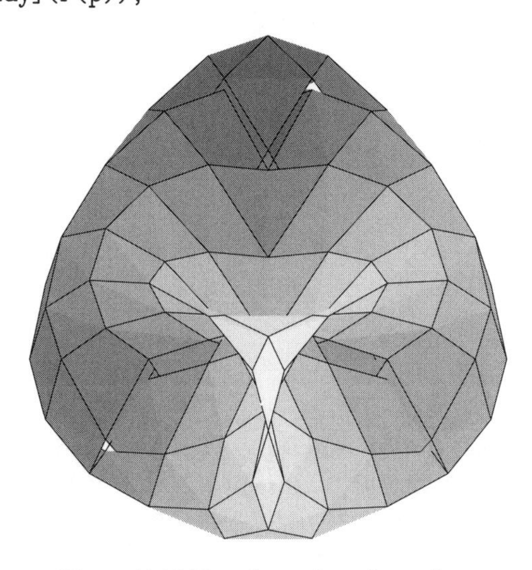

Figure 14.20 Transformation of a surface.

translate

The **translate** function is used to translate two- and three-dimensional plot objects. For two-dimensional plot objects p, the syntax is of the form

translate(p,a,b)

This corresponds to the translation $(x, y) \mapsto (x+a, y+b)$. For three-dimensional plot objects p, the syntax is of the form

translate(p,a,b,c)

This corresponds to the translation $(x, y, z) \mapsto (x+a, y+b, z+c)$. We translate a fixed line $y = x$ by a sequence of rotating vectors and give the result in Figure 14.21.

```
> with(plottools):
> line := plot(x,x=0..1):
> line_seq:=seq(translate(line,cos(Pi*t/32),sin(Pi*t/32)),
   t=0..16):
> plots[display](line_seq,scaling=constrained);
```

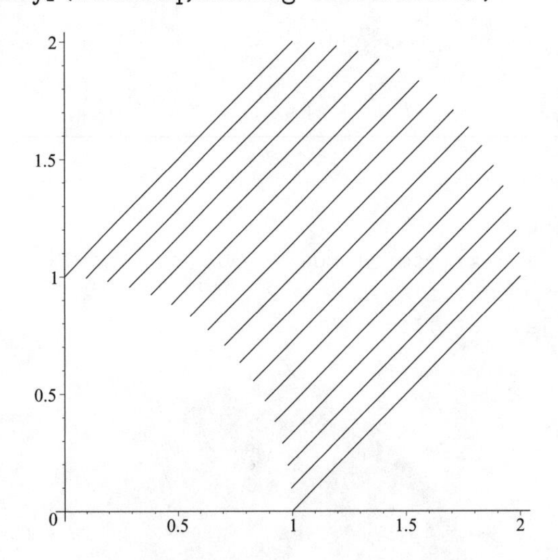

Figure 14.21 A sequence of translations.

14.2 The *geometry* package

The *geometry* package is used for doing two-dimensional Euclidean geometry. An overview of this package will be given in Section 17.7.8. There are two functions in the package for defining regular polygons and regular star polygons. The function `RegularPolygon(` p, n, `pt`, r`)` defines p as a regular n-gon with center `pt` and radius r of the circumscribed circle. Let's plot a regular nonagon. First we use the `point` function to define the center $C = (0, 0)$ of the nonagon.

```
> with(geometry):
> point(C,0,0);
```

$$C$$

We can retrieve the coordinates of C using the `coordinate` function.

```
> coordinates(C);
```

$$[0, 0]$$

We define g as a regular nonagon, centered at C with $r = 1$.

```
> RegularPolygon(g,9,C,1);
```

$$g$$

You can find out a lot about this n-gon using the `detail` function. Try

```
> detail(g);
```

Finally, to draw the nonagon, we use the `draw` function.

```
> draw(g,axes=none);
```

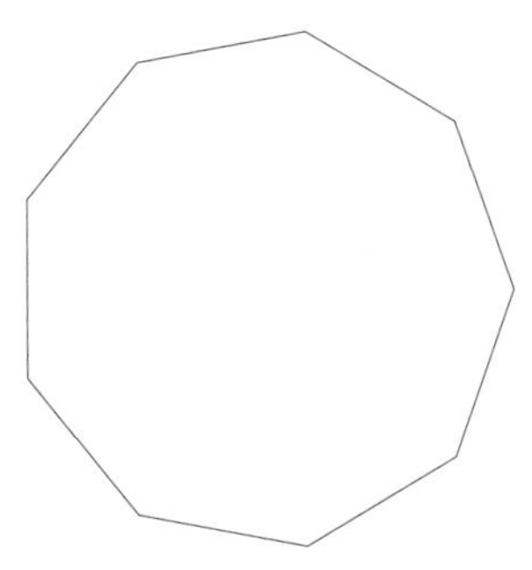

Figure 14.22 A regular nonagon.

Let p, q be positive, relatively prime integers. A star regular p/q-gon is a geometric shape obtained by connecting every qth vertex of a regular p-gon. We use the `StarRegularPolygon` function. The syntax is analogous to the `RegularPolygon` function. We construct a star regular 17/7-gon:

```
> with(geometry):
> RegularStarPolygon(sgon,17/7,point(o,0,0),1):
> draw(sgon,axes=none);
```

Figure 14.23 A star regular 17/7-gon.

14.3 The *geom3d* package

The *geom3d* package is used for doing three-dimensional Euclidean geometry. We used the *geom3d* package briefly in Section 10.2 to compute equations of lines and planes. In this section we mainly concentrate on the plotting capabilities of the *geom3d* package.

14.3.1 Regular polyhedra

The *geom3d* functions for defining regular polyhedra are:

```
GreatDodecahedron              GreatIcosahedron
GreatStellatedDodecahedron     RegularPolyhedron
SmallStellatedDodecahedron     cube
dodecahedron                   hexahedron
icosahedron                    octahedron
tetrahedron
```

Most of these functions are covered by the *plottools* package. We include some examples of the remaining functions. The `RegularPolyhedron` function can be used to define any of the regular polyhedra listed using a Schlafli symbol. See `?geom3d` for more details.

```
> with(geom3d):
> GreatDodecahedron(gdh,point(O,0,0,0),1):
> draw(gdh,orientation=[60,60]);
```

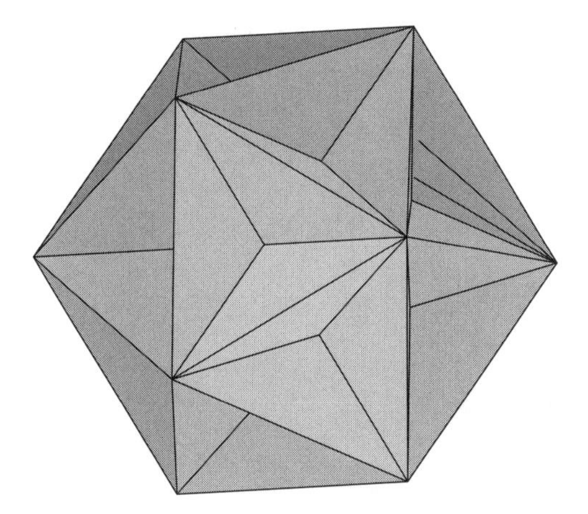

Figure 14.24 A great dodecahedron.

```
> with(geom3d):
> GreatIcosahedron(gih,point(O,0,0,0),1):
> draw(gih,orientation=[75,65]);
```

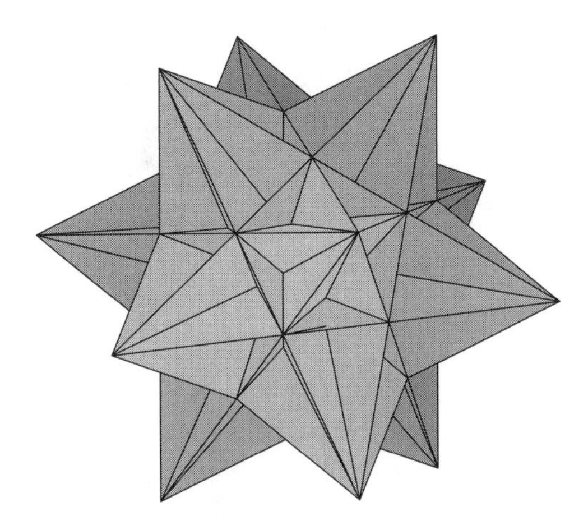

Figure 14.25 A great isocahedron.

```
> with(geom3d):
> GreatStellatedDodecahedron(gsdh,point(O,0,0,0),1):
> draw(gsdh,orientation=[70,20]);
```

Figure 14.26 A great stellated dodecahedron.

```
> with(geom3d):
> SmallStellatedDodecahedron(ssdh,point(O,0,0,0),1):
> draw(ssdh);
> draw(ssdh,orientation=[70,20]);
```

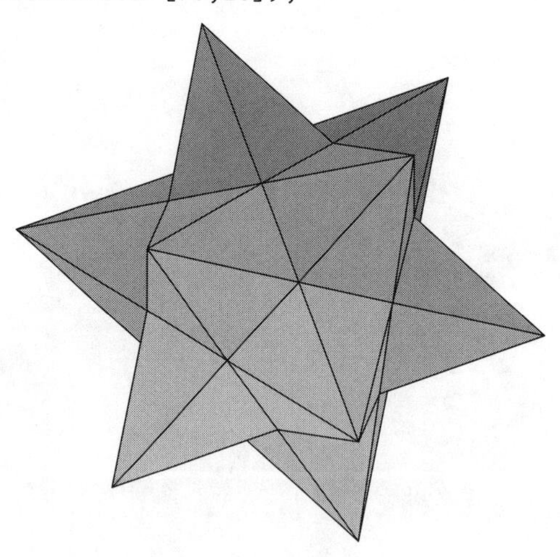

Figure 14.27 A small stellated dodecahedron.

A general tetrahedron can defined using the `gtetrahedron` function. If A, B, C, D are four points, then `gtetrahedron(gt, [`A`,`B`,`C`,`D`])` defines the tetrahedron `gt` with the specified vertices. Try

```
> with(geom3d):
> point(P1,0,0,0), point(P2,0,1,0):
```

```
> point(P3,1,0,0), point(P4,0,1/2,1):
> gtetrahedron(gt, [P1,P2,P3,P4]):
> draw(gt);
```

14.3.2 Quasi-regular polyhedra

There are two quasi-regular polyhedra: the cuboctahedron and the icosidodecahedron. The corresponding MAPLE functions have the same name.

```
> with(geom3d):
> cuboctahedron(coh,point(O,0,0,0),1):
> draw(coh,orientation=[40,25]);
```

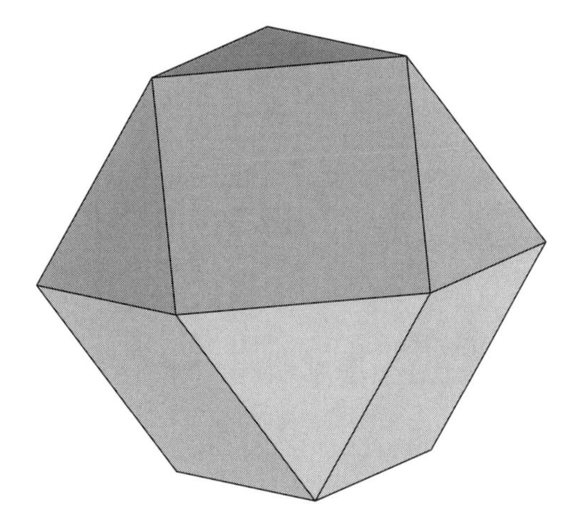

Figure 14.28 A cuboctahedron.

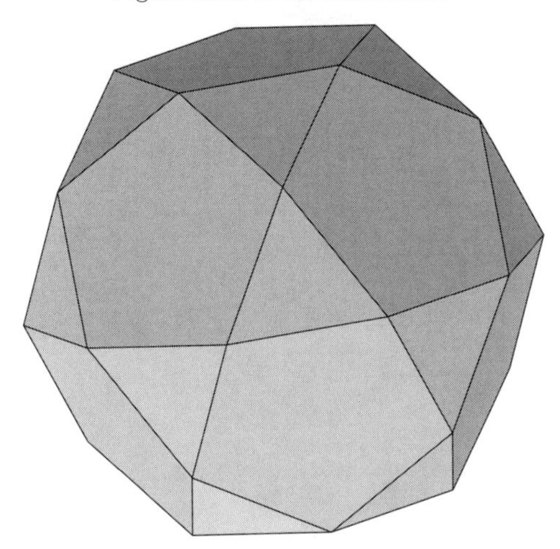

Figure 14.29 An icosidodecahedron.

```
> with(geom3d):
> icosidodecahedron(idh,point(O,0,0,0),1):
> draw(idh,orientation=[80,20]);
```

The resulting plot is given above in Figure 14.29.

14.3.3 The Archimedean solids

The Archimedean solids are convex polyhedra whose faces are regular polygons of at least two types. There are 13 Archimedean solids. Their MAPLE names are given below:

```
GreatRhombicuboctahedron       GreatRhombiicosidodecahedron
SmallRhombicuboctahedron       SmallRhombiicosidodecahedron
SnubCube                       SnubDodecahedron
TruncatedCuboctahedron         TruncatedDodecahedron
TruncatedHexahedron            TruncatedIcosahedron
TruncatedIcosidodecahedron     TruncatedOctahedron
TruncatedTetrahedron
```

We plot a few examples in Figures 14.30 to 14.32.

```
> with(geom3d):
> TruncatedIcosidodecahedron(tid,point(P,0,0,0),1):
> draw(tid);
```

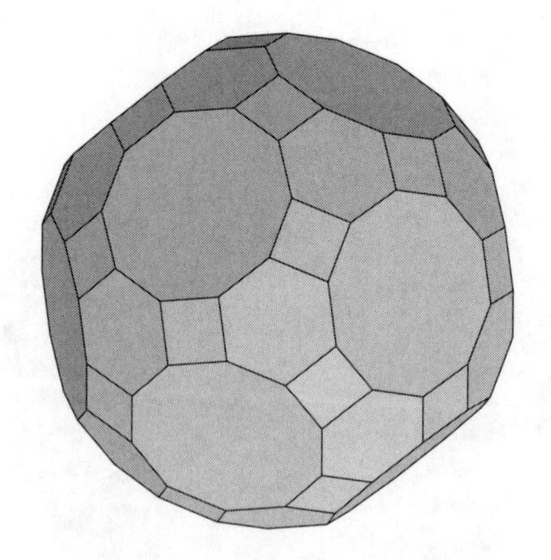

Figure 14.30 A truncated icosidodecahedron.

```
> SmallRhombicuboctahedron(srco,point(P,0,0,0),1):
> draw(srco);
```

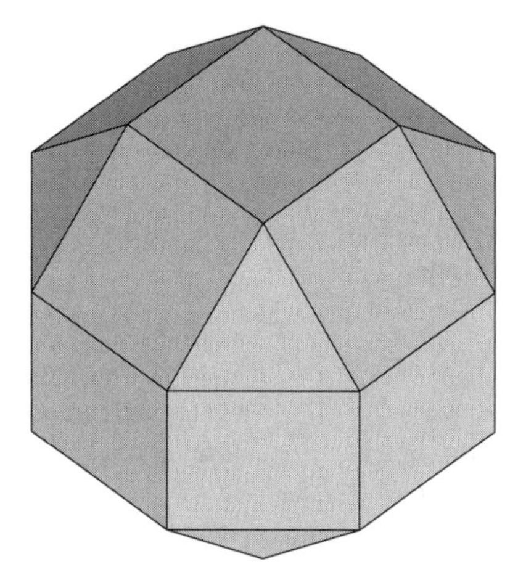

Figure 14.31 A small rhombicuboctahedron.

```
> TruncatedIcosahedron(tic,point(P,0,0,0),1):
> draw(tic);
```

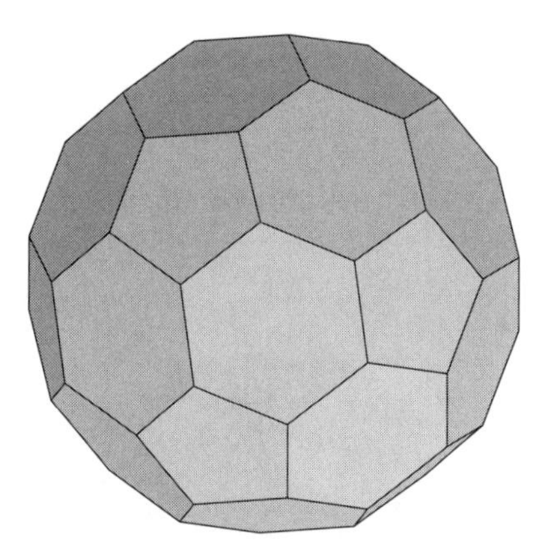

Figure 14.32 A truncated icosahedron.

To plot all the Archimedean solids try

```
> archset:={GreatRhombicuboctahedron,
   GreatRhombiicosidodecahedron,
   SmallRhombicuboctahedron, SmallRhombiicosidodecahedron,
   SnubCube, SnubDodecahedron, TruncatedCuboctahedron,
   TruncatedDodecahedron, TruncatedHexahedron,
   TruncatedIcosahedron, TruncatedIcosidodecahedron,
   TruncatedOctahedron, TruncatedTetrahedron};
> with(geom3d):
> for a in archset do
>   printf("______________________________________________________\n");
>   a(gon,point(P,0,0,0),1):
```

```
>     printf(cat(convert(a,string),"\n"));
>     draw(gon);
>   end do;
```

The duals of the Archimedean solids are also in the *geom3d* package. Their MAPLE names are given below:

```
HexakisIcosahedron            HexakisOctahedron
PentagonalHexacontahedron     PentagonalIcositetrahedron
PentakisDodecahedron          RhombicDodecahedron
RhombicTriacontahedron        TetrakisHexahedron
TrapezoidalHexecontahedron    TrapezoidalIcositetrahedron
TriakisIcosahedron            TriakisOctahedron
TriakisTetrahedron
```

We plot a few examples in Figures 14.33 to 14.35.

```
> with(geom3d):
> TrapezoidalHexecontahedron(thc,point(P,0,0,0),1):
> draw(thc);
```

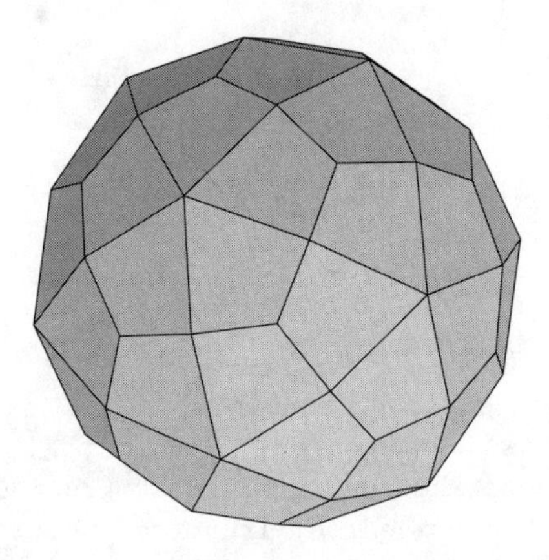

Figure 14.33 A trapezoidal hexecontahedron.

```
> HexakisOctahedron(hoc,point(P,0,0,0),1):
> draw(hoc);
```

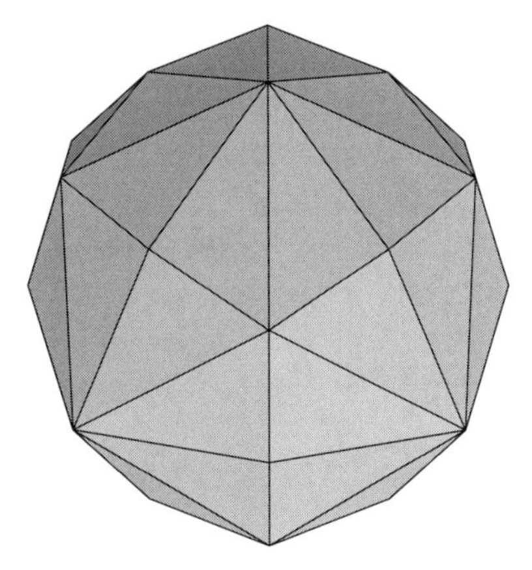

Figure 14.34 A hexakis octahedron.

```
> TrapezoidalIcositetrahedron(tict,point(P,0,0,0),1):
> draw(tict);
```

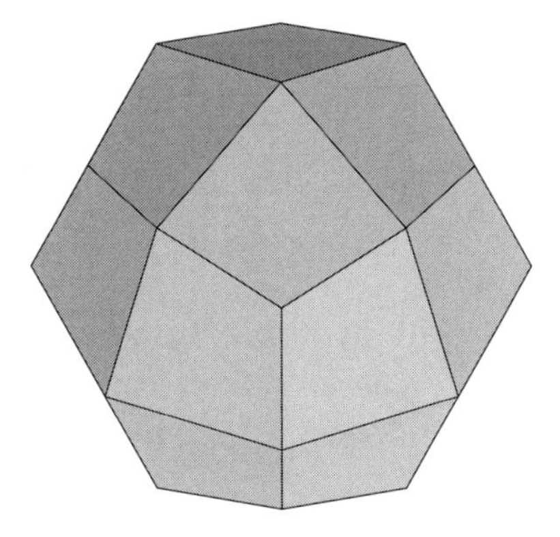

Figure 14.35 A trapezoidal icositetrahedron.

15. SPECIAL FUNCTIONS

To get a list of MAPLE's standard library functions, type

```
> ?index,function
```

There are 152 functions in all.

15.1 Overview of mathematical functions

To obtain a list of the mathematical functions, type

```
> ?inifcn
```

There are 118 functions not including the trigonometric and hyperbolic functions and their inverses. These are the initially known functions. There are additional mathematical functions in various packages.

Below we list these 118 functions:

`abs` - absolute value of real or complex number

`AiryAi`, `AiryAiZeros`, `AiryBi`, `AiryBiZeros` - Airy wave functions and their negative real zeros

`AngerJ` - Anger J function

`argument` - argument of a complex number

`bernoulli` - Bernoulli numbers and polynomials

`BesselI`, `BesselJ` - modified Bessel functions and Bessel functions of the first kind

`BesselJZeros` - nonnegative real zeros of Bessel J

`BesselK`, `BesselY` - modified Bessel functions and Bessel functions of the second kind

`BesselYZeros` - positive real zeros of Bessel Y

`Beta` - Beta function

`binomial` - binomial coefficients

`ceil` - smallest integer greater than or equal to a number

`Chi` - hyperbolic cosine integral

`Ci` - cosine integral

`conjugate` - conjugate of a complex number or expression

`csgn` - complex "half-plane" signum function

`dilog` - dilogarithm function

`Dirac` - Dirac delta function

`Ei` - exponential integrals

`EllipticCE`, `EllipticCK`, `EllipticCPi`, `EllipticE`, `EllipticF`, `EllipticK`, `EllipticModulus`, `EllipticNome`, `EllipticPi` - Complete, incomplete, and complementary elliptic integrals and related functions

`erf` - error function

`erfc` - complementary error function and its iterated integrals

`erfi` - imaginary error function

`euler` - Euler numbers and polynomials

`exp` - exponential function

`factorial` - factorial function

`floor` - greatest integer less than or equal to a number

`frac` - fractional part of a number

`FresnelC`, `Fresnelf`, `Fresnelg`, `FresnelS` - Fresnel integrals and auxiliary functions

`GAMMA` - Gamma and incomplete Gamma functions

`GaussAGM` - Gauss arithmetic geometric mean

`HankelH1`, `HankelH2` - Hankel functions (Bessel functions of the third kind)

`harmonic` - partial sum of the harmonic series

`Heaviside` - Heaviside step function

`hypergeom` - generalized hypergeometric function

`ilog10`, `ilog` - integer logarithms

`Im` - imaginary part of a complex number

`JacobiAM`, `JacobiCN`, `JacobiCD`, `JacobiCS`, `JacobiDN`, `JacobiDC`, `JacobiDS`, `JacobiNC`, `JacobiND`, `JacobiNS`, `JacobiSC`, `JacobiSD`, `JacobiSN` - Jacobi elliptic functions

`JacobiTheta1`, `JacobiTheta2`, `JacobiTheta3`, `JacobiTheta4` - Jacobi theta functions

`JacobiZeta` - Jacobi Zeta function

`KelvinBer`, `KelvinBei`, `KelvinHer`, `KelvinHei`, `KelvinKer`, `KelvinKei` - Kelvin functions

`KummerM`, `KummerU` - Kummer functions

`LegendreP`, `LegendreQ` - Legendre functions

`LerchPhi` - Lerch's Phi function

`Li` - logarithmic integral

`ln` - natural logarithm

`lnGAMMA` - log-Gamma function

`log` - logarithm to arbitrary base

`log10` - log to the base 10

`LommelS1`, `LommelS2` - Lommel functions

`MeijerG` - a modified MeijerG function

`max`, `min` - maximum/minimum of a sequence of real values

`pochhammer` - pochhammer symbol

`polar` - polar representation of complex numbers

`polylog` - polylogarithm function

`Psi` - polygamma function

`Re` - real part of a complex number

`round` - nearest integer to a number

`signum` - sign of a real or complex number

`Shi` - hyperbolic sine integral

`Si` - sine integral

`sqrt` - square root

`Ssi` - shifted sine integral

`StruveH`, `StruveL` - Struve functions

`surd` - nonprincipal root function

`trunc` - nearest integer to a number in the direction of 0

`LambertW` - Lambert W function

`WeberE` - Weber E function

`WeierstrassP` - Weierstrass P-function

`WeierstrassPPrime` - Derivative of Weierstrass P-function

`WeierstrassZeta` - Weierstrass zeta-function

`WeierstrassSigma` - Weierstrass sigma-function

`WhittakerM`, `WhittakerW` - Whittaker functions

`Zeta` - Riemann and Hurwitz zeta functions

15.2 Bessel functions

MAPLE knows the Bessel functions of the first, second, and third kinds. To obtain a list of these functions, type

```
> ?Bessel
```

The Bessel functions of the first kind $J_n(x)$ and the second kind $Y_n(x)$ are given in MAPLE by `BesselJ(n,x)` and `BesselY(n,x)`, respectively. They satisfy Bessel's equation:

$$x^2 y'' + x y' + (x^2 - n^2) y = 0.$$

MAPLE can compute derivatives and series expansions of the Bessel functions and can also compute floating-point approximations:

```
> diff(BesselJ(n,x),x);
```

$$-\mathrm{BesselJ}(n+1,x) + \frac{n\mathrm{BesselJ}(n,x)}{x}$$

```
> evalf(BesselJ(3,0.8));
```

$$0.01024676633$$

```
> series(BesselJ(3,x),x,10);
```

$$\frac{1}{48}x^3 - \frac{1}{768}x^5 + \frac{1}{30720}x^7 - \frac{1}{2211840}x^9 + O\left(x^{10}\right)$$

We see that MAPLE knows the formula

$$\frac{d}{dx} J_n(x) = -J_{n+1}(x) + \frac{n}{x} J_n(x).$$

We found

$$J_3(0.8) \approx .01024676633,$$
$$J_3(x) = \frac{1}{48}x^3 - \frac{1}{768}x^5 + \frac{1}{30720}x^7 - \frac{1}{2211840}x^9 + O\left(x^{10}\right)$$

MAPLE can also compute real zeros of $J_n(x)$ and $Y_n(x)$. `BesselJZeros(n, m)` `BesselYZeros(n, m)` give the mth real positive zero of $J_n(x)$, $Y_n(x)$, respectively. We compute the first real positive zero of $J_3(x)$.

```
> BesselJZeros(3,1);
```

$$\mathrm{BesselJZeros}(3,1)$$

```
> evalf(%);
```

$$6.380161896$$

We found that the first positive real zero of $J_3(x)$ is approximately 6.380161896. Let's plot the first ten zeros of $J_3(x)$ and $J_4(x)$ on the number line. See Figure 15.1.

```
> zpts3 := [seq([evalf(BesselJZeros(3,n)),0],n=1..10)]:
> zpts4 := [seq([evalf(BesselJZeros(4,n)),0],n=1..10)]:
> p1 := plot(zpts3,style=point,symbol=circle,color=blue):
> p2 := plot(zpts4,style=point,symbol=cross,color=red):
> plots[display](p1,p2,axes=none);
```


Figure 15.1 Real zeros of $J_3(x)$ and $J_4(x)$.

The zeros of $J_3(x)$ are marked by a blue circle, and those of $J_4(x)$ are marked by a red cross. What do you notice?

The modified Bessel functions of the first kind and second kind $I_n(x)$ and $K_n(x)$ are given in MAPLE by `BesselI(n,x)` and `BesselK(n,x)`, respectively. They satisfy the modified Bessel equation

$$x^2\,y'' + x\,y' - (x^2 + n^2)\,y = 0.$$

There are Bessel functions of the third kind, usually known as Hankel functions. The Hankel functions of the first and second kinds are defined by

$$\begin{aligned} H_n^{(1)}(x) &= J_n(x) + i\,Y_n(x), \\ H_n^{(2)}(x) &= J_n(x) - i\,Y_n(x). \end{aligned}$$

In MAPLE they are given by `HankelH1(n,x)` and `HankelH2(n,x)` respectively.

15.3 The Gamma function

The Gamma function $\Gamma(z)$ can be defined in terms of a certain infinite product

$$\Gamma(z) = \frac{1}{z}\prod_{n=1}^{\infty}\left\{\left(1+\frac{1}{n}\right)^{z}\left(1+\frac{z}{n}\right)^{-1}\right\},$$

for $z \neq 0, -1, -2, \ldots$. The Gamma function $\Gamma(z)$ is an analytic function (of a complex variable z), except for simple poles at $z = 0, -1, -2, \ldots$. For $\Re z > 0$, the Gamma function is given by

$$\Gamma(z) = \int_0^\infty e^{-t}\,t^{z-1}\,dt.$$

In MAPLE the Gamma function is given by `GAMMA(z)`. Let's plot a graph of the Gamma function $\Gamma(x)$ for real x. See Figure 15.2.

```
> plot(GAMMA(x),x=-4..4,y=-10..10,discont=true);
```

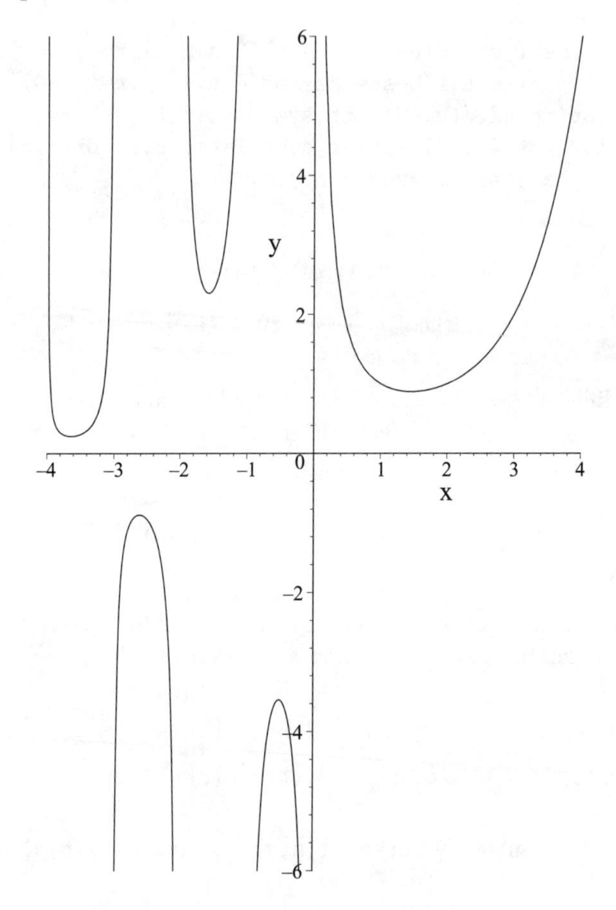

Figure 15.2 The graph of the Gamma function.

The Gamma function interpolates the factorial function

$$\Gamma(n) = (n-1)!,$$

when n is a positive integer. We compute some values of the Gamma function.

```
> GAMMA(10) = 9!;
```

$$362880 = 362880$$

```
> GAMMA(1/2);
```

$$\sqrt{\pi}$$

```
> GAMMA(1/3);
```

$$2/3\,\frac{\pi\sqrt{3}}{\Gamma(\frac{2}{3})}$$

```
> evalf(%);
```

$$2.678938537$$

```
> GAMMA(-1+I);
```

$$\Gamma(-1+i)$$

```
> evalf(%);
```

$$-0.1715329199 + 0.3264827482\,I$$

We found that

$$\begin{aligned}
\Gamma(10) &= 362880 = 9!, \\
\Gamma(1/2) &= \sqrt{\pi}, \\
\Gamma(1/3) &= \frac{2\pi\sqrt{3}}{3\Gamma(2/3)} \approx 2.678938537, \\
\Gamma(-1+i) &\approx -.1715329199 + .3264827482\,i
\end{aligned}$$

We can convert factorials and binomial coefficients to values of the Gamma function using the `convert` function.

```
> fn := n!;
```

$$n!$$

```
> bnk:=binomial(n,k);
```

$$bnk := \texttt{binomial}(n,k)$$

```
> gfn := convert(fn,GAMMA);
```

$$\Gamma(n+1)$$

```
> gbnk := convert(bnk,GAMMA);
```

$$\frac{\Gamma(n+1)}{\Gamma(k+1)\Gamma(n-k+1)}$$

MAPLE found that

$$\begin{aligned}
n! &= \Gamma(n+1), \\
\binom{n}{k} &= \frac{\Gamma(n+1)}{\Gamma(k+1)\Gamma(n-k+1)}.
\end{aligned}$$

Of course, here n and k are nonnegative integers satisfying $0 \le k \le n$. We can also `convert` back. Try

```
> convert(gfn,factorial);
> convert(gbnk,binomial);
```

See `?convert[GAMMA]` for more information.

The beta function $\beta(x,y)$ is defined by

$$\beta(x,y) = \int_0^1 t^{x-1}(1-t)^{y-1}\,dt,$$

for $\Re x > 0$, $\Re y > 0$. In MAPLE it is given by `Beta(x,y)`.

```
> Beta(x,y);
```

$$\beta(x,y)$$

```
> convert(%,GAMMA);
```

$$\frac{\Gamma(x)\Gamma(y)}{\Gamma(x+y)}$$

```
> int(t^(x-1)*(1-t)^(y-1),t=0..1);
```

$$\frac{\Gamma(x)\Gamma(y)}{\Gamma(x+y)}$$

MAPLE knows that

$$\beta(x,y) = \frac{\Gamma(x)\Gamma(y)}{\Gamma(x+y)}.$$

In MAPLE the log Gamma function $\operatorname{Log}\Gamma(z)$ is given by `lnGAMMA(z)`. This is the principal value of the log of $\Gamma(z)$.

```
> lnGAMMA(-3/2);
```

$$\mathtt{lnGAMMA}(-\frac{3}{2})$$

```
> evalf(%);
```

$$0.8600470154 - 6.283185307\,I$$

```
> taylor(lnGAMMA(z+1),z=0,6);
```

$$-\gamma\,z + 1/12\,\pi^2 z^2 - \frac{1}{3}\,\zeta(3)z^3 + \frac{1}{360}\,\pi^4 z^4 - \frac{1}{5}\,\zeta(5)z^5 + O\left(z^6\right)$$

We found that $\operatorname{Log}\Gamma(-3/2) \approx .8600470154 - 6.283185307\,i$. We found the first few terms of the Taylor expansion of $\operatorname{Log}\Gamma(z+1)$ near $z=0$.

$$\operatorname{Log}\Gamma(z+1) = -\gamma\,z + 1/12\,\pi^2 z^2 - \frac{1}{3}\,\zeta(3)z^3 + \frac{1}{360}\,\pi^4 z^4 - \frac{1}{5}\,\zeta(5)z^5 + O\left(z^6\right).$$

Here γ is Euler's constant

$$\gamma = \lim_{n\to\infty}\left(\sum_{k=1}^{n}\frac{1}{k}\right) - \ln(n),$$

which is given in MAPLE by `gamma`. Also, $\zeta(n)$ is the Riemann zeta function, given in MAPLE by `Zeta`. See Section 15.9.

The Psi function $\Psi(z)$ is the logarithmic derivative of the Gamma function

$$\Psi(z) = \frac{\Gamma'(z)}{\Gamma(z)}.$$

In MAPLE it is given by `Psi(z)`.

```
> diff(lnGAMMA(z),z);
```

$$\Psi(z)$$

Naturally, MAPLE found that

$$\frac{d}{dz}\operatorname{Log}\Gamma(z) = \Psi(z).$$

MAPLE knows certain values of the Psi function.

```
> Psi(1/6);
```

$$-\gamma - 2\ln(2) - \frac{3}{2}\ln(3) - \frac{1}{2}\pi\sqrt{3}$$

MAPLE knows that

$$\Psi(1/6) = -\gamma - 2\ln(2) - \frac{3}{2}\ln(3) - \frac{1}{2}\pi\sqrt{3}.$$

Try

```
> Psi(1/2);
> Psi(1/4);
> Psi(3/4);
> exp(Psi(1/6)+Psi(5/6)+2*gamma);
> simplify(%);
```

The nth derivative of the Psi function is given in MAPLE by `Psi(z,n)`.

```
> Psi(1,2);
```

$$-1 + \frac{1}{6}\pi^2$$

MAPLE knows that

$$\Psi''(1) = -1 + \frac{1}{6}\pi^2.$$

15.4 Hypergeometric functions

Let p, q be nonnegative integers. The generalized hypergeometric function ${}_pF_q$ is given by

$${}_pF_q\left(\begin{matrix} a_1, a_2, \cdots, a_p \\ b_1, b_2, \cdots, b_q \end{matrix}; \; z\right) = \sum_{n=0}^{\infty} \frac{(a_1)_n (a_2)_n \cdots (a_p)_n}{(b_1)_n (b_2)_n \cdots (b_q)_n} \frac{z^n}{n!},$$

where $(a)_n$ is the Pochhamer symbol

$$(a)_n = a(a+1)\cdots(a+n-1) = \frac{\Gamma(a+n)}{\Gamma(a)}.$$

In MAPLE, this hypergeometric function is given by

`hypergeom([`a_1`,`a_2`, ... ,`a_p`], [`b_1`,`b_2`, ... ,`b_q`], z).`

The Pochhammer symbol $(a)_n$ is given in MAPLE by `pochhammer(a,n)`. The Gaussian hypergeometric function $F(a, b, c; z)$ is given by

$$F(a,b,c;z) = {}_2F_1\left(\begin{matrix} a, b \\ c \end{matrix}; z\right).$$

Most of the special functions used by physicists and engineers are special cases of hypergeometric functions.

```
> simplify( hypergeom([],[],z) );
```

$$e^z$$

```
> simplify( hypergeom([a],[],z) );
```

$$(1-z)^{-a}$$

```
> simplify( hypergeom([],[3/2],-z^2/4) );
```

$$\frac{\sin(z)}{z}$$

```
> simplify( hypergeom([],[1/2],-z^2/4) );
```

$$\cos(z)$$

```
> simplify( hypergeom([],[a],-z^2/4) );
```

$$-(-2\,a\mathrm{BesselJ}(a,z) + \mathrm{BesselJ}(a+1,z)z)\,\Gamma(a)2^{a-1}z^{-a}$$

```
> simplify( hypergeom([1,1],[2],-z) );
```

$$\frac{\ln(1+z)}{z}$$

MAPLE knows the results

$$e^z = {}_0F_0(z),$$

$$(1-z)^{-a} = {}_1F_0\left(\begin{matrix} a \\ - \end{matrix}; z\right),$$

$$\sin z = z\,{}_0F_1\left(\begin{matrix} - \\ \frac{3}{2} \end{matrix}; -\frac{z^2}{4}\right),$$

$$\cos z = {}_0F_1\left(\begin{matrix} - \\ \frac{1}{2} \end{matrix}; -\frac{z^2}{4}\right),$$

$$z^a\,{}_0F_1\left(\begin{matrix} - \\ a \end{matrix}; -\frac{z^2}{4}\right) = -\left(-2\,a\,J_a(z) + z\,J_{a+1}(z)\right)\Gamma(a)2^{a-1}$$

$$\ln(1+z) = z\,{}_2F_1\left(\begin{matrix} 1, 1 \\ 2 \end{matrix}; -z\right).$$

MAPLE knows some hypergeometric summation theorems.

```
> simplify( hypergeom([a,b],[c],1) );
```

$$\frac{\Gamma(c)\Gamma(c-a-b)}{\Gamma(c-a)\Gamma(c-b)}$$

It knows the Gauss summation

$${}_2F_1\left(\begin{matrix}a,b\\c\end{matrix};\ 1\right) = \frac{\Gamma(c)\Gamma(c-a-b)}{\Gamma(c-a)\Gamma(c-b)},$$

for $\Re(c-a-b) > 0$. Try

```
> simplify( hypergeom([a,b,c],[1+a-b,1+a-c],1) );
> simplify( hypergeom([a,1+a/2,b,c,d],
   [a/2,1+a-b,1+a-c,1+a-d],1) );
```

Certain series can be evaluated using `convert` with the `hypergeom` option.

```
> Sum(pochhammer(a,n)*pochhammer(b,n)/pochhammer(c,n)/n!,
   n=0..infinity);
```

$$\sum_{n=0}^{\infty} \frac{\mathrm{pochhammer}(a,n)\mathrm{pochhammer}(b,n)}{\mathrm{pochhammer}(c,n)n!}$$

```
> convert(%,hypergeom);
```

$$\frac{\mathrm{pochhammer}(a,0)\mathrm{pochhammer}(b,0)\Gamma(c)\Gamma(c-a-b)}{\mathrm{pochhammer}(c,0)\Gamma(c-a)\Gamma(c-b)}$$

```
> simplify(%);
```

$$\frac{\Gamma(c)\Gamma(c-a-b)}{\Gamma(c-a)\Gamma(c-b)}$$

MAPLE found (eventually) that

$$\sum_{n=0}^{\infty} \frac{(a)_n(b)_n}{(c)_n\, n!} = \frac{\Gamma(c)\Gamma(c-a-b)}{\Gamma(c-a)\Gamma(c-b)},$$

which is just Gauss's summation of a ${}_2F_1$.

15.5 Elliptic integrals

The incomplete elliptic integral of the first kind $F(z,k)$ is given by

$$F(z,k) = \int_0^z \frac{dt}{\sqrt{(1-t^2)(1-k^2t^2)}}.$$

In MAPLE it is given by `EllipticF(z,k)`. Here $0 < k < 1$ is called the modulus. The incomplete elliptic integral of the second kind $E(z,k)$ is given by

$$E(z,k) = \int_0^z \frac{\sqrt{1-k^2t^2}}{\sqrt{1-t^2}}\,dt.$$

In MAPLE it is given by `EllipticE(z,k)`. The incomplete elliptic integral of the third kind $\Pi(z,\nu,k)$ is given by

$$\Pi(z,\nu,k) = \int_0^z \frac{dt}{(1-\nu t^2)\sqrt{(1-k^2t^2)(1-t^2)}}.$$

In MAPLE it is given by `EllipticPi(z,nu,k)`.

The complete elliptic integrals of the first and second kind, respectively, are given by

$$K(k) = F(1,k),$$
$$E(k) = E(1,k).$$

They are given in MAPLE by `EllipticK` and `EllipticE`, respectively. There are complementary integrals K' and E', which are integrals in the complementary variable $k' = \sqrt{1-k^2}$

$$K'(k) = K(k'),$$
$$E'(k) = E(k').$$

In MAPLE these are given by `EllipticCK` and `EllipticCE`. We compute the first few terms of the Taylor expansion of $K(k)$ near $k = 0$.

```
> taylor(EllipticK(k),k=0,10);
```

$$\frac{1}{2}\pi + \frac{1}{8}\pi k^2 + \frac{9}{128}\pi k^4 + \frac{25}{512}\pi k^6 + \frac{1225}{32768}\pi k^8 + O\left(k^{10}\right)$$

We found that

$$K(k) = \pi\left(\frac{1}{2} + \frac{1}{8}k^2 + \frac{9}{128}k^4 + \frac{25}{512}k^6 + \frac{1225}{32768}k^8 + \cdots\right).$$

Try

```
> taylor(EllipticK(k)-Pi/2*hypergeom([1/2,1/2],[1],k^2),
   k=0,100);
```

MAPLE knows the derivatives of the elliptic integrals. We compute the derivative of $E(k)$.

```
> diff(EllipticE(k),k);
```

$$-\frac{\mathrm{EllipticK}(k)}{k} + \frac{\mathrm{EllipticE}(k)}{k}$$

MAPLE knows that

$$\frac{d}{dk}\,E(k) = \frac{E(k) - K(k)}{k}.$$

Try

```
> diff(EllipticK(k),k);
```

The nome $q = q(k)$ is defined by

$$q = e^{-\pi K'(k)/K(k)}.$$

In MAPLE, it is given by `EllipticNome(k)`. The inverse function is called the modulus and is given in MAPLE by `EllipticModulus(q)`. This function can be given in terms of Jacobi's theta functions

$$k = \frac{\vartheta_2^2(0,q)}{\vartheta_3^2(0,q)}.$$

See Section 15.7 for definition and computation of the theta-functions. We compute an example.

```
> k := 0.35;
```

$$0.35$$

```
> q := EllipticNome(k);
```

$$0.008166668955$$

```
> EllipticModulus(q);
```

$$0.3500000000$$

We found that the nome corresponding to $k = 0.35$ is $q \approx .008166668955$.

15.6 The AGM

The *arithmetic-geometric mean* (AGM) iteration of Gauss is the following two-term recursion:

$$a_{n+1} = \frac{a_n + b_n}{2},$$
$$b_{n+1} = \sqrt{a_n\, b_n}.$$

If $a_0 = a$, $b_0 = b$, then both sequences $\{a_n\}$, $\{b_n\}$, converge to the same limit $M(a,b)$. The function $M(a,b)$ is given in MAPLE by `GaussAGM(a,b)`.

```
> g1 := 1/GaussAGM(1,sqrt(2));
```

$$\left(\mathrm{GaussAGM}(1,\sqrt{2})\right)^{-1}$$

```
> g2 := 2/Pi*int(1/sqrt(1-t^4),t=0..1);
```

$$1/2\,\frac{\beta(\frac{1}{4},1/2)}{\pi}$$

```
> evalf(g1,12);
```

0.834626841678

```
> evalf(g2,12);
```

0.834626841675

This numerical result was observed by Gauss on May 30, 1799. He commented in his diary that this result "will surely open up a whole new field of analysis." See Jon and Peter Borwein's book [3] for a fascinating account of Gauss's AGM and its connection with elliptic integrals, hypergeometric functions, theta functions, and approximations to π. Try

```
> evalf(g1-g2,100);
```

15.7 Jacobi's theta functions

There are four types of theta-functions

$$\vartheta_1(z,q) = 2\,q^{\frac{1}{4}}\sum_{n=0}^{\infty}(-1)^n\,q^{n(n+1)}\sin((2\,n+1)\,z),$$

$$\vartheta_2(z,q) = 2\,q^{\frac{1}{4}}\sum_{n=0}^{\infty}q^{n(n+1)}\cos((2\,n+1)\,z),$$

$$\vartheta_3(z,q) = 1+2\sum_{n=1}^{\infty}q^{n^2}\cos(2\,nz),$$

$$\vartheta_4(z,q) = 1+2\sum_{n=1}^{\infty}(-1)^n\,q^{n^2}\cos(2\,nz).$$

A good reference is Whittaker and Watson's book [4]. In MAPLE, these four functions are given by `JacobiTheta1`, `JacobiTheta2`, `JacobiTheta3`, and `JacobiTheta4`, respectively.

```
> Digits:=30:
> z:=rand(0..2000)()/2000.;
```

0.0240000000000000000000000000000

```
> q:=rand(0..1000)()/1001.;
```

0.366633366633366633366633366633

```
> f1 := JacobiTheta1(z,q)^2*JacobiTheta4(0,q)^2:
> f2 := JacobiTheta3(z,q)^2*JacobiTheta2(0,q)^2:
```

```
> f3 := JacobiTheta2(z,q)^2*JacobiTheta3(0,q)^2:
> f1 - f2 + f3;
```

$$8.0 \times 10^{-29}$$

This result suggests that

$$\vartheta_1^2(z,q)\vartheta_4^2(0,q) = \vartheta_3^2(z,q)\vartheta_2^2(0,q) - \vartheta_2^2(z,q)\vartheta_3^2(0,q).$$

Try verifying the result for other values of z, q. Unfortunately, at this point MAPLE is unable to compute series expansions of the theta-functions.

15.8 Elliptic functions

There are three basic Jacobi elliptic functions: $\mathrm{sn}\,(u,k)$, $\mathrm{dn}\,(u,k)$, and $\mathrm{cn}\,(u,k)$. These are given in MAPLE by `JacobiSN(u,k)`, `JacobiCN(u,k)`, and `JacobiDN(u,k)`, respectively. They arise as inverse functions of incomplete elliptic integrals and are doubly periodic (elliptic) functions. For example,

$$u = \int_0^{\mathrm{sn}\,(u,k)} \frac{dt}{(1-t^2)(1-k^2t^2)}.$$

Let's verify this result for $u = 1.0$ and $k = 0.5$:

```
> u:=1.0:
> k:=0.5:
> s:=JacobiSN(u,k);
```

$$0.8226355779$$

```
> EllipticF(s,k);
```

$$0.9999999996$$

The value of `EllipticF(JacobiSN(`u`,k),k)` should be u. Here we obtained 0.9999999996, which is close enough to the correct value 1.0.

The other Jacobi elliptic functions in MAPLE are `JacobiAM`, `JacobiCD`, `JacobiCS`, `JacobiDC`, `JacobiDS`, `JacobiNC`, `JacobiND`, `JacobiNS`, `JacobiSC`, and `JacobiSD`.

An alternative approach to elliptic functions is due to Weierstrass. The Weierstrass $\wp$-function is defined by

$$\wp(z) = \wp(z;\Omega) = \frac{1}{z^2} + \sum_{\substack{\omega\in\Omega \\ \omega\neq 0}} \left(\frac{1}{(z-\omega)^2} - \frac{1}{\omega^2} \right).$$

Here Ω is a lattice in $\mathbb{C}$

$$\Omega = \{m\omega_1 + n\omega_2 \,:\, m,n \in \mathbb{Z}\},$$

where $\omega_2/\omega_1 \notin \mathbb{R}$. There are two important invariants

$$g_2 = g_2(\Omega) = 60 \sum_{\substack{\omega\in\Omega \\ \omega\neq 0}} \frac{1}{\omega^4},$$

$$g_3 = g_3(\Omega) = 140 \sum_{\substack{\omega\in\Omega \\ \omega\neq 0}} \frac{1}{\omega^6}.$$

The invariants g_2, g_3 uniquely determine the lattice Ω. In MAPLE the Weierstrass $\wp$ is given in terms of g_2, g_3. The function `WeierstrassP(z, g2, g3)` corresponds to $\wp(z;\Omega)$. We compute the Laurent series of $\wp(z)$ near $z = 0$.

```
> series(WeierstrassP(z,g2,g3),z,10);
```

$$z^{-2} + 1/20\,\mathrm{g2}\,z^2 + 1/28\,\mathrm{g3}\,z^4 + \frac{1}{1200}\,\mathrm{g2}^2 z^6 + \frac{3}{6160}\,\mathrm{g2}\,\mathrm{g3}\,z^8 + O\left(z^{10}\right)$$

We found that

$$\wp(z) = \frac{1}{z^2} + \frac{1}{20}\,g_2 z^2 + \frac{1}{28}\,g_3 z^4 + \frac{1}{1200}\,{g_2}^2 z^6 + \frac{3}{6160}\,g_2 g_3 z^8 + \cdots$$

In MAPLE the derivative of $\wp$-function $\wp'(z)$ is given by `WeierstrassPPrime`.

```
> diff(WeierstrassP(z,g2,g3),z);
```

$$\mathrm{WeierstrassPPrime}(z, \mathrm{g2}, \mathrm{g3})$$

Try

```
> WeierstrassPPrime(z,g2,g3)^2 - 4*WeierstrassP(z,g2,g3)^3
    + g2*WeierstrassP(z,g2,g3):
> series(%,z,20):
> normal(%);
```

What did you get?

The Weierstrass zeta-function $\zeta(z)$ is defined by

$$\zeta(z) = \zeta(z;\Omega) = \frac{1}{z} + \sum_{\substack{\omega\in\Omega \\ \omega\neq 0}} \left(\frac{1}{(z-\omega)} + \frac{1}{\omega} + \frac{z}{\omega^2} \right).$$

This is not to be confused with the Riemann zeta-function of Section 15.9.

```
> diff(WeierstrassZeta(z,g2,g3),z);
```

$$-\mathrm{WeierstrassP}(z, \mathrm{g2}, \mathrm{g3})$$

MAPLE knows that

$$\zeta'(z) = -\wp(z).$$

The Weierstrass σ-function is defined by

$$\sigma(z) = \sigma(z;\Omega) = z \prod_{\substack{\omega\in\Omega \\ \omega\neq 0}} \left\{ \left(1 - \frac{z}{\omega}\right) \exp\left(\frac{z}{\omega} + \frac{z^2}{2\omega^2}\right) \right\}.$$

```
> diff(WeierstrassSigma(z,g2,g3),z);
```

$$\mathrm{WeierstrassZeta}(z, \mathrm{g2}, \mathrm{g3})\,\mathrm{WeierstrassSigma}(z, \mathrm{g2}, \mathrm{g3})$$

MAPLE knows that

$$\frac{\sigma'(z)}{\sigma(z)} = \zeta(z).$$

15.9 The Riemann zeta-function

The Riemann zeta-function $\zeta(z)$ is given by

$$\zeta(z) = \sum_{n=1}^{\infty} \frac{1}{n^z},$$

for $\Re(z) > 1$. The zeta-function has analytic continuation to the whole complex plane except for a simple pole at $z = 1$. For even integers n, it is known that $\zeta(n)$ is a rational multiple of π^n:

```
> for n from 2 by 2 to 10 do
>   print(zeta(n)=Zeta(n));
> end do;
```

$$\zeta(2) = \frac{1}{6}\,\pi^2$$
$$\zeta(4) = \frac{1}{90}\,\pi^4$$
$$\zeta(6) = \frac{1}{945}\,\pi^6$$
$$\zeta(8) = \frac{1}{9450}\,\pi^8$$
$$\zeta(10) = \frac{1}{93555}\,\pi^{10}$$

The function `Zeta(n,z)` gives the nth derivative of the zeta-function, and `Zeta(n,z,a)` gives the nth derivative of the Hurwitz zeta function

$$\zeta(z, a) = \sum_{n=0}^{\infty} \frac{1}{(n+a)^z}.$$

We plot the absolute value of the zeta-function on the line $\Re(z) = 1/2$. See Figure 15.3.

```
> plot(abs(Zeta(1/2+t*I)), t= 0..40);
```

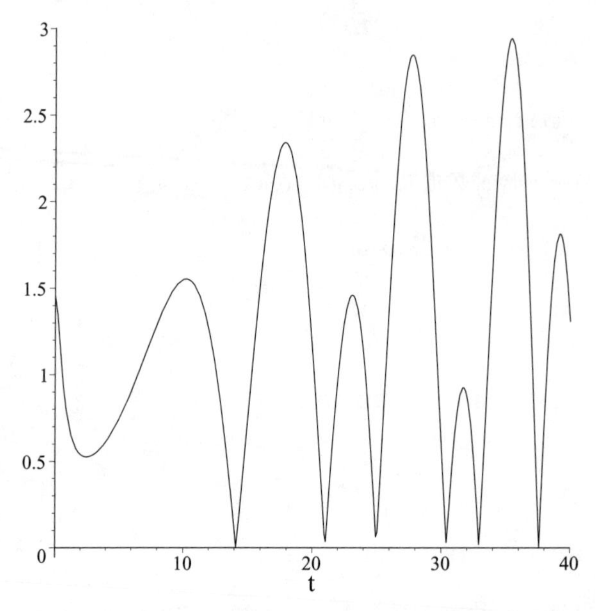

Figure 15.3 The graph of $|\zeta(1/2 + i\,t)|$, $0 \le t \le 40$.

15.10 Orthogonal polynomials

The orthogonal polynomial package is *orthopoly*.

```
> with(orthopoly);
```

$$[G, H, L, P, T, U]$$

The functions in the package are used to define eight orthogonal polynomials:

`G(n,a,x)`	Gegenbauer polynomial
`H(n,x)`	Hermite polynomial
`L(n,x)`	Laguerre polynomial
`L(n,a,x)`	generalized Laguerre polynomial
`P(n,x)`	Legendre polynomial
`P(n,a,b,x)`	Jacobi polynomial
`T(n,x)`	Chebyshev polynomial (first kind)
`U(n,x)`	Chebyshev polynomial (second kind)

As an example, we compute the general second-degree Jacobi polynomial. The Jacobi polynomials $P_n^{(a,b)}(x)$ are orthogonal on the interval $[-1, 1]$ with respect to the weight function $w(x) = (1-x)^a(1+x)^b$, where a, b are constants greater than -1.

```
> with(orthopoly):
> P(2,a,b,x);
```

$$\frac{1}{8}a^2 - \frac{1}{8}a - \frac{1}{4}ab - \frac{1}{2} - \frac{1}{8}b + \frac{1}{8}b^2 + \frac{1}{4}(3+a+b)(a-b)x + \frac{1}{8}(4+a+b)(3+a+b)x^2$$

We found that

$$P_2^{(a,b)}(x) = \frac{1}{8}a^2 - \frac{1}{8}a - \frac{1}{4}ab - \frac{1}{2} - \frac{1}{8}b + \frac{1}{8}b^2 + \frac{1}{4}(3+a+b)(a-b)x + \frac{1}{8}(4+a+b)(3+a+b)x^2.$$

15.11 Integral transforms

The *inttrans* package contains many functions for computing integral transforms.

```
> with(inttrans);
```

[addtable, fourier, fouriercos, fouriersin, hankel, hilbert, invfourier, invhilbert, invlaplace, invmellin, laplace, mellin, savetable]

We have seen the `laplace`, `invlaplace` functions in Section 8.6. These two functions compute Laplace and inverse Laplace transforms. In Section 8.6 we also saw how to use the `addtable` function.

15.11.1 Fourier transforms

Suppose $f(x)$ is a function defined on $(-\infty,\infty)$. The Fourier transform of $f(x)$ is given by

$$F(x) = \int_{-\infty}^{\infty} f(t)\, e^{-itx}\, dt.$$

In MAPLE it is given by `fourier(f(t),t,x)`. As an example, we compute the Fourier transform of $f(t) = e^{-t^2/2}$.

```
> with(inttrans):
> f := exp(-t^2/2);
```

$$f := e^{-\frac{1}{2}t^2}$$

```
> F := fourier(f,t,x);
```

$$F := \sqrt{2}\sqrt{\pi}e^{-\frac{1}{2}x^2}$$

We find that the Fourier transform of $f(t) = e^{-t^2}$ is

$$F(x) = \sqrt{2\pi}\, e^{-x^2/2}.$$

The inverse Fourier transform of $F(x)$ is

$$f(t) = \frac{1}{2\pi} \int_{-\infty}^{\infty} F(x)\, e^{itx}\, dx.$$

In MAPLE it is given by `invfourier(F(t),t,x)`. We check our work by computing the inverse Fourier transform.

```
> with(inttrans):
> F := sqrt(2*Pi)*exp(-x^2/2);
```

$$F := \sqrt{2\pi} e^{-\frac{1}{2}x^2}$$

```
> f := invfourier(F,x,t);
```

$$f := e^{-\frac{1}{2}t^2}$$

We found that the inverse Fourier transform of $F(x) = \sqrt{2\pi}e^{-x^2/2}$ is $f(t) = e^{-t^2/2}$ as expected.

The function

$$F(x) = \sqrt{\frac{2}{\pi}} \int_0^{\infty} f(t)\, \cos xt\, dt,$$

is called the Fourier cosine transform of $f(t)$, and the function

$$\Phi(x) = \sqrt{\frac{2}{\pi}} \int_0^{\infty} f(t)\, \sin xt\, dt,$$

is called the Fourier sine transform of $f(t)$. In MAPLE these transforms are given by `fouriercos(f(t),t,x)` and `fouriersin(f(t),t,x)`, respectively. For $a > 0$ we compute the Fourier cosine and sine transforms of $f(t) = e^{-at}$.

```
> with(inttrans):
> assume(a>0):
> f := exp(-a*t);
```

$$f := e^{-a\sim t}$$

```
> F := fouriercos(f,t,x);
```

$$F := \frac{\sqrt{2}a\sim}{\sqrt{\pi}\,(a\sim^2 + x^2)}$$

```
> Phi := fouriersin(f,t,x);
```

$$\Phi := \frac{\sqrt{2}x}{\sqrt{\pi}\,(a\sim^2 + x^2)}$$

We found that the Fourier cosine and sine transforms of $f(t) = e^{-at}$ are

$$F(x) = \sqrt{\frac{2}{\pi}} \frac{a}{a^2 + x^2},$$
$$\Phi(x) = \sqrt{\frac{2}{\pi}} \frac{x}{a^2 + x^2},$$

respectively. There is no need to define `invfouriercos` and `invfouriersin`, since the Fourier cosine and sine transforms are inverses of each other. Confirm this by trying

```
> fouriercos(F,x,t);
> radsimp(%);
> fouriersin(Phi,x,t);
> radsimp(%);
```

15.11.2 Hilbert transform

The Hilbert transform of a function $f(x)$ is defined as the principal value integral

$$F(x) = \frac{1}{\pi} \mathrm{PV} \int_{-\infty}^{\infty} \frac{f(t)}{t - x}\, dt = \frac{1}{\pi} \lim_{y \to \infty} \int_{-y}^{y} \frac{f(t)}{t - x}\, dt.$$

This is given in MAPLE by `hilbert(f(t),t,x)`. We compute the Hilbert transform of

$$f(t) = \frac{1}{1 + t^2}.$$

```
> with(inttrans):
> f := 1/(1+t^2);
```

$$f := \frac{1}{1 + t^2}$$

```
> hilbert(f,t,x);
```

$$-\frac{x}{x^2 + 1}$$

We found that the Hilbert transform is

$$F(x) = -\frac{x}{1 + x^2}.$$

The inverse Hilbert transform is simply the negative of the Hilbert transform. We confirm this for our example.

```
> with(inttrans):
> F := -x/(1 + x^2);
```

$$F := -\frac{x}{1 + x^2}$$

```
> invhilbert(F,x,t);
```

$$\frac{1}{1+t^2}$$

```
> hilbert(F,x,t);
```

$$-\frac{1}{1+t^2}$$

15.11.3 Mellin transform

The Mellin transform of a function $f(t)$ is

$$F(s) = \int_0^\infty f(t)\, t^s \, \frac{dt}{t}.$$

This is given in MAPLE by `mellin(f(t),t,s)`. We compute the Mellin transform of $f(t) = \sin t$.

```
> with(inttrans):
> mellin(sin(t),t,s);
```

$$\Gamma(s)\sin(\frac{1}{2}\,\pi\, s)$$

We found that the Mellin transform of $f(t) = \sin t$ is

$$F(s) = \Gamma(s)\,\sin(\pi s/2).$$

The inverse Mellin transform is given by

$$f(t) = \frac{1}{2\pi i}\int_{c-i\infty}^{c+i\infty} F(s)t^{-s}\, ds.$$

Here c is a sufficiently large real constant.

```
> with(inttrans):
> F := GAMMA(s)*sin(Pi*s/2);
```

$$F := \Gamma(s)\sin(\frac{1}{2}\,\pi\, s)$$

```
> invmellin(F,s,t);
```

$$\mathit{invmellin}(\Gamma(s)\sin(\frac{1}{2}\,\pi\, s), s, t)$$

We see that MAPLE was unable to recognize the inverse Mellin transform of $F(s)$. We try computing the inverse Mellin transform of $\Gamma(s)$.

```
> with(inttrans):
> F := GAMMA(s);
```

$$\Gamma(s)$$

```
> invmellin(F,s,t);
Invmellin can transform GAMMA(t) if Re(t)<>0, Re(t) > -1
```

MAPLE needs to know a range for the constant c before it can compute the integral in the inverse Mellin transform.

```
> with(inttrans):
> F := GAMMA(s);
```

$$\Gamma(s)$$

```
> invmellin(F,s,t,0..infinity);
```

$$e^{-t}$$

The command `invmellin(F,s,t,0..infinity)` tells MAPLE to assume that $c > 0$. We found that the inverse Mellin transform of $\Gamma(s)$ is $f(t) = e^{-t}$. Try checking this by computing the Mellin transform of $f(t) = e^{-t}$:

```
> with(inttrans):
> f := exp(-t);
> mellin(f,t,s);
```

15.12 Fast Fourier transform

Let $n + 1 = 2^m$, and $\omega = \exp(2\pi i/(n + 1))$, so that ω is a primitive $(n + 1)$th root of unity. Suppose we are given a sequence $S = \{a_k\}_{k=0}^{n}\}$ of complex numbers. The fast fourier transform (FFT) of S is the sequence $\{\alpha_j\}_{j=0}^{n}$, where

$$\alpha_k = \sum_{k=1}^{n} a_j\, \omega^j.$$

In MAPLE the sequence S is entered as two arrays of real and imaginary parts `x` and `y`. The FFT of S is computed using the MAPLE command `FFT(m,x,y)`. Let S be the sequence

$$S = \{1, 2, 3, 4, 5, 6, 7, 8\}.$$

We compute the FFT of S. Here $m = 3$ because our sequence has length $8 = 2^3$.

```
> x := array([seq(k,k=1..8)]);
```

$$x := [1, 2, 3, 4, 5, 6, 7, 8]$$

```
> y := array([seq(0,k=1..8)]);
```

$$y := [0, 0, 0, 0, 0, 0, 0, 0]$$

```
> FFT(3,x,y);
```

$$8$$

```
> print(x);
```

$$[36, -4.000000002, -4., -3.999999998, -4, -3.999999998, -4., -4.000000002]$$

```
> print(y);
```

$$[0, 9.656854244, 4., 1.656854248, 0, -1.656854248, -4., -9.656854244]$$

The arrays x and y now correspond to the real and imaginary parts of the output sequence. The FFT of S is the sequence

$$\{36, -4.000000002 + 9.656854244\,i, -4.0 + 4.0\,i, -3.999999998 + 1.656854248\,i,$$
$$-4, -3.999999998 - 1.656854248\,i, -4.0 - 4.0\,i,$$
$$-4.000000002 - 9.656854244\,i\}$$

To check this calculation let

$$p(z) = \sum_{k=0}^{n} a_k z^k.$$

Then $p(\omega^j) = \alpha_j$, for $0 \le j \le n$. Check this for our example sequence S:

```
> omega := exp(2*Pi*I/8);
> sum((k+1)*z^k,k=0..7):
> p := unapply(%,z);
> for k from 0 to 7 do
        k, x[k+1]+I*y[k+1], evalf(p(omega^k));
    end do:
```

Did it check out?

In MAPLE the inverse FFT is iFFT. Let's compute the inverse FFT of the output sequence found above.

```
> x := array([seq(k,k=1..8)]):
> y := array([seq(0,k=1..8)]):
> FFT(3,x,y):
> iFFT(3,x,y);
```

$$8$$

```
> print(x);
```

$$[1.000000000, 2.000000002, 3.000000001, 4.000000004, 5.000000000,$$
$$5.999999998, 6.999999999, 7.999999996]$$

```
> print(y);
```

$$[0., .6250000000\,10^{-9}, 0., -.6250000000\,10^{-9}, 0., -.6250000000\,10^{-9}, 0.,$$
$$.6250000000\,10^{-9}]$$

Taking into account floating point error, this gives our original sequence as expected.

15.13 Asymptotic expansion

To find the first n terms of the asymptotic expansion of the function $f(z)$, we use the command `asympt(`$f(z)$`, `z`, `n`)`. For example, below we find the first few terms of the asymptotic expansion of the Psi function (which you should recall as the logarithmic derivative of the gamma function).

```
> z:='z':
> asympt(Psi(z),z,3);
```

$$\ln(z) - \frac{1}{2}\frac{1}{z} - \frac{1}{12}\frac{1}{z^2} + O(z^{-4})$$

Try finding the first few terms of the asymptotic expansion of the gamma function $\Gamma(z)$.

```
> asympt(GAMMA(z),z,3);
```

16. STATISTICS

16.1 Introduction

The *stats* package provides basic data analysis and plotting functions. We load the package:

```
> with(stats);
```

$$[anova, describe, fit, importdata, random, statevalf, statplots, transform]$$

As you see, there are eight subpackages: *anova*, *describe*, *fit*, *importdata*, *random*, *statevalf*, *statplots*, and *transform*.

To illustrate various MAPLE statistics applications, we will invoke the following "Lamp Example" due to William E. Wilson, engineer and inventor.

Lamp Example

The proper operation of a fluorescent lamp depends on depositing an adequate amount of electron emitter on the filaments located at each end of the glass tube during the manufacturing process. The life of the fluorescent lamp is directly proportional to the amount of deposited electron emitter. Other factors that affect the life of a lamp include the type of gas used to fill the lamp tube and the type of fixture into which the lamp is inserted. Two gas types are typically used in the manufacture of household fluorescent lamps: argon gas and a neon-argon gas mixture. Three common types of fixture into which a fluorescent lamp is inserted are the instant-start, rapid-start, and preheat fixtures.

The following three data sets have been constructed with simulated data:

Data Set 1 contains data for 200 lamps: the life of each lamp (measured in hours) and the amount of electron emitter deposited on its filaments (measured in milligrams).

Data Set 2 contains data for 500 lamps: the life of a fluorescent lamp (measured in hours) and the type of gas (argon or neon-argon) used to fill the lamp.

Data Set 3 contains data for 300 lamps: the life of a fluorescent lamp (measured in hours) and the type of fixture (rapid-start, preheat, instant-start) used to test the lamp.

These data sets and more details about the Lamp Example are available on the Web at
`http://www.math.ufl.edu/~frank/maple-book/lamp/index.html`
Go to this page using your favorite Web browser and download the files: *data1.dat*, *data2.dat*, *data3.dat*, and *lamp.dat*.

16.2 Data sets

The first step in a MAPLE data analysis is to create a MAPLE data set. This can be done by importing a preexisting data set or constructing one with MAPLE commands.

Let's import Data Set 1 from the Lamp Example.

```
> with(stats):
> data1 := importdata("data1.dat",2):
```

Note that the first argument for the importdata function is the data file name and the second is the number of columns or streams into which the data file is split. We assign the names `life` and `amount` to the data elements.

```
> life := data1[1]:
> amount := data1[2]:
```

We can construct a data set using a statistical list:

```
> example_dataset := [10, 20, 30, 30, 30, missing];
```

$$example\_dataset := [10, 20, 30, 30, 30, missing]$$

The value *missing* indicates a missing data point. We can also use the weight function to create the same data set.

```
> example_dataset:=[10,20, Weight(30,3), missing];
```

$$example\_dataset := [10, 20, Weight(30, 3), missing]$$

We can count the number of nonmissing data points in a data stream or data set using the **describe[count]** function. Try some examples.

```
> describe[count](life);
```

$$200$$

```
> describe[count](example_dataset);
```

$$3$$

We can also count the number of missing data points.

```
> describe[countmissing](life);
```

$$0$$

```
> describe[countmissing](example_dataset);
```

$$1$$

16.3 Numerical methods for describing data

A statistic is any quantity calculated from data. Statistics are single numbers which estimate population characteristics as well as summarize information in a data set. The following sections show how statistics can describe various features of a data set.

16.3.1 Describing the center of a data set

The center of a data set gives information about its location. The most common ways of describing the center of a data set are by reporting the mean, median, and mode, statistics which measure central tendency. The harmonic mean, geometric mean, and quadratic mean are other statistical measures of central tendency.

The mean ($\overline{X}$ or arithmetic average) can be calculated using the `mean` function in the *describe* subpackage. Let's compute the mean lamp life using data from the Lamp Example.

```
> with(stats):
> data1:=importdata("data1.dat",2):
> life:=data1[1]:
> amount:=data1[2]:
> describe[mean](life);
```

$$7524.607000$$

```
> describe[mean](amount);
```

$$10.43450000$$

Try

```
> describe[mean]([10, 20, 30, 30, 30, missing]);
```

$$24$$

Note that the missing value was ignored!

The median is the middle of ordered data when N (the number of data points) is odd and the average of the two middle values when N is even. Try

```
> describe[median](life);
```

$$7575.850000$$

```
> describe[median](amount);
```

$$10.12000000$$

```
> describe[median]([10, 20, 30]);
```

$$20$$

```
> describe[median]([10, 20, 30, 40]);
```

$$25$$

The median is not as sensitive as the mean to extreme values in the data set, as the following examples illustrate:

```
> describe[mean]([10, 20, 30, 40]),
   describe[mean]([10, 20, 30, 1000]);
```

$$25, 265$$

```
> describe[median]([10, 20, 30, 40]),
   describe[median]([10, 20, 30, 1000]);
```

$$25, 25$$

The mode is the most frequently occuring value in a data set.

```
> with(stats):
> describe[mode]([10, 20, 30, 30, 30, missing]);
```

$$30$$

The harmonic mean is defined to be the reciprocal of the mean of the reciprocals of the data. The formula for harmonic mean, H, is given by

$$H = \frac{N}{\sum_{i=1}^{N} \frac{1}{x_i}}.$$

The geometric mean of a set of N numbers is the Nth root of the product of those numbers. The formula for geometric mean, G, is given by

$$G = \left(\prod_{i=1}^{N} x_i \right)^{\frac{1}{N}}.$$

The quadratic mean, or root mean square, is the square root of the mean of the squares of the data. The formula for quadratic mean, Q, is given by

$$Q = \sqrt{\frac{1}{N} \sum_{i=1}^{N} x_i^2}.$$

Try

```
> with(stats):
> H := describe[harmonicmean]([10, 20, 30, 30, 30, missing]);
```

$$H := 20$$

```
> G := describe[geometricmean]([10, 20, 30, 30, 30, missing]);
```

$$G := 5400000^{1/5}$$

```
> Q := describe[quadraticmean]([10, 20, 30, 30, 30, missing]);
```

$$Q := 8\sqrt{10}$$

16.3.2 Describing the dispersion of a data set

Measures of the dispersion of a data set give information about the spread or variability of the data. The most common ways of describing the dispersion of a data set are by reporting the range, variance, and standard deviation statistics.

The range function,`describe[range]`, finds the minimum and the maximum values in a data set.

```
> with(stats):
> describe[range]([10, 20, 30, 30, 30, missing]);
```

$$10 \ldots 30$$

A deviation is the difference between a data point and the sample mean. Sample variance, S^2, is the sum of squared deviations divided by $N-1$,

$$S^2 = \frac{\sum_{i=1}^{N}(x_i - \overline{x})^2}{N-1}.$$

The function variance, `describe[variance[1]](data)`, computes the sample variance of the given data.

Try

```
> with(stats):
> describe[variance[1]]([10, 20, 30, 30, 30, missing]);
```

$$80$$

The function `describe[variance[0]](data)` or `describe[variance](data)`, the default, is the sum of squared deviations divided by N, i.e.,

$$\frac{\sum_{i=1}^{N}(x_i - \overline{x})^2}{N}.$$

When a data set represents the entire population, `describe[variance](data)` computes the population variance. Try

```
> with(stats):
> describe[variance [0]]([1, 0,-1]),
    describe[variance[1]]([1,0,-1]);
```

$$2/3, 1$$

Sample standard deviation, S, is the square root of variance,

$$S = \sqrt{\left(\frac{\sum_{i=1}^{N}(x_i - \overline{x})^2}{N-1}\right)}.$$

The `describe[standarddeviation[1]](data)` function computes the sample standard deviation of the given data. Try

```
> with(stats):
> describe[standarddeviation[1]]([10, 20, 30, 30, 30, missing]);
```

$$4\sqrt{5}$$

The function `describe[standarddeviation[0]](data)` or `describe[standarddeviation](data)`, the default, is the square root of the sum of squared deviations divided by N, i.e.,

$$\sqrt{\left(\frac{\sum_{i=1}^{N}(x_i - \overline{x})^2}{N}\right)}.$$

Compare the following

```
> with(stats):
> describe[standarddeviation[0]]([1, 0,-1]),
    describe[standarddeviation[1]]([1,0,-1]);
```

$$1/3\sqrt{6}, \quad 1$$

Using the data from the Lamp Example, compute the range, sample variance, and sample standard deviation of the lamp life.

```
> with(stats):
> data1:=importdata("data1.dat",2):
> life:=data1[1]:
> describe[range](life);
> describe[variance[1]](life);
> describe[standarddeviation[1]](life);
```

The mean deviation is another measure of dispersion. The mean deviation is the average of the absolute value of deviations, i.e.,

$$\frac{1}{N}\sum_{i=1}^{N}|x_i - \overline{x}|.$$

The `describe[meandeviation](data)` function computes the mean deviation of the given data.

Try

```
> with(stats):
> describe[meandeviation]([10, 20, 30, 30, 30, missing]);
```

$$\frac{36}{5}$$

The statistics percentile, decile, quantile, and quartile give information about how an ordered data set is partitioned.

The `describe[percentile[p]](data)` function returns the pth percentile of a data set. Try

```
> with(stats):
> describe[percentile[37]]([seq(i,i=1..100)]);
```

$$37$$

```
> describe[percentile[50]]([seq(i,i=1..100)]),
  describe[median]([seq(i,i=1..100)]);
```

$$50, \frac{101}{2}$$

The `describe[decile[d]](data)` function returns the dth decile of a data set. Try

```
> with(stats):
> describe[decile[6]]([70,10,80,20,30,40,100,50,60,90]);
```

$$60$$

```
> describe[percentile[2]]([seq(i,i=1..100)]),
  describe[percentile[20]]([seq(i,i=1..100)]),
  describe[decile[2]]([seq(i,i=1..100)]);
```

$$2, 20, 20$$

The `describe[quantile[r, offset]](data)` $(0 < r < 1)$ function generalizes the concept of median, quartile, percentile, etc. Select a fraction r between

0 and 1. Sort the data. The quantile function returns the value of the precise position $rN + offset$, where N is the number of data points in the data set. Try

```
> with(stats):
> describe[quantile[1/2]]([1,2,3]);
```

$$3/2$$

```
> describe[quantile[1/2,1/2]]([1,2,3])=
    describe[median]([1,2,3]);
```

$$2 = 2$$

```
> describe[percentile[20]]([seq(i,i=1..100)])=
    describe[quantile[20/100]]([seq(i,i=1..100)]);
```

$$20 = 20$$

```
> describe[percentile[21]]([seq(i,i=1..100)]),
    describe[quantile[20/100,.9]]([seq(i,i=1..100)]);
```

$$21, 20.9$$

Quartiles divide the data set into four portions and are the 25th, 50th, and 75th percentiles of a data set. Try

```
> with(stats):
> describe[quartile[1]]([seq(i,i=1..100)])=
    describe[percentile[25]]([seq(i,i=1..100)]);
```

$$25 = 25$$

```
> describe[quartile[2]]([seq(i,i=1..100)])=
    describe[percentile[50]]([seq(i,i=1..100)]);
```

$$50 = 50$$

```
> describe[median]([seq(i,i=1..100)]);
```

$$\frac{101}{2}$$

```
> describe[quartile[3]]([seq(i,i=1..100)])=
    describe[percentile[75]]([seq(i,i=1..100)]);
```

$$75 = 75$$

The interquartile range, IQR, is a measure of dispersion which is not sensitive to extreme values. The IQR is the difference between the upper quartile (the 75th percentile) and the lower quartile (the 25th percentile).

```
> with(stats):
> IQR1 := describe[quartile[3]]([10, 20, 30, 40])
    - describe[quartile[1]]([10, 20,30, 40]);
```

$$20$$

```
> IQR2 := describe[quartile[3]]([10, 20, 30, 1000])
    - describe[quartile[1]]([10, 20, 30, 1000]);
```

$$20$$

16.3.3 Describing characteristics of a data set

The `describe[moment[r,origin,1]](data)` function computes the various moments of the given data about any origin. The formula for the rth moment, M_r, about an origin is given by

$$M_r = \frac{1}{N-1}\sum_{i=1}^{N}(x_i - origin)^r.$$

Try

```
> with(stats):
> describe[moment[3,0,1]]([10,20,30])=(1/2)*(10^3+20^3+30^3);
```

$$18000 = 18000$$

```
> describe[moment[4,20,1]]([10,20,30])=
   describe[moment[4,mean,1]]([10,20,30]);
```

$$10000 = 10000$$

The `describe[moment[r,origin,0]](data)` (or `describe[moment[r,origin]](data)`) function uses the formula

$$M_r = \frac{1}{N}\sum_{i=1}^{N}(x_i - origin)^r.$$

Try

```
> with(stats):
> describe[moment[3]]([10,20,30])=(1/3)*(10^3+20^3+30^3);
```

$$12000 = 12000$$

The formula for the `describe[sumdata[r,origin]](data)` (the default values for r and $origin$ are 1 and 0, respectively) function is given by

$$M_r = \sum_{i=1}^{N} (x_i - origin)^r.$$

Try

```
> with(stats):
> describe[sumdata[3]]([10,20,30])=(10^3+20^3+30^3);
```

$$36000 = 36000$$

```
> describe[sumdata]([10,20,30])=(10+20+30);
```

$$60 = 60$$

```
> 3*describe[moment[4,1,0]]([10,20,30])
   = describe[sumdata[4,1]]([10,20,30]);
```

$$844163 = 844163$$

```
> '' = ( (10-1)^4+(20-1)^4+(30-1)^4 );
```

$$= 844163$$

Skewness is defined to be the third moment about the sample mean, divided by the third power of the standard deviation, and it measures the degree of symmetry of a data set. A perfectly symmetric data set has a skewness of zero. If the data set has some extremely small values, then the skewness will be negative. If the data set has some extremely large values, then the skewness will be positive. The formula for `describe[skewness[1]](data)` is given by

$$\frac{M_3}{S^3} = \frac{\frac{1}{N-1}\sum_{i=1}^{N}(x_i - \overline{X})^3}{\left[\frac{1}{N-1}\sum_{i=1}^{N}(x_i - \overline{X})^2\right]^3}.$$

Try the following

```
> with(stats):
> describe[skewness[1]]([-1,0,1]);
```

$$0$$

```
> describe[skewness[1]]([-1,0,1000]):
> evalf(%);
```

$$0.5773483226$$

The formula for describe[skewness[0]](data) or default is given by

$$\frac{M_3}{S^3} = \frac{\frac{1}{N}\sum_{i=1}^{N}(x_i - \overline{X})^3}{\left[\frac{1}{N}\sum_{i=1}^{N}(x_i - \overline{X})^2\right]^3}.$$

Try

```
> with(stats):
> describe[skewness[1]]([-1000,0,1]):
> sk1 := evalf(%);
```

$$sk1 := -0.5773483226$$

```
> describe[skewness[0]]([-1000,0,1]):
> sk2 := evalf(%);
```

$$sk2 := -0.7071043969$$

```
> describe[skewness]([-1000,0,1]):
> sk3 := evalf(%);
```

$$sk3 := -0.7071043969$$

```
> sk1 <> sk2;
```

$$-0.5773483226 \neq -0.7071043969$$

```
> `` = sk3;
```

$$= -0.7071043969$$

Kurtosis is defined to be the fourth moment about the sample mean divided by the fourth power of the standard deviation, and it measures the degree of flatness or peakedness of a data set. For the normal distribution, the kurtosis is 3. If the distribution has a flatter top, the kurtosis is less than 3. If the distribution has a high peak, the kurtosis is greater than 3. Refer to the formulas for the skewness function for the definitions of describe[kurtosis[1]](data) and describe[kurtosis[0]](data).

Using the data from the Lamp Example, find the skewness and kurtosis of the *lamp life* and *emitter amount* variables.

```
> with(stats):
> data1:=importdata("data1.dat",2):
> life:=data1[1]:
> amount:=data1[2]:
> describe[skewness[1]](life),describe[skewness[1]](amount);
```

$$0.04407219456, 1.500624764$$

Note that the *lamp life* has a nearly symmetric distribution and *emitter amount* is positively skewed.

```
> describe[kurtosis[1]](life), describe[kurtosis[1]](amount);
```

$$2.493837420,\ 7.077100768$$

Note that the *lamp life* is more normally distributed than *emitter amount.*

The `describe[coefficientofvariation[1]](data)` function computes the coefficient of variation of the given data, which is the standard deviation divided by the mean. The coefficient of variation expresses the standard deviation as a percent of the mean. When means are not equal to zero, the dispersion in data sets with different units of measure can be compared by computing the coefficient of variation for each data set.

Note that the standard deviation is the same for the following two data sets.

```
> with(stats):
> describe[standarddeviation[1]]([10,20,30])
    = describe[standarddeviation[1]]([110,120,130]);
```

$$10 = 10$$

But the coefficient of variations are quite different!

```
> with(stats):
> data1:=importdata("data1.dat",2):
> life:=data1[1]:
> amount:=data1[2]:
> describe[coefficientofvariation[1]]([10,20,30])
    <> describe[coefficientofvariation[1]]([110,120,130]);
```

$$1/2 \neq 1/12$$

```
> describe[coefficientofvariation[1]](life),
    describe[coefficientofvariation[1]](amount);
```

$$0.1320181761,\ 0.1802995240$$

A bivariate data set contains two measurements (say "X" and "Y") made on a single subject and consists of ordered (X,Y) pairs. The functions `describe[covariance](X,Y)` and `describe[linearcorrelation](X,Y)` compute the covariance and correlation of X and Y, respectively. The formula for `describe[covariance](X,Y)` is given by

$$\frac{1}{N}\sum_{i=1}^{N}(x_i - \overline{X})(y_i - \overline{Y}).$$

The formula for describe[linearcorrelation](X,Y) is

$$\frac{\sum_{i=1}^{N}(x_i - \overline{X})(y_i - \overline{Y})}{\sqrt{\sum_{i=1}^{N}(x_i - \overline{X})^2}\sqrt{\sum_{i=1}^{N}(y_i - \overline{Y})^2}}.$$

Try

```
> with(stats):
> describe[covariance]([-1,0,1],[-1,0,1]);
```

$$\frac{2}{3}$$

```
> describe[covariance]([-1,0,1],[1,0,-1]);
```

$$-\frac{2}{3}$$

Consider a bivariate data set that consists of the ordered pairs: $(-10, -20)$, $(-9, -18)$, ..., $(0, 0)$, ... $(10, 20)$. To plot these data points, we use the scatterplot function in the *statplots* subpackage. A graph is given below in Figure 16.1.

```
> with(stats):
> x1:=[seq(i,i=-10..10)]:
> y1:=[seq(i*2,i=-10..10)]:
> statplots[scatterplot](x1,y1);
> describe[linearcorrelation](x1,y1);
```

$$1$$

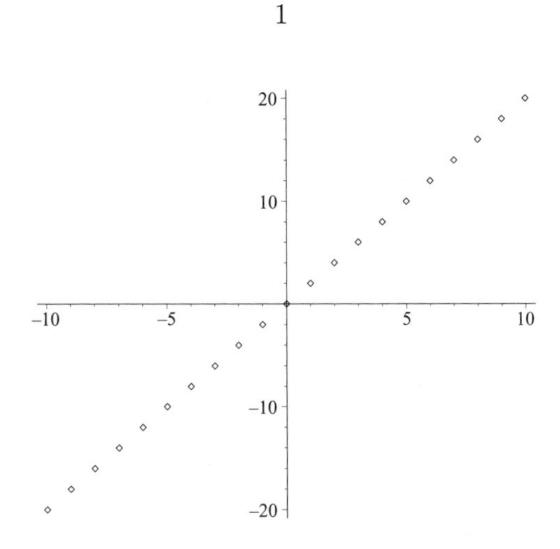

Figure 16.1 MAPLE plot of data points.

Note that the value of the linear correlation is 1. This means that the X_1-Y_1 values fall perfectly on a straight line and that an increase in X_1 corresponds to an increase in Y_1.

Now consider a bivariate data set consisting of the ordered pairs $(-10, 10)$, $(-9, 9)$, ... , $(0, 0)$, ... $(10, 10)$. Try graphing these data points and finding the correlation coefficient.

```
> x2:=[seq(i,i=-10..10)]:
> y2:=[seq(-i,i=-10..10)]:
> statplots[scatterplot](x2,y2);
> describe[linearcorrelation](x2,y2);
```

Finally, consider a bivariate data set consisting of the ordered pairs $(-2, 2)$, $(-2, -2)$, $(-1, 1)$, $(-1, -1)$, $(0, 0)$, $(1, -1)$, $(1, 1)$, $(2, -2)$, $(2, 2)$. Try graphing these data points.

```
> x3:=[-2,-2,-1,-1,0,1,1,2,2]:
> y3:=[2,-2,1,-1, 0,-1,1,2,-2]:
> scatterplot(x3,y3, symbol=circle, symbolsize=20, color=red);
> describe[linearcorrelation](x3,y3);
```

Find the linear correlation between *lamp life* and *emitter amount.*

```
> describe[linearcorrelation](life, amount);
```

$$0.7739070274$$

The lamp life is positively correlated with the amount of emitter deposited on the filaments during the manufacturing process.

16.4 Transforming data

The subpackage *transform* provides various tools for transforming lists of statistical data.

<u>`transform[statsort]`</u>

The function `transform[statsort]` sorts the statistical data.

```
> with(stats):
> transform[statsort]([15..17, 4, Weight(3,10),missing,
  Weight(11..12,3),missing]);
```

$$[\mathit{Weight}(3, 10), 4, \mathit{Weight}(11 \ldots 12, 3), 15 \ldots 17, \mathit{missing}, \mathit{missing}]$$

<u>`transform[split[n]]`</u>

The function `transform[split[n]]` splits the data into n data lists of the same weight.

```
> transform[split[3]]([15..17, 4, Weight(3,10),missing,
   Weight(11..12,3),missing]);
```

$$[[15\ldots 17, 4, \mathit{Weight}(3, 11/3)], [\mathit{Weight}(3, \frac{17}{3})], [\mathit{Weight}(3, \frac{2}{3}), \mathit{missing}, \mathit{Weight}(11\ldots 12, 3), \mathit{missing}]]$$

transform[frequency]

The `transform[frequency]` function computes the frequencies in the given data.

```
> transform[frequency]([Weight(3,10),missing, 4,
   Weight(11..12,3), 15..17,missing]);
```

$$[10, 1, 1, 3, 1, 1]$$

`transform[cumulativefrequency]`

The `transform[cumulativefrequency]` function computes the partial sums of the frequencies.

```
> transform[cumulativefrequency]([Weight(3,10), missing, 4,
   Weight(11..12,3), 15..17, missing])
   =[10, 10+1, 10+1+1,10+1+1+3, 10+1+1+3+1, 10+1+1+3+1+1];
```

$$[10, 11, 12, 15, 16, 17] = [10, 11, 12, 15, 16, 17]$$

`transform[deletemissing]`

The function `transform[deletemissing]` removes missing data.

```
> transform[deletemissing]([10, missing, 20, missing,
   Weight(30,3), missing]);
```

$$[10, 20, \mathit{Weight}(30, 3)]$$

`transform[subtractfrom]`

The function `transform[subtractfrom]` subtracts a number or the value of a statistic.

```
> transform[subtractfrom[25]]([10, 20, 30, 40,missing])
   =transform[subtractfrom[mean]]([10, 20, 30, 40,missing]);
```

$$[-15, -5, 5, 15, \mathit{missing}] = [-15, -5, 5, 15, \mathit{missing}]$$

transform[divideby]

The function transform[divideby] divides the data by the given divisor.

```
> transform[divideby[25]]([10, 20, 30, 40, missing])
   =transform[divideby[mean]]([10, 20, 30, 40,missing]);
```

$$[\frac{2}{5}, \frac{4}{5}, \frac{6}{5}, \frac{8}{5}, missing] = [\frac{2}{5}, \frac{4}{5}, \frac{6}{5}, \frac{8}{5}, missing]$$

transform[standardscore[n_constraints]]

The function transform[standardscore[n_constraints]] replaces each data value by its standard score (z-score). The standard score of a data point x is (x-mean)/standarddeviation. For more details about n_constraints, refer to ?describe[standarddeviation].

```
> transform[standardscore[1]]([1,2,3])
   =[(1-2)/1, (2-2)/1, (3-2)/1];
```

$$[-1, 0, 1] = [-1, 0, 1]$$

```
> data1:=importdata("data1.dat", 2):
> life:=data1[1]:
> transform[standardscore[1]](life);
```

The transform[apply] function applies the requested function to the given data.

```
> transform[apply[x->sqrt(x)]]([100,36,30,49]);
```

$$[10, 6, \sqrt{30}, 7]$$

transform[multiapply]

The function transform[multiapply] applies the requested function across the given data.

```
> transform[multiapply[(x,y)->5*x+y^2]]([[1,2,3],[4,5,6]])
   =[5*1+4^2 , 5*2+5^2 , 5*3+6^2];
```

$$[21, 35, 51] = [21, 35, 51]$$

transform[moving[size, func]]

The function transform[moving[size, func]] is used to smooth the data. The function transform[moving[n]] replaces each data point by the mean of itself and its next $n-1$ neighbors to the right.

```
> with(describe):
> L := [12,3,5,2,7,20];
```

$$[12, 3, 5, 2, 7, 20]$$

```
> transform[moving[3]](L);
```

$$[\frac{20}{3}, \frac{10}{3}, \frac{14}{3}, \frac{29}{3}]$$

```
> [mean([12,3,5]), mean([3,5,2]), mean([5,2,7]),
   mean([2,7,20])];
```

$$[\frac{20}{3}, \frac{10}{3}, \frac{14}{3}, \frac{29}{3}]$$

To use the mean function, we loaded the *describe* subpackage. To do the same thing with the median, try

```
> L := [12,3,5,2,7,20];
```

$$[12, 3, 5, 2, 7, 20]$$

```
> transform[moving[3,median]](L);
```

$$[5, 3, 5, 7]$$

```
> [median([12,3,5]), median([3,5,2]), median([5,2,7]),
   median([2,7,20])];
```

$$[5, 3, 5, 7]$$

transform[statvalue]

The function transform[statvalue] sets each data point's weight to 1.

```
> transform[statvalue]([Weight(3,10), missing, 4,
   Weight(11..12,3), 15..17, missing]);
```

$$[3, \mathit{missing}, 4, 11 \ldots 12, 15 \ldots 17, \mathit{missing}]$$

transform[scaleweight]

The function transform[scaleweight] multiplies the weights of the data by the given amount.

```
> with(stats):
> transform[scaleweight[1/2]]([Weight(3,10),missing, 4,
   Weight(11..12,3), 15..17, missing]);
```

$$[\mathit{Weight}(3, 5), \mathit{Weight}(\mathit{missing}, \frac{1}{2}), \mathit{Weight}(4, \frac{1}{2}), \mathit{Weight}(11 \ldots 12, \frac{3}{2}),$$
$$\mathit{Weight}(15 \ldots 17, \frac{1}{2}), \mathit{Weight}(\mathit{missing}, \frac{1}{2})]$$

`transform[tally]`

The function `transform[tally]` tallies each data item.

```
> with(stats):
> transform[tally]([3, 3, 3, 3, 3, missing, 4, 11..12,
    11..12, 11..12, 15..17, missing]);
```

$$[Weight(3,5), 4, Weight(missing,2), Weight(11\ldots12,3), 15\ldots17]$$

`transform[tallyinto](data, partition)`

The function `transform[tallyinto](data, partition)` tallies each item into the pattern given by partition.

```
> transform[tallyinto]([3, 3, 3, 3, 3, missing, 4, 11..12,
    11..12, 11..12, 15..17, missing],[3..5, 11..17]);
```

$$[Weight(11\ldots17,4), Weight(missing,2), Weight(3\ldots5,6)]$$

`transform[classmark]`

The `transform[classmark]` function replaces classes by their midpoint:

```
> transform[classmark]([1 ..  3, 4 ..  5, Weight(11..12,3)]);
```

$$[2, \frac{9}{2}, Weight(23/2,3)]$$

```
> [(1+3)/2, (4+5)/2, (11+12)/2];
```

$$[2, \frac{9}{2}, 23/2]$$

16.5 Graphical methods for describing data

The subpackage *statplots* provides the capability to create various statistical plots. We load the packages:

```
> with(stats):
> with(stats[statplots]);
```

$$[boxplot, histogram, scatterplot, xscale, xshift, xyexchange, xzexchange, yscale, yshift, yzexchange, zscale, zshift]$$

Before we go to the specific statistical plots, we will introduce three types of utility functions (scale, shift, exchange) that can be used to modify any plot. See Chapter 6 for more information on plotting.

The `xscale(`*amount*`, plot)` function changes the scale of the x-coordinate by multiplying every x-coordinate by the value *amount*. The `yscale(`*amount*`, plot)` and `zscale(`*amount*`, plot)` functions work in the same way. Try

```
> pred:=plot(statevalf[pdf,normald], -3..3,color=red):
> pblue:=plot(statevalf[pdf,normald], -3..3,color=blue):
> pblue2:=xscale(2,pblue):
> plots[display](pred,pblue2);
```

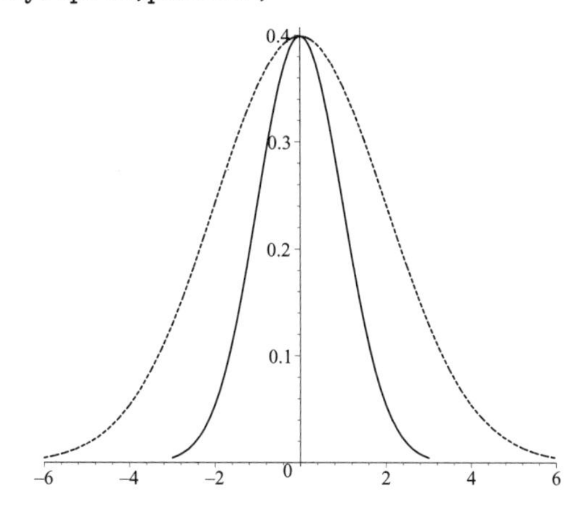

Figure 16.2 MAPLE plot of two normal curves.

The red curve is a graph of the standard normal distribution and the blue curve is a scaled version. In Figure 16.2 red appears as a solid line and blue is a dashed line.

The `xshift(`*amount*`, plot)` function shifts the x-coordinates by adding the value *amount* to every x coordinate. The `yshift(`*amount*`, plot)` and `zshift(`*amount*`, plot)` functions work in the same way. Try

```
> pblue3:=xshift(3,pblue):
> plots[display](pred,pblue3);
```

The resulting plot is given below in Figure 16.3. The red (solid) curve is a graph of the standard normal distribution and the blue (dashed) curve is a shifted version.

The `xyexchange(plot)` function exchanges the x- and y-coordinates. The `xzexchange(plot)` and `yzexchange(plot)` functions work in the same way.

```
> with(stats):
> with(stats[statplots]):
> p1:=plot('exp(-abs(x))','x'=-2..2, color=red):
> p2:=plot('exp(-abs(x))','x'=-2..2, color=blue):
> xp2:=xyexchange(p2):
> plots[display](p1,xp2);
```

The resulting plot is given below in Figure 16.4. The red (solid) curve is a graph of the double exponential distribution, and the blue (dashed) curve is an exchanged version.

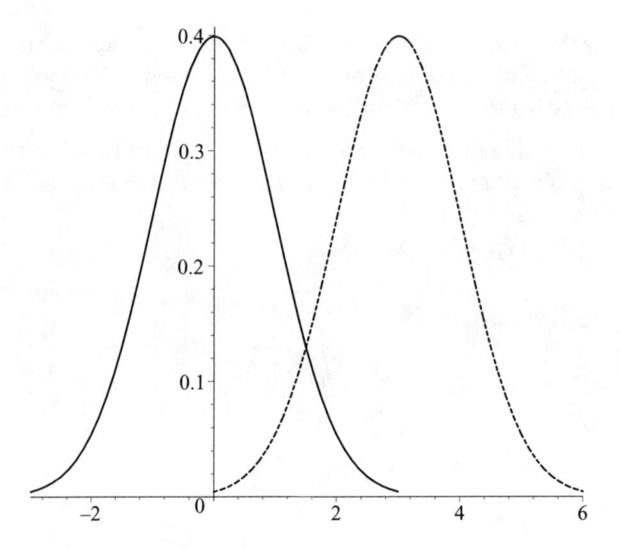

Figure 16.3 MAPLE plot of a normal curve and an x-shifted one.

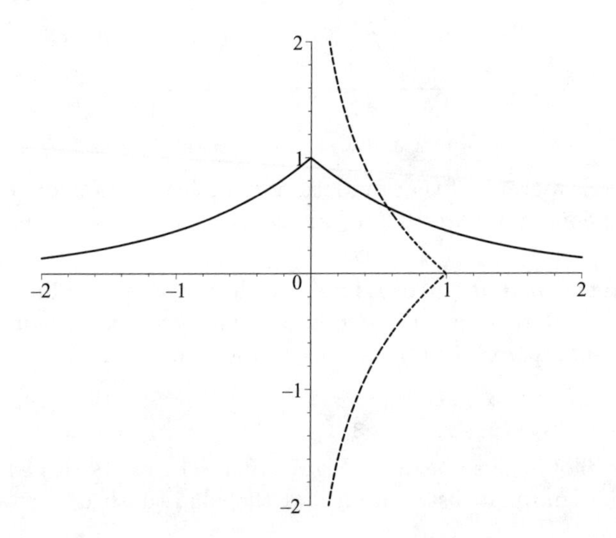

Figure 16.4 Double exponential distribution and an xy-exchanged one.

16.5.1 Histogram

The `histogram(data, area=`a`, numbars=`n`)` function will plot a histogram for a given data set. The parameter `numbars` allows the user to specify how many divisions into which the data should be separated. If the data are spread uniformly, then `numbars=`n should produce a histogram with n columns. When `area = `a is included in the function syntax, the histogram bars are forced to have equal width and have a total height equal to a. To make the total area of the bars equal to the total weight of the data, use `area = count`.

Using Data Set 1 from the Lamp Example, try

```
> with(stats):
> with(stats[statplots]):
> data1:=importdata("data1.dat",2):
> life:=data1[1]:
> histogram(life,color=green);
> histogram(life,color=green,numbars=20, area=count);
> histogram(life,color=green,numbars=10, area=1);
```

The resulting plots are given below in Figures 16.5, 16.6, and 16.7.

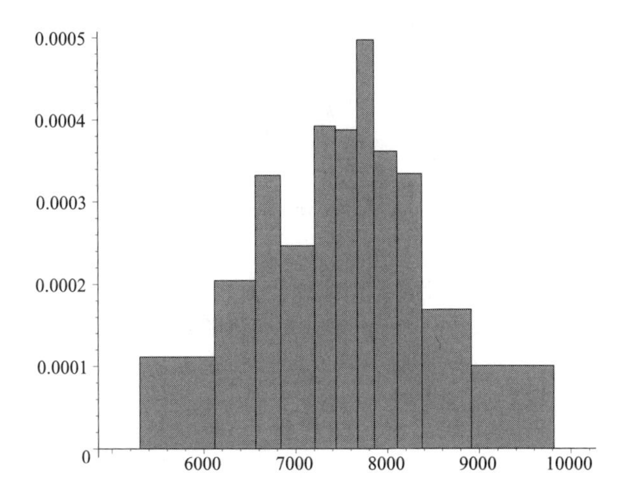

Figure 16.5 MAPLE plot of a histogram.

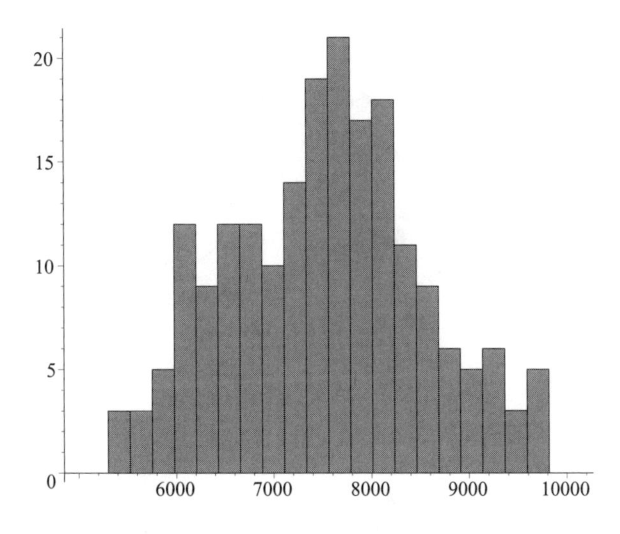

Figure 16.6 MAPLE plot of a histogram with area=count.

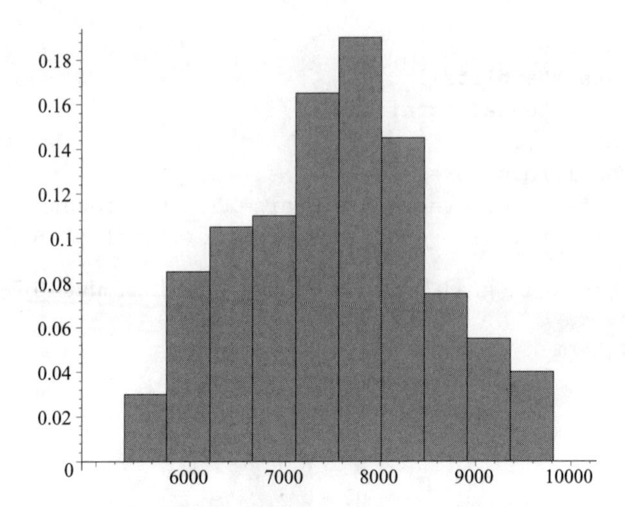

Figure 16.7 MAPLE plot of a histogram with area=1.

When `histogram` with two data sets is used, two three-dimensional histograms are plotted. Using Data Set 2 from the Lamp Example, we obtain a 3-D plot of two histograms, which is given in Figure 16.8.

```
> data2:=importdata("data2.dat",2):
> argon:=data2[1]:
> neon_argon:=data2[2]:
> histogram(argon,neon_argon,color=green,numbars=20, area=count,
    axes=boxed);
```

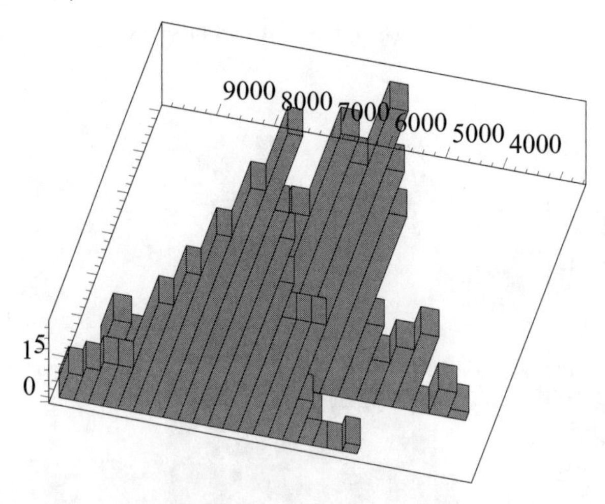

Figure 16.8 3-D plot of two histograms.

16.5.2 Box plot

A *box plot* is a compact graph providing information about the center, spread, and symmetry or skewness of the data. The `boxplot(data, shift=`s`,` `width=`w`, format=notched)` function will plot a box plot for a given data set. The parameter `shift=`s centers the box plot at value s. The parameter `width=`w creates a box plot with a width of w. The center line of a box plot shows the location of the median, while the lower and upper edges of the box indicate the first and third quartiles, respectively. Two lines extend from the central box to the data values, which are within a distance of up to 1.5 times the interquartile range. Box plots are quite useful for comparing data sets. Using Data Set 2 from the Lamp Example, we plot side-by-side box plots. See Figure 16.9.

```
> with(stats):
> with(stats[statplots]):
> life:=importdata("data2.dat",2):
> argon:=life[1]:
> neon_argon:=life[2]:
> boxplot(argon,neon_argon,width=1/2, shift=1);
```

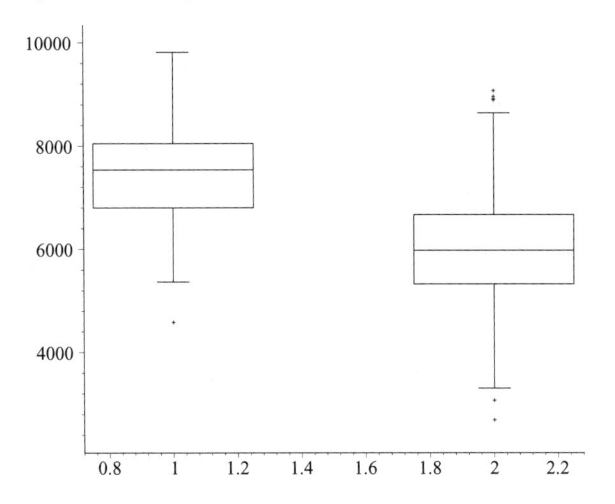

Figure 16.9 Box plot of lamp life by gas type.

Note that the lamps filled with argon gas have a longer life.

Using the parameter `format=notched` will create a box plot with one additional feature. The sides of the box are indented, or notched, at the median line. Using Data Set 3 from the Lamp Example, try plotting notched box plots.

```
> life3:=importdata("data3.dat",3):
> instant:=life3[1]:
> preheat:=life3[2]:
> rapid:=life3[3]:
> boxplot(rapid,preheat,instant,format=notched,width=1/4,
    shift=1);
```

You should find that the lamps tested in the rapid-start fixture have the longest life.

16.5.3 Scatter plot

The `scatterplot` function will produce a scatter plot of the points in a given data set. There are some formats that are valid only for one-dimensional scatter plots: `jittered`,`projected`, `stacked`, and `symmetry`.

The `format=projected` option for one-dimensional plots is the default. Points are plotted at their x-value along the line $y = 1$. Try

```
> with(stats):
> with(stats[statplots]):
> scatterplot([10, 20, Weight(30,3), 40, 50, 60, Weight(70,5),
   missing]);
```

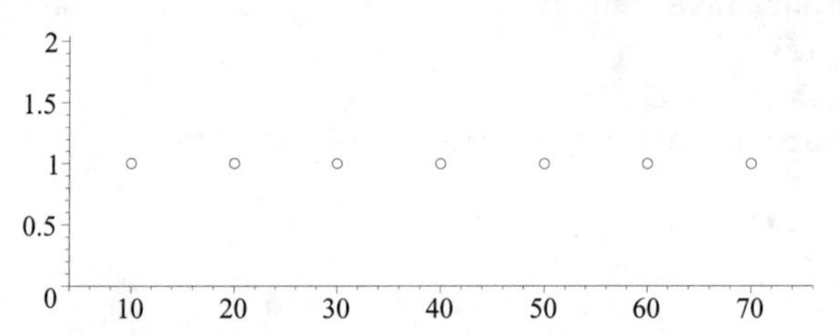

Figure 16.10 One-dimensional scatter plot.

Note that repeated x-values are plotted only once!

The `format=jittered` option for one-dimensional plots causes the points corresponding to a particular x-value to be randomly scattered along the vertical line at that x-value. Try

```
> scatterplot([10, 20, Weight(30,3), 40, 50, 60, Weight(70,5),
   missing],format=jittered,symbol=circle);
```

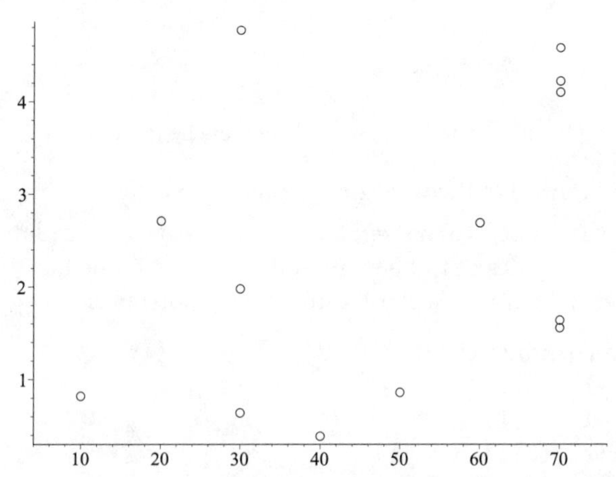

Figure 16.11 One-dimensional scatter plot with `jittered` format option.

The `format=stacked` option is the same as the `jittered` except that the points are equally spaced. Repeated values are stacked vertically so that the height of the stack gives the number of repeated values. Try

```
> scatterplot([10, 20, Weight(30,3), 40, 50, 60, Weight(70,5),
    missing],format=stacked,symbol=circle);
```

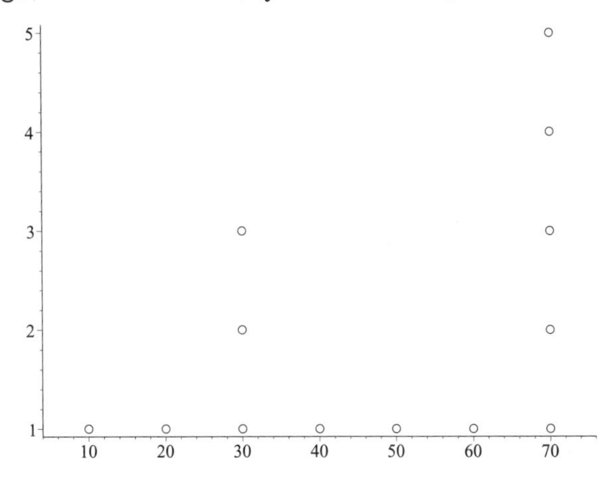

Figure 16.12 One-dimensional scatter plot with `stacked` format option.

The `format=symmetry` option for one-dimensional plots produces a symmetry plot of the data. In this type of plot, the data are ordered and divided into two halves along the value of the median, $\widetilde{X}$. Let $X_{(1)}$, $X_{(2)}$,$X_{(3)}$, ... , $\widetilde{X}$, ... , $X_{(N-1)}$, $X_{(N)}$ denote the ordered data. Next, ordered pairs are formed from the split data set where the abscissa is taken from the lower half and the ordinate is taken from the upper half. The ordered pairs look like $(|X_1 - \widetilde{X}|, |X_N - \widetilde{X}|)$, $(|X_2 - \widetilde{X}|, |X_{N-1} - \widetilde{X}|)$, $(|X_3 - \widetilde{X}|, |X_{N-2} - \widetilde{X}|)$, etc. If the data are symmetric (with respect to the median), then the plot will produce points on the straight line $y = x$. Departure from this line indicates deviation from symmetry. Here is a simple example:

```
> scatterplot([10, 20, 30, 40, 50, 60, 70],format=symmetry,
    symbol=circle);
```

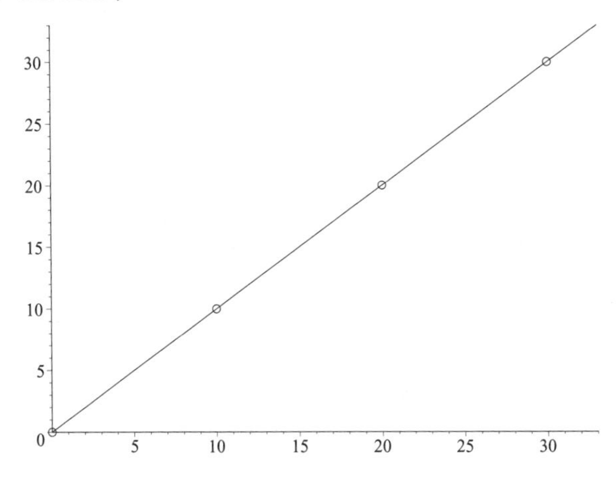

Figure 16.13 One-dimensional scatter plot with `symmetry` format option.

Since the median of the data set $[10, 20, 30, 40, 50, 60, 70]$ is 40, the `scatterplot` function above the ordered pairs $(|10-40|, |70-40|)$, $(|20-40|, |60-40|)$, $(|30-40|, |50-40|)$. The symmetry plot above shows a data set which is perfectly symmetric. The following examples show data sets that show skewness.

```
> scatterplot([10, 20, Weight(30,3), 40, 50, 60, Weight(70,5),
   missing],format=symmetry);
> scatterplot([10, 20, Weight(30,7), 40, 50, 60, Weight(70,3)],
   format=symmetry);
```

In two or three dimensions the `scatterplot(`*data*`1,`*data*`2,`*data*`3)` function will produce a two- or three-dimensional scatter plot with *data1* plotted on the x-axis, *data2* plotted on the y-axis, and *data3* plotted on the z-axis. A simple example can be obtained by using Data Set 1 from the Lamp Example.

```
> data1:=importdata("data1.dat",2):
> life:=data1[1]:
> amount:=data1[2]:
> scatterplot(amount, life, symbol=circle);
```

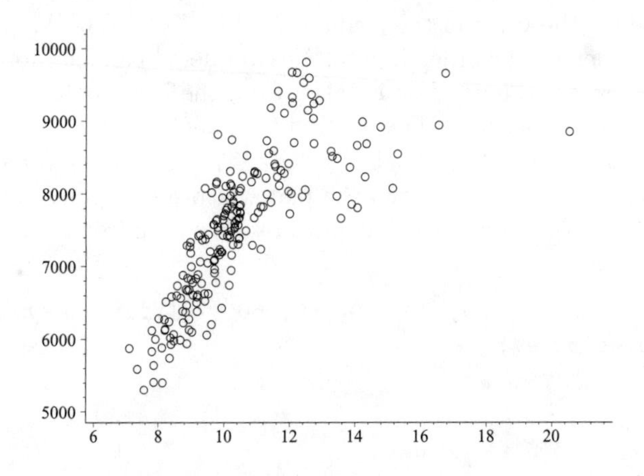

Figure 16.14 A two-dimensional scatter plot.

There are four formats that can be used for one- and two-dimensional scatter plots: `agglomerated`, `excised`, `quantile`, and `sunflower`.

A quantile plot for one-dimensional data is a graph of the ordered pairs with the observed data value as the ordinate and the quantile of the observed data value as the abscissa. The `scatterplot(data, format=quantile` function generates a quantile plot. Try using Data Set 1 from the Lamp Example to generate a quantile plot.

```
> scatterplot(life, format=quantile);
```

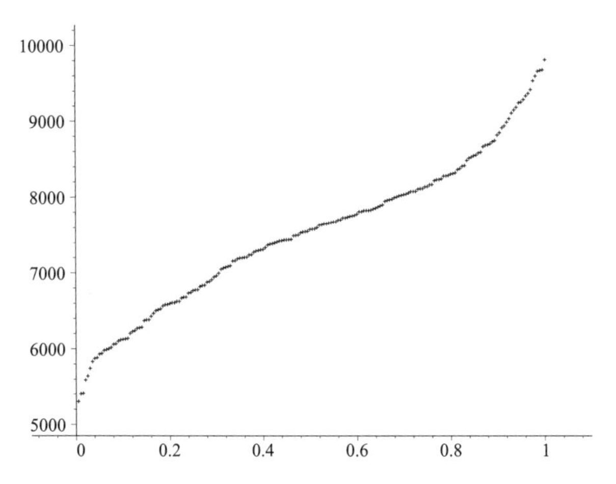

Figure 16.15 Quantile plot.

A quantile-quantile plot, or q-q plot, for two-dimensional data is a graph of the data paired by quantile value. Try using Data Set 2 from the Lamp Example to generate a quantile plot.

```
> data2:=importdata("data2.dat",2):
> argon:=data2[1]:
> neon_argon:=data2[2]:
> scatterplot(argon, neon_argon, format=quantile);
```

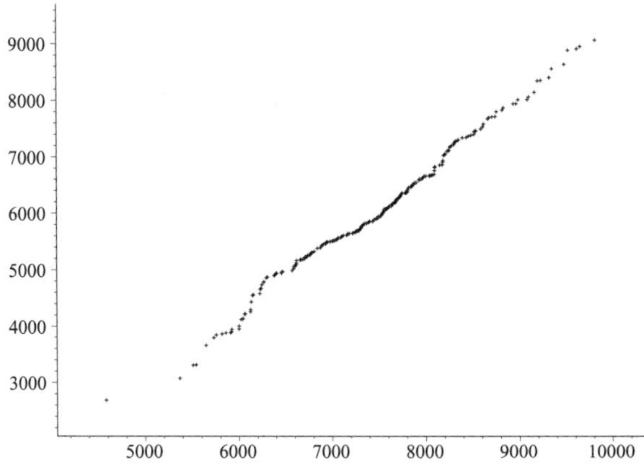

Figure 16.16 A Quantile-Quantile plot.

A q-q plot is often used to compare sample data to data randomly generated from a known distribution. If the sample has the same distribution, the q-q plot will resemble a straight line. The following example shows how to generate a q-q plot comparing *lamp life* data from Data Set 1 (transformed to z-scores) to data randomly generated from a standard normal distribution.

```
> life_zscore:=transform[standardscore](life):
> life_zscore:=transform[statsort](life_zscore):
```

```
> life_zquantile:=seq(life_zscore[i], i=1..199):
> z_score:=seq(statevalf[icdf,normald](i/200),i=1..199):
> scatterplot([life_zquantile],[z_score],axes=frame);
```

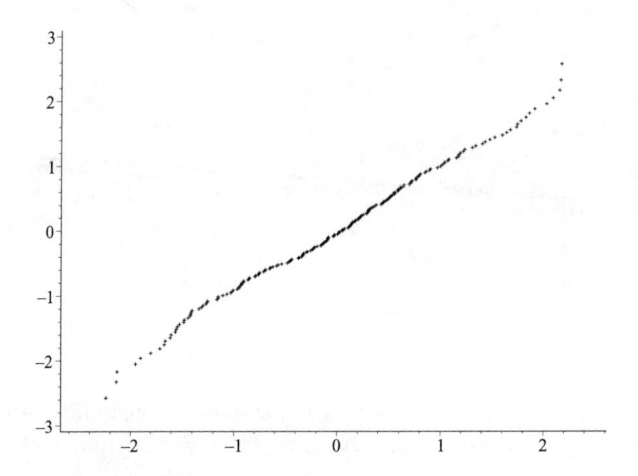

Figure 16.17 Quantile-Quantile plot of lamp life and normal.

The `format=sunflower[l]` option replaces points with *sunflowers*. Each sunflower has one arm for every data point. The value l specifies the maximum length of the arms; the default value is one tenth the range of the data plotted on the x-axis. Try

```
> scatterplot([10, 20, Weight(30,3), 40, 50, 60, Weight(70,5),
   missing],format=sunflower[1]);
> scatterplot(life, format=sunflower);
```

The resulting plots are given below in Figures 16.18 and 16.19.

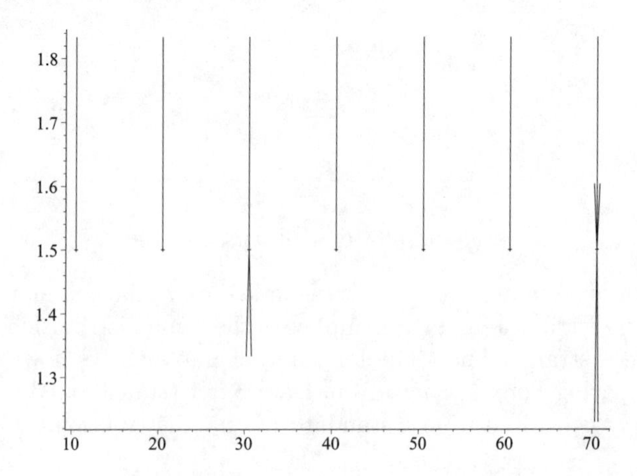

Figure 16.18 Sunflower plot of sample data.

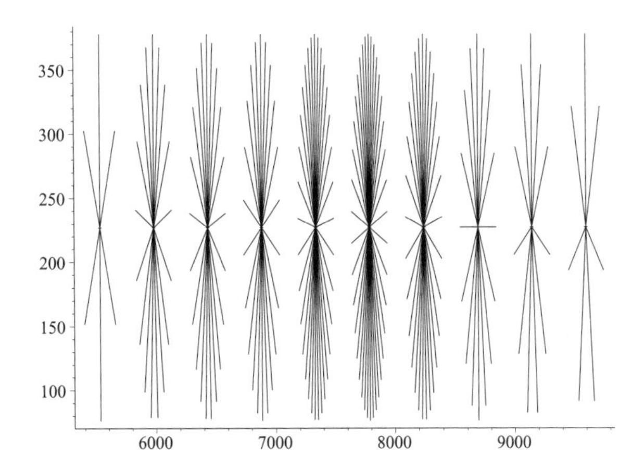

Figure 16.19 Sunflower plot of lamp life data.

MAPLE plotting functions can be combined to powerfully represent in a single graph the information conveyed in several separate graphs. As an example, the graph below combines a scatter plot and box plots for the *lamp life* and *emitter amount* data.

```
> data1:=importdata("data1.dat",2):
> life:=data1[1]:
> amount:=data1[2]:
> sp:=scatterplot(amount,life):
> bp1:=boxplot(amount,width=400,color=red):
> bp1:=yshift(10500,xyexchange(bp1)):
> bp2:=boxplot(life,width=1,shift=22,color=cyan):
> plots[display]({sp,bp1,bp2},view=[6..23,5000..11000]);
```

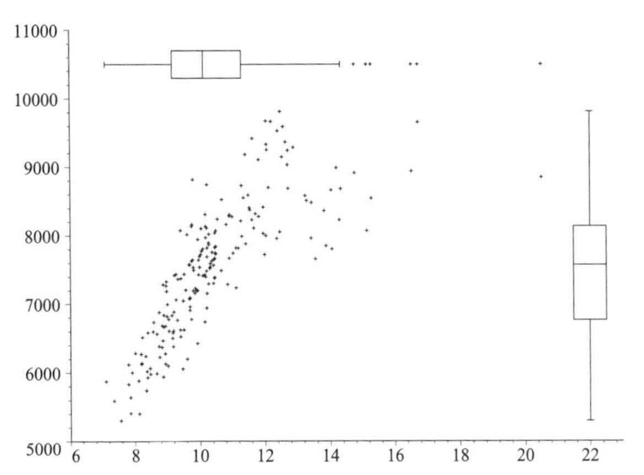

Figure 16.20 Plot of lamp life and emitter amount.

The `agglomerated[n, l]` and `excised[n, l]` are other format options for the `scatterplot` function. See `?scatterplot` for more information.

16.6 Linear regression

The *fit* subpackage provides a tool for fitting curves to bivariate statistical data. The `fit[leastsquare[[x,y],y=curve,`$b_0, b_1, \ldots, b_k$`]]([x-data, y-data])` fits the specified `curve` to the given data using the method of least squares, where `curve` is a linear function in the unknown parameters $b_0, b_1, \ldots, b_k$. The default of `curve` is $y = ax + b$. Try

```
> with(stats):
> fit[leastsquare[[x,y]]]([[10,20,30,40],[22,42,62,82]]);
```

$$y = 2 + 2\,x$$

```
> with(stats):
> data1:=importdata("lamp.dat",2):
> life:=data1[1]:
> amount:=data1[2]:
> simple_fit:=fit[leastsquare[[x,y]]]([amount, life]);
```

$$y = 3260.660333 + 408.6392896\,x$$

```
> lamp_fit:=fit[leastsquare[[x,y], y=a*x^2+b*x+c, {a,b,c}]]
    ([amount,life]);
```

$$y = -54.29005649\,x^2 + 1694.889037\,x - 4058.482288$$

The `leastmediansquare` is another format option for the `fit` function. See `?fit` for more information.

16.7 ANOVA

The *anova* subpackage provides a tool for conducting an analysis of variance, a method for comparing two or more population means. The `anova[oneway]` function result contains the information presented in a standard ANOVA table in the following format:

[[treatment df, treatment sum of squares, treatment mean sum of squares],
[error df, error sum of squares, error mean sum of squares],
[total df, total sum of squares]],
[treatment df, error df, F-ratio, Prob (F < F-ratio)]

A simple example of a one-way ANOVA is given below.

```
> with(stats):
> data3:=importdata("data3.dat",3):
> instant:=data3[1]:
> preheat:=data3[2]:
> rapid:=data3[3]:
```

```
> life3:=[instant, preheat, rapid]:
> anova[oneway](life3);
```

$$[[2, 640574782.49, 320287391.245], [297, 295982552.72, 996574.251582],\\ [299, 936557335.21]][2, 297, 321.388387003, 1.0]$$

Because the p-value (Prob(F > F-ratio)) is clearly less than 0.05, we conclude that the choice of lamp fixture significantly affects lamp life.

16.8 Distributions

MAPLE has a number of well-known distributions for discrete and continuous random variables. The following discrete distributions are available:

```
binomiald[n,p]           discreteuniform[a,b]
empirical[list_prob]     hypergeometric[N1,N2,n]
negativebinomial[n,p]    poisson[mu]
```

The following continuous distributions are available:

```
beta[nu1,nu2]            cauchy[a,b]        chisquare[nu]
exponential[alpha,a]     fratio[nu1,nu2]    gamma[a,b]
laplaced[a,b]            logistic[a,b]      lognormal[mu,sigma]
normald[mu,sigma]        studentst[nu]      uniform[a,b]
weibull[a,b]
```

See `?stats,distributions` for the definitions of distribution parameters.

16.8.1 Evaluating distributions

The subpackage *statevalf* provides numerical evaluations of statistical functions.

The functions available for discrete distributions are

`dcdf`	discrete cumulative probability function
`idcdf`	inverse discrete cumulative probability function
`pf`	probability function

The functions available for continuous distributions are

`cdf`	cumulative density function
`icdf`	inverse cumulative density function
`pdf`	probability density function

The various distributions take their parameters as indices to the distributions. See `?stats[distributions]` for information on each available distribution. The following examples show how the binomial distribution can be evaluated.

```
> with(stats):
> statevalf[pf,binomiald[10,0.3]](0);
```

$$0.0282475249$$

```
> statevalf[dcdf,binomiald[10,0.3]](3);
```

$$0.6496107184$$

```
> statevalf[idcdf,binomiald[10,0.3]](0.6496);
```

$$2.0$$

Plot the probability density function and the cumulative density function of the binomial distribution with $n = 10$ and $p = 0.3$.

```
> x:=seq(i,i=0..10):
> Binomial_pdf:=seq(statevalf[pf,binomiald[10,0.3]](i),
   i=0..10):
> Binomial_cdf:=seq(statevalf[dcdf,binomiald[10,0.3]](i),
   i=0..10):
> sp1:=scatterplot([x],[Binomial_pdf],color=red,symbol=circle):
> sp2:=scatterplot([x],[Binomial_cdf],color=blue,symbol=cross):
> plots[display]({sp1,sp2}, view=[0..10,0..1]);
```

The red circles are the points of the PDF function and the blue crosses are the points of the CDF function.

The following examples show how the normal distribution can be evaluated:

```
> with(stats):
> statevalf[cdf,normald[0,1]](1.96);
```

$$0.9750021049$$

```
> statevalf[icdf,normald[0,1]](0.975);
```

$$1.959963985$$

```
> statevalf[cdf,normald](1.96);
```

$$0.9750021049$$

Plot the probability, cumulative, and inverse cumulative density functions of the standard normal distribution.

```
> p1:= plot(statevalf[pdf,normald], -3..3, colour=green):
> p2:=plot(statevalf[cdf,normald], -3..3, colour=blue):
```

```
> p3:=plot(statevalf[icdf,normald], 0.1..0.9, colour=red):
> plots[display]({p1,p2,p3},view=[-3..3,-1..1]);
```

16.8.2 Generating random distributions

The `random[`*distribution*`](`*n*`)` function generates n random numbers with a given *distribution*. To generate a random number between 0 and 1 try

```
> stats[random,uniform[0,1]](1);
```

$$0.4274196691$$

To generate 100 random numbers from a normal distribution with mean 3 and standard deviation 2, and a histogram of the random numbers, try

```
> normal_rando:=stats[random,normald[3,2]](20):
> stats[statplots,histogram]([normal_rando], area=1, numbars=10,
    color=blue);
```

17. Overview of Other Packages

We have already seen many MAPLE packages including *Units*, *student*, *LinearAlgebra*, *codegen*, *geom3d*, *geometry*, *inttrans*, *linalg*, *orthopoly*, *plots*, *plottools*, and *stats*. In this chapter we give an overview of the remaining packages. To see a list of the available packages try

```
> ?index[packages]
```

The resulting table of package names, with their descriptions, is given below.

algcurves	Algebraic curves
codegen	Code generation
combinat	Combinatorial functions
combstruct	Combinatorial structures
context	Context-sensitive menus
CurveFitting	Fitting curves to data points
DEtools	Differential equations tools
diffalg	Differential algebra
difforms	Differential forms
Domains	Create domains of computation
ExternalCalling	Link to external functions
finance	Financial mathematics
GaussInt	Gaussian integers
genfunc	Rational generating functions
geom3d	Euclidean three-dimensional geometry
geometry	Euclidean geometry
Groebner	Groebner basis calculations in skew algebras
group	Permutation and finitely-presented groups
inttrans	Integral transforms
liesymm	Lie symmetries
linalg	Linear algebra based on array data structures
LinearAlgebra	Linear algebra based on rtable data structures
LinearFunctionalSystems	Solving linear functional systems of equations
LinearOperators	Solving operator equations, building annihilators
ListTools	Manipulating lists
LREtools	Manipulate linear recurrence relations
Matlab	Matlab link
MathML	Convert MAPLE expressions to MathML
networks	Graph networks
numapprox	Numerical approximation
numtheory	Number theory
Ore_algebra	Basic calculations in algebras of linear operators
OrthogonalSeries	Series of orthogonal polynomials
orthopoly	Orthogonal polynomials
padic	p-Adic numbers

PDEtools	Tools for solving partial differential equations
plots	Graphics package
plottools	Basic graphical objects
PolynomialTools	New polynomial tool package
polytools	Polynomial tools
powseries	Formal power series
process	(Unix)-multiprocessing
RandomTools	Random objects
RationalNormalForms	Representing hypergeometric terms
RealDomain	Restricting domain to real numbers
simplex	Linear optimization
Sockets	Network connection tools
SolveTools	Solving systems of algebraic equations
Spread	Spreadsheets
stats	Statistics
StringTools	Manipulating strings
student	Student calculus
sumtools	Indefinite and definite sums
tensor	Tensor computations and general relativity
Units	Unit conversion
XMLTools	Extensible markup language tools

The Galois Fields package *GF* is listed as a package in MAPLE 6 but not in MAPLE 7. The new packages for MAPLE 7 are *CurveFitting*, *ExternalCalling*, *LinearFunctionalSystems*, *LinearOperators*, *ListTools*, *MathML*, *OrthogonalSeries*, *PolynomialTools*, *RandomTools*, *RationalNormalForms*, *RealDomain*, *Sockets*, *SolveTools*, *StringTools*, *Units*, and *XMLTools*.

17.1 Finite fields

Any finite field $\mathbb{F}$ must have p^k elements for some prime p. First we consider the case when $k = 1$. Arithmetic in the field of p elements, $\mathbb{Z}_p$, coincides with arithmetic over the integers modulo p. This is handled in MAPLE by the functions `mod` and `modp`. For example, let's compute the following in $\mathbb{Z}_{17} = \{0, 1, 2, \ldots, 16\}$:

$$13 + 15,$$
$$13 - 15,$$
$$13/15,$$
$$13^{-1}.$$

```
> 13 + 15 mod 17;
```

$$11$$

```
> 13 - 15 mod 17;
```

$$15$$

```
> 13/15 mod 17;
```

$$2$$

```
> 13^(-1) mod 17;
```

$$4$$

Thus, in $\mathbb{Z}_{17}$ we see that

$$\begin{aligned} 13 + 15 &= 11, \\ 13 - 15 &= 15, \\ 13/15 &= 2, \\ 13^{-1} &= 4. \end{aligned}$$

MAPLE can do calculations in the polynomial ring $\mathbb{Z}_p[x]$. For example, to factorize a polynomial, we use the `Factor` function.

```
> P := x^4+3*x+1;
```

$$x^4 + 3\,x + 1$$

```
> factor(P);
```

$$x^4 + 3\,x + 1$$

```
> Factor(P) mod 43;
```

$$(x + 16)\left(x^3 + 27\,x^2 + 41\,x + 35\right)$$

Although the polynomial $x^4 + 3\,x + 1$ is irreducible over $\mathbb{Q}$, it does factor in $\mathbb{Z}_{43}$ as

$$x^4 + 3\,x + 1 = (x + 16)\left(x^3 + 27\,x^2 + 41\,x + 35\right).$$

Besides the `Factor` function there are many other functions compatible with the `mod` function for doing polynomial and linear algebra calculations over $\mathbb{Z}_p$:

```
Content     Det         DistDeg     Divide      Eval
Expand      Factor      Factors     Frobenius   Gausselim
Gaussjord   Gcd         Gcdex       Hermite     Interp
Inverse     Issimilar   Lcm         Normal      Nullspace
Power       Powmod      Prem        Primitive   Primpart
Quo         Randpoly    Randprime   Rem         Resultant
Roots       Smith       Sprem       Sqrfree     taylor
```

MAPLE can also handle calculations in a field $\mathbb{Z}_p[\alpha]$, where α is the root of an irreducible polynomial over $\mathbb{Z}_p$. For example, the polynomial x^2+1 is irreducible over $\mathbb{Z}_5$, so we let α be a root and let $\mathbb{F} = \mathbb{Z}_5[\alpha]$. We compute $1/(1+\alpha)$ in $\mathbb{F}$.

```
> alias(alpha=RootOf(y^2+1)):
> Normal(1/(1+alpha)) mod 5;
```

$$2\,\alpha + 3$$

We see that in $\mathbb{F}$,

$$\frac{1}{1+\alpha} = 2\alpha + 3.$$

To handle a finite field with p^k elements for $k > 1$, we use the *GF* package. This package differs from other packages in that it is not loaded with the `with` function. For p a prime and k a positive integer, the function `GF(p,k)` creates a table of functions and constants for doing arithmetic in the finite field $\text{GF}(p^k)$. The finite field $\text{GF}(p^k)$ can be constructed as a finite extension $\mathbb{Z}_p[\alpha]$, where α is the root of an irreducible polynomial $P(x)$ of degree k over $\mathbb{Z}_p$. This can be created in MAPLE using the function `GF(p,k,P(α))`. As an example, we consider the field $\text{GF}(2^3)$.

```
> Factor(x^3+x+1) mod 2;
```

$$x^3 + x + 1$$

Since the polynomial $x^3 + x + 1$ is irreducible over $\mathbb{Z}_2$, we can use it to construct $\text{GF}(2^3)$.

```
> G8 := GF(2,3,alpha^3+alpha+1);
G8 := module()
export `+`, `-`, `*`, `/`, `^`, input, output, inverse,
extension, variable, factors, norm, trace, order, random, size,
isPrimitiveElement, PrimitiveElement, ConvertIn, ConvertOut,
zero, one, init;
end module
```

On the `export` line above we see all the functions available. In this example, each function f is called using `G[f]`. We can list the elements in the field $GF(2^3) = \mathbb{Z}_2[\alpha]$ using the `input` function.

```
> seq(G8[input](k),k=1..8);
```

$$1, \alpha, 1+\alpha, \alpha^2, 1+\alpha^2, \alpha+\alpha^2, 1+\alpha+\alpha^2, \alpha^3$$

`G8[input](k)` gives the kth element of the field. We can assign names to field elements using the `ConvertIn` function. We let $a = 1 + \alpha$ and $b = \alpha^2$:

```
> a := G8[ConvertIn](1+alpha);
```

$$a := 1 + \alpha$$

```
> b := G8[ConvertIn](alpha^2);
```

$$b := \alpha^2$$

We can perform field operations in the obvious way.

```
> G8[`^`](a,4);
```

$$1 + \alpha + \alpha^2$$

```
> G8[`/`](a,b);
```

$$\alpha$$

```
> G8[`*`](a,b);
```

$$1 + \alpha + \alpha^2$$

In the field $\mathbb{Z}_2[\alpha]$ we found that

$$\begin{aligned}(1+\alpha)^4 &= 1 + \alpha + \alpha^2,\\ \frac{1+\alpha}{\alpha^2} &= \alpha,\\ (1+\alpha)\alpha^2 &= 1 + \alpha + \alpha^2.\end{aligned}$$

For more examples see `?GF`.

17.2 Polynomials

In this section we discuss the `factor` function and the *polytools* and *PolynomialTools* packages. *PolynomialTools* is a MAPLE 7 package. It is an updated version of the *polytools* package. Back in Chapter 3 we saw how to `factor` polynomials over $\mathbb{Z}$ or $\mathbb{Q}$. The `factor` function can also be used to factor over an extension field of $\mathbb{Q}$. In the previous section we used the `Factor` function to factor polynomials over field extensions of $\mathbb{Z}_p$. As an example, we factor the polynomial

$$P(x,y) = x^4 - 4\,x^2y^2 + 4\,y^4 - 6\,x^2 - 12\,y^2 + 9$$

over the fields $\mathbb{Q}$, $\mathbb{Q}[\sqrt{2}]$, and $\mathbb{Q}[\sqrt{2}, \sqrt{3}]$:

```
> p:= x^4-4*x^2*y^2+4*y^4-6*x^2-12*y^2+9;
```

$$p := x^4 - 4\,x^2y^2 + 4\,y^4 - 6\,x^2 - 12\,y^2 + 9$$

```
> factor(p);
```

$$x^4 - 4\,x^2y^2 + 4\,y^4 - 6\,x^2 - 12\,y^2 + 9$$

```
> factor(p,sqrt(2));
```

$$\left(x^2 - 2\,xy\sqrt{2} + 2\,y^2 - 3\right)\left(x^2 - 3 + 2\,xy\sqrt{2} + 2\,y^2\right)$$

```
> factor(p,sqrt(2),sqrt(3));
```

$$\left(x - \sqrt{3} - y\sqrt{2}\right)\left(x + \sqrt{3} - y\sqrt{2}\right)\left(x + \sqrt{3} + y\sqrt{2}\right)\left(x - \sqrt{3} + y\sqrt{2}\right)$$

We see that $P(x,y)$ is irreducible over $\mathbb{Q}$. Over $\mathbb{Q}[\sqrt{2}]$, it has the factorization

$$P(x,y) = \left(x^2 - 2\,xy\sqrt{2} + 2\,y^2 - 3\right)\left(x^2 - 3 + 2\,xy\sqrt{2} + 2\,y^2\right),$$

and over $\mathbb{Q}[\sqrt{2}, \sqrt{3}]$, it factors completely into linear factors

$$P(x,y) = \left(x - \sqrt{3} - y\sqrt{2}\right)\left(x + \sqrt{3} - y\sqrt{2}\right)\left(x + \sqrt{3} + y\sqrt{2}\right)\left(x - \sqrt{3} + y\sqrt{2}\right).$$

The syntax of `factor` has the form `factor(p,K)`, where is p is a polynomial and K is a set of radicals or `RootOf`s that generate an extension field. Let β satisfy $\beta^3 + \beta + 1$. We factor the polynomial

$$Q(x) = x^6 - 2x^4 - x^3 + 2x^2 - 1,$$

over $\mathbb{Q}[\beta]$. We use `RootOf` to define β.

```
> beta := RootOf(Z^3+Z+1);
```

$$\beta := \text{RootOf}(\_Z^3 + \_Z + 1)$$

```
> Q := x^6-2*x^4-x^3+2*x^2-1;
```

$$Q := x^6 - 2\,x^4 - x^3 + 2\,x^2 - 1$$

```
> factor(Q,beta);
```

$$(x^4 - x^3\%1 - x^2 + x^2\%1^2 + x\%1 + 1)(x^2 + x\%1 - 1)$$
$$\%1 := \texttt{RootOf}(\_Z^3 + \_Z + 1)$$

Over $\mathbb{Q}[\beta]$, MAPLE found the factorization

$$Q(x) = (x^4 - \beta x + (\beta^2 - 1)x^2 + \beta x + 1)(x^2 + \beta x - 1).$$

To factor over $\mathbb{C}$, we use the `complex` option.

```
> factor(Q,complex);
```

$$(x + 1.010648068 + 0.6987007022\,I)\,(x + 1.010648068 - 0.6987007022\,I)$$
$$(x + 0.7154310100)\,(x - 0.6694841663 + 0.4628406978\,I)$$
$$(x - 0.6694841663 - 0.4628406978\,I)\,(x - 1.397758814)$$

Of course, the complex constants in this factorization have been approximated.

The *polytools* package contains five functions that operate on polynomials:

`minpoly` `recipoly` `split` `splits` `translate`

The *PolynomialTools* package is new to MAPLE 7, but it is just an updated version of the *polytools* package. It contains three additional functions:

`Shorten` `PolynomialToPDE` `PDEToPolynomial`

For a floating-point constant c and a positive integer n, the minpoly(c,n) function returns a polynomial of degree n with small integer coefficients, one of whose roots agrees with c. As an example, suppose we suspect that

$$c \approx 0.31783724519578224 47$$

is the root of a nice quartic polynomial.

```
> c := 0.31783724519578224 47:
> p := polytools[minpoly](c,4);
```

$$p := 1 - 10\,\_X^2 + \_X^4$$

```
> r := solve(p);
```

$$r := -\sqrt{3} - \sqrt{2}, \sqrt{3} + \sqrt{2}, -\sqrt{3} + \sqrt{2}, \sqrt{3} - \sqrt{2}$$

```
> map(evalf,[r]);
```

$$[-3.146264370, 3.146264370, -0.317837246, 0.317837246]$$

MAPLE found that c is an approximate root of the polynomial $p(x) = 1 - 10x^2 + x^4$. We used solve to obtain the roots $x = \pm\sqrt{3} \pm \sqrt{2}$, which we approximated using evalf. We are now led to conjecture that

$$c = \sqrt{3} - \sqrt{2}.$$

For a polynomial $p(x)$, the function recipoly($p(x)$,x) determines whether $p(x)$ is self-reciprocal; i.e., whether

$$p(x) = x^d\, p(1/x),$$

where d is the degree of $p(x)$.

For a polynomial $p(x)$, the function split($p(x)$,x) computes a complete factorization of $p(x)$ in some splitting field. We find the complete factorization of the polynomial $p(x) = 1 + x^2 + x^4$:

```
> P := 1 + x^2 + x^4;
```

$$P := 1 + x^2 + x^4$$

```
> polytools[split](P,x);
```

$$(x - 1 + \%1)(x + \%1)(x + 1 - \%1)(x - \%1)$$
$$\%1 := \mathrm{RootOf}(\_Z^2 - \_Z + 1)$$

This means that the polynomial $p(x)$ splits in $\mathbb{Q}(\omega)$ as

$$p(x) = (x - 1 + \omega)(x + \omega)(x + 1 - \omega)(x - \omega),$$

where ω satisfies $\omega^2 - \omega + 1$. The function `splits` is the same as `split` except that it returns the factorization in list form.

For a polynomial $p(x)$ and a constant a, `translate(`$p(x)$`,`x`,`a`)` returns $p(x+a)$. See `?polynomial` for a list of all MAPLE operations that are defined on polynomials.

17.3 Group theory

The group theory package is *group*. It contains the following functions:

```
DerivedS       LCS          NormalClosure   RandElement
SnConjugates   Sylow        areconjugate    center
centralizer    convert      core            cosets
cosrep         derived      elements        grelgroup
groupmember    grouporder   inter           invperm
isabelian      isnormal     issubgroup      mulperms
normalizer     orbit        parity          permgroup
hpermrep       pres         subgrel         transgroup
transnames     type
```

In the *group* package, groups are represented either as permutation groups or by sets of generators and relations.

There are two ways to represent permutations. The MAPLE list

`[`i_1`, `i_2`, ..., `i_r`]`

corresponds to the permutation $\sigma(k) = i_k$, for $1 \le k \le r$. Permutations can also be given as products of disjoint cycles. A cycle is represented by a list. For example, the cycle $\sigma = (a_1\, a_2\, \ldots\, a_r)$ is represented in MAPLE by the list `[`$a_1, a_2, \ldots, a_r$`]`. Then products of cycles are represented by a list of lists. For example, the permutation $\sigma = (1\,3\,4)(2\,7)(5\,6)$ is represented in MAPLE by `[[1,2,3],[2,7],[5,6]]`. We can use the `convert` function to convert between the two ways of representing permutations. Suppose we are given the permutation

$$\sigma = \begin{pmatrix} 1 & 2 & 3 & 4 & 5 & 6 \\ 3 & 4 & 6 & 2 & 5 & 1 \end{pmatrix}.$$

We enter this into MAPLE.

```
> sigma := [3,4,6,2,5,1];
```

$$\sigma := [3, 4, 6, 2, 5, 1]$$

We can convert this to a product of disjoint cycles by using the `convert` function with the `disjcyc` option.

```
> dcp := convert(sigma,'disjcyc');
```

$$dcp := [[1,3,6],[2,4]]$$

We found that

$$\sigma = (1\,3\,6)\,(2\,4).$$

To convert from a product of disjoint cycles to the list form, we use the `convert` function with the `permlist` option.

```
> convert(dcp,'permlist',6);
```

$$[3,4,6,2,5,1]$$

In general, the syntax takes the form

convert(dcp , 'permlist', n)

where `dcp` is a product of disjoint cycles (given as a list of lists) and n is the degree of the permutation. In our example we needed $n = 6$.

A permutation group is defined in MAPLE using the `permgroup` function. The syntax has the form

permgroup(n, {dcp1, dcp2, ... , dcpk})
permgroup(n, {a1=dcp1, a2=dcp2, ... , ak=dcpk})

Here n is the degree of the permutations, and the second argument is a set of generators. Each generator is given as a product of disjoint cycles. In the second form, names are assigned to the generators. We define the group $G < S_4$ generated by $\sigma = (1\,2\,3\,4)$, and $\tau = (1\,3)\,(2\,4)$.

```
> with(group):
> G:=permgroup(4,sigma=[[1,2,3,4]],tau=[[1,3],[2,4]]);
```

$$G := \texttt{permgroup}(4, \{\sigma = [[1,2,3,4]], \tau = [[1,3],[2,4]]\})$$

We can compute the order of this group using `groupord`.

```
> grouporder(G);
```

$$4$$

We see that $|G| = 4$. We can list the elements of the group using the `elements` function.

```
> elements(G);
```

$$\{[], [[1,2,3,4]], [[1,3],[2,4]], [[1,4,3,2]]\}$$

We see that

$$G = \{e, \sigma, \tau, (1\,4\,3\,2)\},$$

where e is the identity permutation.

Groups can also be defined in terms of generators and relations using the `grelgroup` function. The syntax has the form `grelgroup(`S`,`R`)`, where S is a set of generators and R is a set of relations. Each relation is given as a list of certain generators or their inverses. The relation corresponds to setting the product of the elements equal to the identity. For example, we consider the group G with generators a, b and relations $a^5 = e$, $b^2 = e$, $aba = b$.

```
> with(group):
> G := grelgroup(a,b,[a,a,a,a,a],[b,b],[a,b,a,1/b]);
```

$$G := \texttt{grelgroup}(\{a, b\}, \{[a, a, a, a, a], [b, b], [a, b, a, b^{-1}]\})$$

```
> grouporder(G);
```

$$10$$

We found that $|G| = 10$.

We summarize the remaining functions in the package.

`areconjugate(P,` g_1`,` g_2`)`

Determines whether g_1, g_2 are conjugate in the permutation group P.

`center(`P`)`

Returns the center $C(P)$ of the permutation group P.

`centralizer(`P`,`g`)`

Returns the centralizer $C_P(g)$ of the permutation g in the permutation group P.

`core(`S`,`P`)`

Computes the core of a subgroup S of a permutation group P (i.e., the largest normal subgroup of P that is contained in S).

`cosets(`S`,`P`)`

Computes a complete list of right coset representatives for a subgroup S of a group G. The groups S, G may be given as permutation groups or in terms of generators and relations using `grelgroup` and `subgrel`.

`cosrep`

Expresses a given group element as an element of some right coset.

`derived(`P`)`

Returns the derived subgroup $P' = P^{(1)}$ of the permutation group P, also known as the commutator subgroup.

`DerivedS(`P`)`

Computes the derived series

$$P > P^{(1)} > P^{(2)} > \cdots$$

of a permutation group P. The series is returned as a sequence of permutation groups.

`groupmember(`p`,`P`)`

Determines whether the permutation p is an element of the permutation group P.

`inter(`P_1`, `P_2`)`

Returns the intersection of two permutation groups P_1 and P_2.

`invperm(`p`)`

Computes the inverse of the permutation p given as a product of disjoint cycles.

`isabelian(`P`)`

Determines whether the permutation group P is abelian.

`isnormal`

Determines whether a subgroup is a normal subgroup of a given group.

`issubgroup(`P_1`, `P_2`)`

Determines whether the permutation group P_1 is a subgroup of the permutation group P_2.

`LCS(`P`)`

Computes the lower central series of a permutation group P.

`mulperms(`p_1`, `p_2`)`

Computes the product $p_1 p_2$ of two permutations p_1, p_2 given as products of disjoint cycles.

`NormalClosure(`S`, `P`)`

Computes the smallest normal subgroup of the permutation group P containing the subgroup S.

`normalizer(`P`, `S`)`

Computes the normalizer $N_P(S)$ of S in the permutation group P.

`orbit(`P`, `k`)`

Computes the orbit

$$\{\sigma(k) : \sigma \in P\},$$

for a given integer k and permutation group P.

`parity`

Determines the parity of a permutation group, an individual permutation, or a permutation with a cycle type given by a partition. The function returns 1 if the parity is even and -1 if the parity is odd. The parity of a permutation is also called the sign of a permutation. The parity of a permutation group is even if all of its elements are even, otherwise it is odd.

`permrep`

Computes a permutation representation of a group in a certain sense. See `?group[permrep]` for more details.

`pres`

Finds a set of relations for the generators of a subgroup.

`subgrel`

Defines a subgroup of a given group in terms of generators.

`transgroup`

Returns certain information for a given transitive permutation group. See `?group[transgroup]`.

`?transnames`

Gives a page of information describing the group naming scheme used by `transgroup`.

`type(L, 'disjcyc'(n))`

Checks whether `L` is a valid MAPLE expression that describes a permutation in S_n as a product of disjoint cycles.

The function `galois` computes the Galois group of a polynomial. It is not in the *group* package. We compute the Galois group of the polynomial

$$p(x) = 2\left(x^2 + 6x - 3\right)^3 - 27(x+1)^3\left(x^2 - 1\right).$$

```
> p:= 2*(x^2+6*x-3)^3-27*(x+1)^3*(x^2-1);
```

$$p := 2\left(x^2 + 6x - 3\right)^3 - 27(x+1)^3\left(x^2 - 1\right)$$

```
> galois(p);
```

$$\text{"}6T11\text{"}, \{\text{"}2S_4(6)\text{"}, \text{"}[2^3]S(3)\text{"}, \text{"}2 \texttt{ wr } S(3)\text{"}\}, \text{"} - \text{"}, 48,$$
$$\{\text{"}(36)\text{"}, \text{"}(246)(135)\text{"}, \text{"}(15)(24)\text{"}\}$$

This means that the Galois group G is a group of order 48,

$$G \cong \mathbb{Z}_2 \times S_4 \cong \mathbb{Z}_2 \wr S_3,$$

and as a permutation group G has generators $(3\,4)$, $(2\,4\,6)\,(1\,3\,5)$ and $(1\,5)\,(2\,4)$. For an explanation of how the output of `galois` is interpreted, see `?group[transnames]` and `?group[transgroup]`.

17.4 Combinatorics

There are three packages for doing combinatorics: *combinat*, *combstruct*, and *networks*. The *combinat* package contains functions for counting and listing combinatorial objects such as permutations, combinations, and partitions. The *networks* package is for drawing graphs and doing graph theory calculations. The *combstruct* package is used to define more abstract combinatorial structures.

17.4.1 The *combinat* package

The *combinat* package contains the following functions:

```
Chi          bell         binomial  cartprod    character
choose       composition  conjpart  decodepart  encodepart
fibonacci    firstpart    graycode  inttovec    lastpart
multinomial  nextpart     numbcomb  numbcomp    numbpart
numbperm     partition    permute   powerset    prevpart
randcomb     randpart     randperm  stirling1   stirling2
subsets      vectoint
```

Cartesian products

The `cartprod` function is used to define Cartesian products. The call `cartprod(`[S_1, S_2, ... , S_k]`)` defines the cartesian product

$$S_1 \times S_2 \times \cdots \times S_k,$$

where the S_j are sets or lists. We define the Cartesian product

$$\{1, 2, 3\} \times \{a, b\} \times \{A, B\}.$$

```
> with(combinat):
> T := cartprod([{1,2,3},{a,b},{A,B}]);
T := table([nextvalue = (proc() ...  end proc),
    finished = false])
```

It is possible to iterate through the Cartesian product using `nextvalue`. We use a while loop to generate the elements of the cartesian product.

```
> with(combinat):
> C := {}:
> T := cartprod([1,2,3,a,b,A,B]):
> while not T[finished] do
      C := C union T[nextvalue]():
   end do:
> C;
```

$$\{[1, a, B], [1, a, A], [1, b, B], [1, b, A], [2, a, B], [2, a, A], [2, b, B], [2, b, A], [3, a, B], [3, a, A], [3, b, B], [3, b, A]\}$$

Permutations

The `permute` function is used to generate permutations. When n is a positive integer, `permute(`n`)` generates all permutations of $1, 2, \ldots, n$. When S is a set or list, `permute(S)` generates all permutations of the elements of S. The functions `permute(`n`, `r`)`, and `permute(S, `r`)` generate permutations taken r at a time. We generate the permutations of $1, 2, 3, 4$ taken 2 at a time.

```
> with(combinat):
> permute(4,2);
```

$$[[1, 2], [1, 3], [1, 4], [2, 1], [2, 3], [2, 4], [3, 1], [3, 2], [3, 4], [4, 1], [4, 2], [4, 3]]$$

Try

```
> permute(4);
> permute([a,b,c,d],3);
> permute([a,a,c,d],3);
```

The `randperm` function generates a random permutation. Try

```
> randperm([a,b,c,d]);
```

The number of permutations of n objects taken r at a time is given by `numbperm(`n`,`r`)`. Try

```
> numbperm(12,5);
> 12!/7!;
```

Combinations

The `choose` function is used to generate combinations. Its usage is similar to the `permute` function. We generate the combinations of $1, 2, 3, 4$ taken two at a time.

```
> with(combinat):
> choose(4,2);
```

$$[[1, 2], [1, 3], [1, 4], [2, 3], [2, 4], [3, 4]]$$

We see that there are six combinations. Try

```
> choose([a,b,c,d,e,f],3);
> choose([a,a,c,d,d,d],3);
```

The number of combinations of n objects taken r at a time is given by `numbcomb(n, r)`. The function `randcomb` is used to generate random combinations.

```
> with(combinat):
> numbcomb(12,5);
> 12!/5!/7!;
> randcomb(12,5);
> randcomb([a,b,c,d,e],3);
```

Partitions and compositions

A partition of a positive integer n is a representation of n as a sum of positive integers disregarding order. The function `partition(n)` generates the partitions of n. Each partition is represented by a list of its summands. We compute the partitions of 6.

```
> with(combinat):
> P:=partition(6);
```

$$P := [[1,1,1,1,1,1],[1,1,1,1,2],[1,1,2,2],[2,2,2],[1,1,1,3],[1,2,3],[3,3],\\ [1,1,4],[2,4],[1,5],[6]]$$

```
> nops(P);
```

$$11$$

There are 11 partitions of 6. They are

$$1+1+1+1+1+1,\ 1+1+1+1+2,\ 1+1+2+2,\ 2+2+2,\\ 1+1+1+3,\ 1+2+3,\ 3+3,\ 1+1+4,\ 2+4,\ 1+5,\ 6.$$

The `numbpart` function computes the number of partitions of an integer. We compute the first few terms of the generating function $P(q)$ for $p(n)$, the number of partitions of n.

```
> P := sum(numbpart(n)*q^n,n=0..20);
```

$$1+q+2\,q^2+3\,q^3+5\,q^4+7\,q^5+11\,q^6+15\,q^7+22\,q^8+30\,q^9+42\,q^{10}\\ +56\,q^{11}+77\,q^{12}+101\,q^{13}+135\,q^{14}+176\,q^{15}+231\,q^{16}+297\,q^{17}\\ +385\,q^{18}+490\,q^{19}+627\,q^{20}$$

```
> series(1/P,q,21);
```

$$1-q-q^2+q^5+q^7-q^{12}-q^{15}+O\left(q^{21}\right)$$

What do you notice about the series expansion for the reciprocal of the generating function?

It is possible to iterate through the set of partitions. See ?decodepart, ?encodepart, ?firstpart, ?nextpart, ?prevpart, and ?lastpart for more information. The function randpart(n) generates a random partition of n. For a partition p, conjpart(p) gives the conjugate partition of p. For example, we compute the conjugate partition of $\pi : 1+1+1+3+5+5+12+21+21$.

```
> p:= [1,1,1,3,5,5,12,21,21]:
> conjpart(p);
```

$$[2, 2, 2, 2, 2, 2, 2, 2, 2, 3, 3, 3, 3, 3, 3, 3, 5, 5, 6, 6, 9]$$

The conjugate partition is

$$\pi' : 2+2+2+2+2+2+2+2+2+3+3+3+3+3+3+3+5+5+6 \\ +6+9.$$

A composition of a positive integer n is a like a partition of n except that order counts. The function composition(n,k) generates the compositions of n with k parts. We compute the compositions of six with three parts:

```
> with(combinat):
> composition(6,3);
```

$$\{[4, 1, 1], [3, 2, 1], [2, 3, 1], [1, 4, 1], [3, 1, 2], [2, 2, 2], [1, 3, 2], [2, 1, 3], \\ [1, 2, 3], [1, 1, 4]\}$$

```
> nops(%);
```

$$10$$

There are ten compositions of six with three parts. They are

$$4+1+1,\ 3+2+1,\ 2+3+1,\ 1+4+1,\ 3+1+2,\ 2+2+2,\ 1+3+2, \\ 2+1+3,\ 1+2+3,\ 1+1+4.$$

The function numbcomp(n,k) returns that number of compositions of n with k parts. Try

```
> numbcomp(6,3);
```

Sets

For a positive integer n, the function powerset(n) generates all subsets of the integers 1, 2, ... , n. Let's try $n = 4$.

```
> with(combinat):
> powerset(4);
```

$$\{\{\}, \{1\}, \{1,2,3,4\}, \{2,3,4\}, \{3,4\}, \{1,3,4\}, \{4\}, \{1,4\}, \{2,4\}, \{1,2,4\}, \{2\}, \{1,2\}, \{3\}, \{1,3\}, \{2,3\}, \{1,2,3\}\}$$

```
> nops(%);
```

$$16$$

The `subsets` function is similar to the `cartprod` function. The `cartprod` allows iteration through the elements of a Cartesian product. The `subsets` function allows iteration through the powerset of a given set. Try

```
> PS := subsets({a,b,c,d});
> while not PS[finished] do
      PS[nextvalue]();
  end do;
```

Lists

Let $\mathbb{N} = \{0, 1, 2, \dots\}$ be the set of natural numbers. There is a canonical bijection between lists of natural numbers of a fixed length and $\mathbb{N}$. The function `inttovec(`m`,`n`)` returns the mth vector of length n. For a list `L`, `vectoint(L)` returns the corresponding natural number. We calculate the first six lists of length 3.

```
> with(combinat):
> for j from 1 to 6 do
      inttovec(j,3);
  end do;
```

$$[1, 0, 0]$$
$$[0, 1, 0]$$
$$[0, 0, 1]$$
$$[2, 0, 0]$$
$$[1, 1, 0]$$
$$[1, 0, 1]$$

Other functions

<u>`Chi`</u>

Let n be a positive integer and suppose λ and ρ are partitions of n. `Chi(`λ`,`ρ`)` computes the trace on any matrix in the conjugacy class corresponding to the partition ρ in the irreducible representation of S_n corresponding to the partition λ. Try

```
> with(combinat):
> lambda := [1,1,3]:
> rho := [1,1,1,1,1]:
> Chi(lambda,rho);
```

bell(n)

Returns the nth Bell number B_n.

binomial

For integers $0 \le r \le n$, binomial(n,r) returns the binomial coefficient $\binom{n}{r}$.

character

character(n) returns the character table for the symmetric group S_n. Try

```
> with(combinat):
> character(5);
```

fibonacci

fibonacci(n) gives the nth Fibonacci number. There is a related polynomial given by fibonacci(n,x).

graycode

Let n be a positive integer. graycode(n) returns a list of all 2^n n-bit integers in Gray code order, i.e., consecutive integers in the list differ in only one place in their binary form. Try

```
> with(combinat):
> g := graycode(4);
> printf(cat(` %.4d`$16,`\n`), op(map(convert, g, binary)));
```

multinomial

For natural numbers n, n_1, n_2, ... , n_k, where $n_1+\cdots+n_k = n$, multinomial(n, n_1, n_2, ... , n_k) returns the multinomial coefficient

$$\binom{n}{n_1, n_2, \ldots, n_k} = \frac{n!}{n_1!n_2!\ldots!n_k!}.$$

Try

```
> with(combinat):
> multinomial(12,3,4,5);
> 12!/3!/4!/5!;
```

stirling1(n,k)
stirling2(n,k)

Returns the Stirling number of the first and second kind, usually denoted by $s(n,k)$, and $S(n,k)$, respectively. Here n, k are integers satisfying $0 \le k \le n$.

17.4.2 The *networks* package

The *networks* package is used for for drawing graphs and doing graph theory calculations. The package contains the following functions:

```
acycpoly       addedge       addvertex       adjacency
allpairs       ancestor      arrivals        bicomponents
charpoly       chrompoly     complement      complete
components     connect       connectivity    contract
countcuts      counttrees    cube            cycle
cyclebase      daughter      degreeseq       delete
departures     diameter      dinic           djspantree
dodecahedron   draw          duplicate       edges
ends           eweight       flow            flowpoly
fundcyc        getlabel      girth           graph
graphical      gsimp         gunion          head
icosahedron    incidence     incident        indegree
induce         isplanar      maxdegree       mincut
mindegree      neighbors     new             octahedron
outdegree      path          petersen        random
rank           rankpoly      shortpathtree   show
shrink         span          spanpoly        spantree
tail           tetrahedron   tuttepoly       vdegree
vertices       void          vweight
```

Undirected graphs

We define a new graph G using the `new` function. We use the `addvertex` function to define six vertices in the graph G.

```
> with(networks):
> G:=new():
> addvertex({A,B,C,D,E,F},G);
```

$$E, F, A, B, C, D$$

In MAPLE, undirected edges correspond to two element sets of vertices. The `addedge` function is used to define edges in the graph. We define seven edges: AB, BC, DF, EF, AD, CE, and AC:

```
> addedge([{A,B},{B,C},{D,F},{E,F},{A,D},{C,E},{A,C}],G);
```

$$e1, e2, e3, e4, e5, e6, e7$$

We can draw the graph using the `draw` function. See Figure 17.1.

```
> draw(G);
```

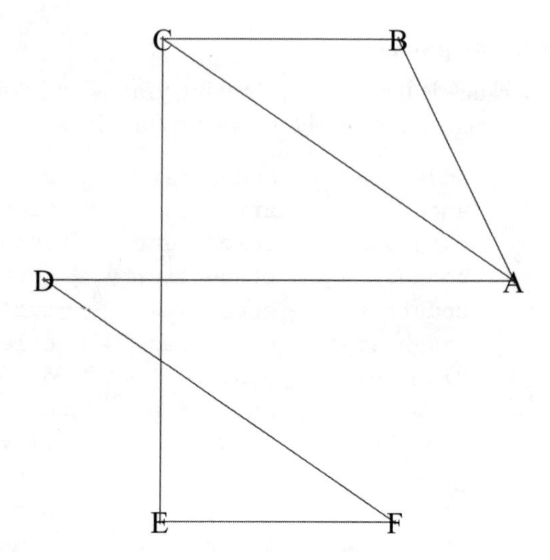

Figure 17.1 An undirected graph with six vertices.

To obtain detailed information about a graph, try the `show` function:

```
> show(G);
```

This information is not completely user-friendly, but it can be deciphered with not too much trouble. To find the ends of an edge e in a graph G use `ends(`e`,`G`)`.

```
> ends(e1,G);
```

$$\{A, B\}$$

For example, `ends(e1,G)` returned $\{A, B\}$. This means that edge `e1` is joined by the vertices A and B. To obtain all the vertices in a graph G, use `vertices(`G`)`; to get the edges, use `end(`G`)`. Try

```
> vertices(G);
> ends(G);
```

We can make a copy of a graph using the `duplicate` function.

```
> H := duplicate(G):
```

We can delete edges and vertices using the `delete` function. We delete the vertex A from the graph G and the edge `e1` from the graph H.

```
> delete(A,G):
> draw(G);
> delete(e1,H):
> draw(H);
```

Because the edge `e1` joins vertices A and B, we could have deleted it from H using `delete({A,B},H)`. There are many other functions. See the table at the beginning of this section.

Directed graphs

A directed graph is a graph wherein each edge has a head and a tail. The directed edge from vertex A to vertex B is represented by the list `[`A`,`B`]`. We construct a graph G with four vertices A, B, C, D and directed edges AB, BA, BC, and CD:

```
> with(networks):
> G:=new():
> addvertex({A,B,C,D},G):
> addedge([[A,B],[B,A],[B,C],[C,D]],G);
```

$$e1, e2, e3, e4$$

For an edge e in a directed graph G, `head(`e`,`G`)`, `tail(`e`,`G`)` return the head and tail of e, respectively. Try

```
> head(e4,G);
> tail(e4,G);
```

Weighted graphs and flows

A weighted graph is a graph wherein weights are attached to edges. We construct a weighted directed graph G with edges AB, AC, AD, BC, BE, CE, DE and corresponding weights 7, 3, 10, 5, 6, 4, 12 using the `addedge` function.

```
> with(networks):
> G := new():
> addvertex({A,B,C,D,E},G):
> addedge([[A,B],[A,C],[A,D],[B,C],[B,E],[C,E],[D,E]],
    names=[AB,AC,AD,BC,BE,CE,DE],
    weights=[7,3,10,5,6,4,12],G);
```

$$AB, AC, AD, BC, BE, CE, DE$$

Notice how we used the `names` option to name the edges. We can obtain the weights by using the `eweight` function.

```
> eweight(G);
```

$$\texttt{table}([BE = 6, BC = 5, CE = 4, AB = 7, DE = 12, AC = 3, AD = 10])$$

In general, to add weights w_1, w_2, ..., w_k to e_1, e_2, ..., e_k to a graph G, use the command

`addedge([`e_1`, `e_2`, ..., `e_k`], weights=[`w_1`, `w_2`, ..., `w_k`]).`

Here the edges e_j are either two elements sets (undirected) or lists (directed).

Edge weights can be interpreted as capacities in a network flow problem. The weight of a directed edge corresponds to the capacity or maximum flow along that edge. In a flow problem the source is the vertex where the flow starts, and the sink is the vertex where the flow ends. We calculate the maximum flow of the network corresponding to the weighted directed graph given above, with source A and sink E.

```
> flow(G,A,E);
```

$$20$$

The maximum flow is 20; i.e., the total flow out of the source (or into the sink) is 20. For a weighted directed graph G with source s and sink t, the maximum flow is given by `flow(`G`,`s`,`t`)`. Saturated edges are edges wherein flow has reached capacity. These can be found by adding an extra name to the argument of the `flow` function:

```
> flow(G,A,E,sateds):
> sateds;
```

$$\{\{A,B\},\{A,C\},\{A,D\},\{E,B\},\{E,C\}\}$$

We see that all edges are saturated except BC and DE. The call `flow(G,A,E, sateds)` assigned the name `sateds` to the saturated edges.

It also possible to add weights to vertices using the addvertex function in a similar way. As an example, we add the weights 0, 1, 2 to the vertices A, B, C of a graph G. Try

```
> with(networks):
> G := new():
> addvertex({A,B,C},weights=[0,1,2],G);
> addedge([[A,B],[B,C],[A,C]],G);
> vweight(G);
```

There are many other functions for doing graph and network calculations. They are given at the beginning of this section. Use the help facility to find more information.

17.4.3 The *combstruct* package

The *combstruct* package is used to define and manipulate more abstract and general combinatorial structures than are available in the *combinat* and *networks* package. The package contains the following functions:

```
allstructs   count     draw          finished     gfeqns
gfseries     gfsolve   iterstructs   nextstruct
```

The `allstructs` function lists all the combinatorial objects with given specifications of a given size. The `count` function counts objects of a given size. The `draw` function returns a random combinatorial object of a given class. The functions `gfeqns`, `gfseries`, `gfsolve` are used to define and solve the associated generating function problem. The functions `finished`, `iterstructs`, `nextstruct` are used to iterate through combinatorial structures. For a nice introduction to this package see the Web site:

http://algo.inria.fr/libraries/autocomb/autocomb.html
which is the *Studies in Automatic Combinatorics* page of the *Algorithms Project* at the Institut National de Recherche en Informatique et en Automatique (INRIA), Le Chesnay, France. Click on Introductory worksheets.

17.5 Number theory

The main package for doing number theory is *numtheory*. Some other packages are *GaussInt* and *padic*.

17.5.1 The *numtheory* package

The *numtheory* package contains the following functions:

```
GIgcd        L           bigomega         cfrac       cfracpol
cyclotomic   divisors    factorEQ         factorset   fermat
imagunit     index       integral_basis   invcfrac    invphi
issqrfree    jacobi      kronecker        lambda      legendre
mcombine     mersenne    minkowski        mipolys     mlog
mobius       mroot       msqrt            nearestp    nthconver
nthdenom     nthnumer    nthpow           order       pdexpand
phi          pi          pprimroot        primroot    quadres
rootsunity   safeprime   sigma            sq2factor   sum2sqr
tau          thue
```

Divisors and factors

For an integer n, `ifactor(`n`)` gives the prime factorization of n. We find the prime factorization of 144.

```
> n := 144;
```

$$144$$

```
> ifactor(n);
```

$$(2)^4\,(3)^2$$

The `ifactor` function is a standard MAPLE function and is not part of the *numtheory* package. The *numtheory* functions `divisors(`n`)` and `factorset(`n`)` return the set of positive divisors and the set of prime divisors of n, respectively. The function `tau(`n`)` gives the number of (positive) divisors of n. We return to our example with $n = 144$.

```
> with(numtheory):
> n := 144;
```

$$144$$

```
> divs := divisors(n);
```

$$\{1, 2, 3, 4, 6, 8, 9, 12, 16, 18, 24, 36, 48, 72, 144\}$$

```
> factorset(n);
```

$$\{2,3\}$$

```
> tau(n);
```

$$15$$

```
> nops(divs);
```

$$15$$

We see that 144 has 15 divisors: 1 2, 3, 4, 6, ... 72, 144, and two prime divisors: 2 and 3.

The function `factorEQ(`n`,`d`)` gives a factorization of n in the ring of integers of the quadratic field $\mathbb{Q}[\sqrt{d}]$ when the ring is norm-Euclidean. This happens when $d = -11, -7, -3, -2, -1, 2, 3, 5, 6, 7, 11, 13, 17, 19, 21, 29, 33, 37, 41, 55, 73$. As an example, we factor 30 in the ring of integers in the field $\mathbb{Q}[\sqrt{-7}]$.

```
> with(numtheory):
> factorEQ(30,-7);
```

$$\left(\frac{1}{2} + 1/2\, I\sqrt{7}\right)\left(\frac{1}{2} + -\frac{1}{2}\, I\sqrt{7}\right)(3)\,(5)$$

In this ring 3 and 5 are primes, and 30 factors

$$30 = z \cdot \overline{z} \cdot 3 \cdot 5,$$

where $z = (1 + i\sqrt{7})/2$. The function `sq2factor(`z`)` gives a factorization of an integer z in the field $\mathbb{Q}[\sqrt{2}]$. The factorization is given as a product of units and primes.

The function `nthpow(`n`, `m`)` returns b^n where b is the largest integer such that $b^n \mid m$.

```
> with(numtheory):
> nthpow(23152500);
```

Primes

See Section 3.3.4 for the functions `ithprime`, `isprime`, `nextprime`, `prevprime`. Above we saw how the function `factorset` gives the prime divisors of an integer. The function `mersenne([`n`])` gives the nth Mersenne prime.

```
> with(numtheory):
> p := mersenne([10]);
```

$$618970019642690137449562111$$

```
> ifactor(p+1);
```

$$(2)^{89}$$

The 10th Mersenne prime is

$$p = 618970019642690137449562111 = 2^{89} - 1.$$

The function `mersenne(`n`)` tests whether the integer $2^n - 1$ is prime.

For an integer n, `pi(`n`)` gives the number of primes less than or equal to n. This is usually denoted by $\pi(n)$. We plot the ratio

$$\frac{\pi(n)}{n/\ln n},$$

for $2 \leq n \leq 1000$. See Figure 17.2.

```
> with(numtheory):
> L := [seq([n,pi(n)/(n/log(n))],n=2..1000)]:
> plot(L, style=point,labels=[n," "]);
```

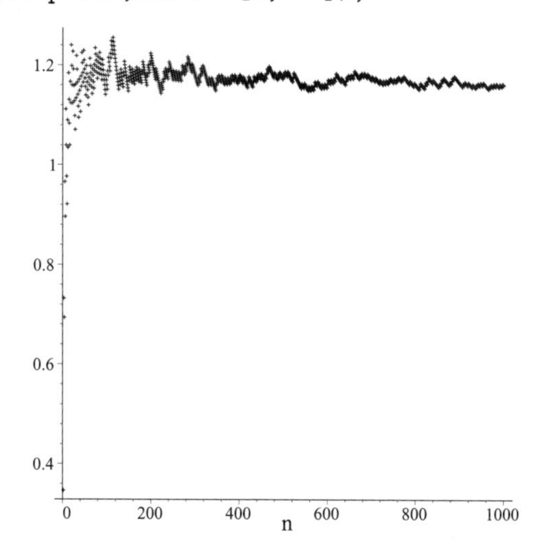

Figure 17.2 A plot of $\pi(n)/(n/\ln n)$.

A safe prime is a prime p such that $(p-1)/2$ is also prime. The function `safeprime(`n`)` returns the smallest safe prime larger than n. Try

```
> with(numtheory):
> safeprime(100);
```

The `bigomega` function computes the number of prime divisors of an integer n counted with multiplicity.

mod n

Calculations modulo n are done using the `mod` and `modp` functions. See sections 3.3.6 and 17.1 for some examples. We describe some functions in the *numtheory* package for doing further calculations mod n.

The residue classes relatively prime to n form a group modulo n. This group is cyclic if $n = p^e$, $2p^e$, 2 or 4, where p is an odd prime. The generator of this cyclic group is called a primitive root modulo n. The function `primroot(`n`)` finds a primitive root modulo n if it exists. The call `primroot(`m`, `n`)` finds the smallest primitive root mod n greater m if it exists. We find a primitive root mod 17.

```
> with(numtheory):
> g := primroot(17);
```

$$3$$

```
> [seq(modp(g^k,17),k=1..16)];
```

$$[3, 9, 10, 13, 5, 15, 11, 16, 14, 8, 7, 4, 12, 2, 6, 1]$$

```
> sort(%);
```

$$[1, 2, 3, 4, 5, 6, 7, 8, 9, 10, 11, 12, 13, 14, 15, 16]$$

3 is a primitive root mod 17. We verified that 3^k gives all the nonzero residue classes mod 17.

Suppose m, n are relatively prime. The function `order(`m`,`n`)` returns the smallest positive integer d such that

$$m^d \equiv 1 \pmod{n}.$$

We compute the order of 2 mod 17.

```
> with(numtheory):
> order(2,17);
```

$$8$$

```
> modp(2^8,17);
```

$$1$$

The order of 2 mod 17 is 8.

Suppose g is a primitive root mod n, and a, n are relatively prime. The function `index(`a`,`g`,`n`)` returns the smallest natural number j such that

$$g^j \equiv a \pmod{n}.$$

This is called the index of a mod n (relative to g). We consider an example mod 47.

```
> with(numtheory):
> g := primroot(47);
```

$$5$$

```
> d := index(6,5,47);
```

$$38$$

```
> modp(g^d,47);
```

$$6$$

5 is a primitive root mod 47. The index of 6 mod 47 with respect to this primitive root is 38.

The function `mroot(`r`,`a`,`n`)` computes an rth root of a mod m, if it exists, i.e., it tries to find an integer x such that

$$x^r \equiv a \pmod{n}.$$

We compute a 5th root of 2 mod 13.

```
> with(numtheory):
> mroot(2,5,13);
```

$$6$$

```
> modp(6^5,13);
```

$$2$$

6 is a 5th root of 2 mod 13. The call `msqrt(`a`, `n`)` tries to find a square root of a mod n. For a prime p, the function `rootsunity(`p`,`n`)` finds all pth roots of unity mod n. We find the 7th roots of unity mod 43.

```
> with(numtheory):
> rootsunity(7,43);
```

$$1, 4, 11, 16, 21, 35, 41$$

```
> map(`^`,[%],7);
```

$$[1, 16384, 19487171, 268435456, 1801088541, 64339296875, 194754273881]$$

```
> map(modp,%,43);
```

$$[1, 1, 1, 1, 1, 1, 1]$$

They are 1, 4, 11, 16, 21, 35, 41. For each number r, we verified that $r^7 \equiv 1 \pmod{43}$.

For a prime p, the Legendre symbol $\left(\frac{a}{p}\right)$ is given in MAPLE by `legendre(`a`,`p`)`, or `L(`a`,`p`)`. We compute the Legendre symbol of 15 mod 23.

```
> with(numtheory):
> legendre(15,23);
```

$$-1$$

```
> msqrt(15,23);
```

$$FAIL$$

We found that

$$\left(\frac{15}{23}\right) = -1.$$

This is confirmed by the fact that 15 does not have a square root mod 23. For a, b relatively prime and b a positive odd integer, the more general Jacobi symbol $\left(\frac{a}{b}\right)$ is computed in MAPLE using `jacobi(a,b)`. The function `quadres(a,b)` returns 1 if a has a square root mod b and -1 otherwise. The function `imagunit(n)` tries to find a square root of -1 mod n. We consider $n = 41$.

```
> with(numtheory):
> legendre(-1,41);
```

$$1$$

```
> imagunit(41);
```

$$9$$

```
> modp(%^2,41);
```

$$40$$

We see that $\left(\frac{-1}{41}\right) = 1$, so that -1 has a square root mod 41. 9 is a square root of -1 mod 41.

Let m_1, m_2 be positive integers and a_1, a_2 be two integers. The command `mcombine(`m_1`,`a_1`,`m_2`,`a_2`)` attempts to find a solution x to the congruences

$$x \equiv a_1 \pmod{m_1},$$
$$x \equiv a_2 \pmod{m_2}.$$

We find a solution to the congruences

$$x \equiv 3 \pmod{4},$$
$$x \equiv 4 \pmod{5}.$$

```
> with(numtheory):
> mcombine(4,3,5,4);
```

$$19$$

By the Chinese remainder theorem, the solutions are given by $x \equiv 19 \pmod{20}$. A more general use of the Chinese remainder theorem can be done using the `chrem` function, which is not part of the *numtheory* package. To solve the simultaneous congruences

$$x \equiv a_i \quad \pmod{m_i},$$

for $1 \le i \le k$, use the command `chrem(`$[a_1, a_2, \ldots, a_k], [m_1, m_2, \ldots, m_k]$`)`.

Continued fractions

For a constant c, `cfrac(c,n)` computes the first $n + 1$ quotients in the simple continued fraction. We compute the first seven quotients of the continued fraction expansion of e.

```
> with(numtheory):
> cfrac(exp(1),6);
```

$$2+\cfrac{1}{1+\cfrac{1}{2+\cfrac{1}{1+\cfrac{1}{1+\cfrac{1}{4+\cfrac{1}{1+\cdots}}}}}}$$

We can get this in list form.

```
> with(numtheory):
> cfrac(exp(1),30,'quotients');
```

$$[2,1,2,1,1,4,1,1,6,1,1,8,1,1,10,1,1,12,1,1,14,1,1,16,1,1,18,1,1,20,1,\ldots]$$

Do you see a pattern? The `cfrac` function can also handle rational functions in one variable. Try out the following example.

```
> with(numtheory):
> y := product( 1-q^(5*n-1))*(1-q^(5*n-4))/(1-q^(5*n-2))
    /(1-q^(5*n-3)),n=1..10):
> cfrac(y,q,6);
```

$$1+\cfrac{1}{-q-1+\cfrac{1}{-q+\cfrac{1}{-q^2+\cfrac{1}{-q^2+\cfrac{1}{-q^3+\cfrac{1}{-q^3+\cdots}}}}}}$$

Do you see a pattern? See `?cfrac` for more options when using the `cfrac` function. The `cfracpol` function will return a continued fraction expansion for each real root of a specified polynomial. Also see the functions `nthconv`, `nthdenom`, `nthnumer`, and `invcfrac`.

Other functions

<u>`cyclotomic`</u>

`cyclotomic(`n`,`x`)` returns the nth cyclotomic polynomial.

fermat

`fermat(`n`)` returns the nth Fermat number $2^{2^n}+1$, for $n < 20$. The form `fermat(`n`,w)` gives further information about a particular Fermat number.

GIgcd

Computes the gcd of Gaussian integers.

invphi

This is the inverse totient function. `invphi(`m`)` will a list of integers n such that $\phi(n) = m$.

issqrfree

Tests whether an integer is square-free.

lambda

This is Carmichael's lambda function. `lambda(`n`)` returns the order of the largest cyclic subgroup of $\mathbb{Z}_n$.

minkowski

The `minkowski` function is used to solve certain Diophantine inequalities.

mipolys

For a prime p, `mipolys(`n`,`p`,`m`)` returns the number of monic irreducible polynomials of degree n over the finite field with p^m elements.

mobius

`mobius(`n`)` returns the value of the Möbius function $\mu(n)$.

nearestp

Returns the nearest lattice point to a specified point for a given n-dimensional real lattice.

pdexpand

For a rational number q, `pdexpand(`q`)` returns the periodic decimal expansion of q.

phi

`phi(`n`)` is the value of Euler's totient function $\phi(n)$.

sigma

`sigma[`k`](`n`)` returns the sum of the kth power of the positive divisors of n:

$$\sigma_k(n) = \sum_{d|n} d^k.$$

sum2sqr

`sum2sqr(n)` returns pairs of integers $[a, b]$ such that

$$n = a^2 + b^2.$$

thue

The `thue` function is used to find integer solutions to equations $p(x, y) = m$, or inequalities of the form $p(x, y) \leq m$, where $p(x, y)$ is an irreducible homogeneous polynomial.

17.5.2 The *GaussInt* package

Gaussian integers have the form $a + b\,i$, where a, b are integers. The set of Gaussian integers is denoted by $\mathbb{Z}[i]$. The *GaussInt* package is used for doing calculations in this ring. It contains the following functions:

```
GIbasis     GIchrem     GIdivisor   GIfacpoly     GIfacset
GIfactor    GIfactors   GIgcd       GIgcdex       GIhermite
GIissqr     GIlcm       GImcmbine   GInearest     GInodiv
GInorm      GInormal    GIorder     GIphi         GIprime
GIquadres   GIquo       GIrem       GIroots       GIsieve
GIsmith     GIsqrfree   GIsqrt      GIunitnormal
```

See `?GaussInt` for more information.

17.5.3 p-adic numbers

The *padic* package is used for doing calculations with p-adic numbers. It contains the following functions

```
arccoshp    arccosp     arccothp    arccotp     arccschp
arccscp     arcsechp    arcsecp     arcsinhp    arcsinp
arctanhp    arctanp     coshp       cosp        cothp
cotp        cschp       cscp        evalp       expansion
expp        lcoeffp     logp        orderp      ordp
ratvaluep   rootp       sechp       secp        sinhp
sinp        sqrtp       tanhp       tanp        valuep
```

Many of these functions are p-adic counterparts of the corresponding real-valued functions. For example, for a prime p, `sinp(x,p)` is the p-adic analog of `sin(x)`. For p prime and z a nonzero integer, define $\text{ord}_p(z)$ as the largest integer a such that $p^a \mid z$. For a nonzero rational $x = r/s$, where r, s are integers, define

$$\text{ord}_p(x) = \text{ord}_p(r) - \text{ord}_p(s).$$

The p-adic absolute value is defined by

$$|x|_p = \frac{1}{p^{\text{ord}_p(x)}}.$$

Define $|0|_p = 0$. Then $|\ |_p$ defines a non-Archimedean metric on $\mathbb{Q}$. The completion of $\mathbb{Q}$ is denoted by $\mathbb{Q}_p$ and is called the set of p-adic numbers. Any p-adic number x has an expansion

$$x = \sum_{j=m}^{\infty} a_j p^j,$$

where $m \in \mathbb{Z}$, each $0 \le a_j < p$, and the convergence is relative to $|\ |_p$. In MAPLE, $\text{ord}_p(x)$ is given by `ordp(x,p)`.

```
> with(padic):
> r := 1705725;
```

$$1705725$$

```
> s := 5561328861;
```

$$5561328861$$

```
> ifactor(r);
```

$$(3)^3\,(5)^2\,(7)\,(19)^2$$

```
> ifactor(s);
```

$$(3)^7\,(11)\,(19)\,(23)^3$$

```
> ordp(r,3);
```

$$3$$

```
> ordp(s,3);
```

$$7$$

```
> x:=r/s;
```

$$\frac{3325}{10840797}$$

```
> ordp(x,3);
```

$$-4$$

For $x = 1705725/5561328861 = 3325/10840797$, we see that $\text{ord}_3(x) = -4$. In MAPLE, $|x|_p$ is given by `valuep(x,p)`.

```
> valuep(x,3);
```

$$3^4$$

We see that $|x|_3 = 3^4$. To compute the first T terms in the p-adic expansion of a rational number x, $j < T$, we use the command `evalp(x,p,T)`. We compute the 3-adic expansion of our rational number x:

```
> x;
```

$$\frac{3325}{10840797}$$

```
> evalp(x,3);
```

$$3^{-4} + 2\,3^{-3} + 2\,3^{-2} + 3^{-1} + 2 + 3 + 2\,3^2 + 2\,3^3 + 3^4 + O(3^6)$$

We see that

$$\frac{3325}{10840797} = (1)3^{-4} + (2)\,3^{-3} + (2)\,3^{-2} + (1)\,3^{-1} + (2)\,3^{0} + (1)\,3^{1} + (2)\,3^{2} + (2)\,3^{3} + (1)\,3^{4} + (0)\,3^{5} + \ldots.$$

By default, MAPLE returns the first ten terms in the expansion. To compute more terms try

```
> evalp(x,3,20);
```

In MAPLE, a p-adic number may be defined in terms of the coefficients in its p-adic expansion. The p-adic number

$$x = a_m p^m + a_{m+1} p^{m+1} + a_n p^n + O(p^{n+1})$$

is represented in MAPLE by

p_adic(p, m, [a_m, a_{m+1}, ..., a_n]).

As an example, we consider the 3-adic expansion we found earlier.

```
> with(padic):
> px := p_adic(3,-4,[1,2,2,1,2,1,2,2,1,0]);
```

$$\text{p\_adic}(3, -4, [1, 2, 2, 1, 2, 1, 2, 2, 1, 0])$$

```
> evalp(px,3);
```

$$3^{-4} + 2\,3^{-3} + 2\,3^{-2} + 3^{-1} + 2 + 3 + 2\,3^{2} + 2\,3^{3} + 3^{4} + O(3^{6})$$

MAPLE is able to compute p-adic expansion of values of p-adic analogues of analytic functions. For example, the square root of p-adic number can be computed using the sqrtp function.

```
> with(padic):
> z := sqrtp(10,3);
```

$$z := 1 + 2\,3^{2} + 3^{3} + 2\,3^{4} + 3^{5} + 3^{8} + O(3^{9})$$

```
> evalp(z^2,3);
```

$$1 + 3^{2} + O(3^{9})$$

```
> sqrtp(2,3);
```

FAIL

We see that 10 has a 3-adic square root and

$$\sqrt{10} = 1 + 2\,3^{2} + 3^{3} + 2\,3^{4} + 3^{5} + 3^{8} + \cdots$$

Since 2 is a quadratic nonresidue mod 3, naturally enough 2 does not have a 3-adic square root. To see a complete list of such p-adic functions type

```
> ?padic,function
```

17.6 Numerical approximation

The *numapprox* package is used for numerical approximation of functions. It contains the following functions

`chebdeg`	`chebmult`	`chebpade`	`chebsort`	`chebyshev`
`confracform`	`hermite_pade`	`hornerform`	`infnorm`	`laurent`
`minimax`	`pade`	`remez`		

Let m, n be positive integers. The $[m/n]$-Padé approximant of a function $f(x)$ is a rational function $R(x) = \frac{p_m(x)}{q_n(x)}$, where $p_m(x)$, $q_n(x)$ are polynomials of degrees at most m, n respectively, such that the Maclaurin expansion of $f(x)$ agrees with that of $R(x)$ as much as possible. It is given in MAPLE by `pade(`$f(x)$`,`x`,[`m`,`n`])`. We compute the $[3,3]$-Padé approximant of e^x.

```
> with(numapprox):
> f := exp(x):
> padf := pade(f,x,[3,3]);
```

$$\frac{1+\frac{1}{2}\,x+\frac{1}{10}\,x^2+\frac{1}{120}\,x^3}{1-\frac{1}{2}\,x+\frac{1}{10}\,x^2-\frac{1}{120}\,x^3}$$

```
> taylor(f-padf,x,10);
```

$$-\frac{1}{100800}x^7-\frac{1}{100800}x^8-\frac{97}{18144000}x^9+O\left(x^{10}\right)$$

We see that the $[3,3]$-Padé approximant of e^x is

$$\frac{1+\frac{1}{2}\,x+\frac{1}{10}\,x^2+\frac{1}{120}\,x^3}{1-\frac{1}{2}\,x+\frac{1}{10}\,x^2-\frac{1}{120}\,x^3},$$

and this agrees with the Maclaurin expansion of e^x to $O(x^7)$.

We give a brief description of other functions in the package:

`chebdeg`	Degree of a polynomial in Chebyshev form
`chebmult`	Multiply two Chebyshev series
`chebpade`	Compute a Chebyshev-Padé approximation
`chebsort`	Sort the terms in a Chebyshev series
`chebyshev`	Chebyshev series expansion
`confracform`	Convert a rational function to continued-fraction form
`hermite_pade`	Compute a Hermite-Padé approximation
`hornerform`	Convert a polynomial to Horner form
`infnorm`	Compute the L-infinity norm of a function
`laurent`	Laurent series expansion
`minimax`	Minimax rational approximation
`remez`	Remez algorithm for minimax rational approximation

17.7 Miscellaneous packages

17.7.1 The *algcurves* package

The *algcurves* package is used for doing algebraic curve calculations. Functions include

`algfun_series_sol`	Find series solutions with nice coefficients
`differentials`	Holomorphic differentials of an algebraic curve
`genus`	The genus of an algebraic curve
`homogeneous`	Make a polynomial in two variables homogeneous in three variables
`homology`	Tretkoff's algorithm for finding a canonical homology basis
`implicitize`	Find an implicit equation for curve or surface
`integral_basis`	Compute an integral basis for an algebraic function field
`is_hyperelliptic`	Test if an algebraic curve is hyperelliptic
`j_invariant`	The j-invariant of an elliptic curve
`monodromy`	Compute the monodromy of an algebraic curve
`parametrization`	Find a parametrization for a curve with genus 0
`periodmatrix`	Compute the periodmatrix of an algebraic curve
`plot_knot`	Make a tubeplot for a singularity knot
`puiseux`	Determine the Puiseux expansions of an algebraic function
`singularities`	The singularities of an algebraic curve
`Weierstrassform`	Normal form for elliptic or hyperelliptic curves

The functions `algfun_series_sol` and `implicitize` are new to MAPLE 7.

17.7.2 The *codegen* package

The code generation package is *codegen*. It contains a collection of tools for creating, manipulating, and translating MAPLE procedures into other programming languages. In Chapter 7, we used the `C` and `fortran` functions for generating C and Fortran code. The package also includes tools for automatic differentiation of MAPLE procedures, code optimization, and an operation count of a MAPLE procedure. Functions include

`C`	Generate C code
`cost`	Operation evaluation count
`declare`	Declare the type of a parameter
`dontreturn`	Don't return a value from a MAPLE procedure
`eqn`	Produce output suitable for troff/eqn printing
`fortran`	Generate Fortran code
`GRADIENT`	Compute the Gradient of a MAPLE procedure
`HESSIAN`	Compute the Hessian matrix of a MAPLE procedure
`horner`	Convert formulae in a procedure to horner form
`intrep2maple`	Convert an abstract syntax tree to a MAPLE procedure

JACOBIAN	Compute the Jacobian matrix of a MAPLE procedure
joinprocs	Join the body of two or more MAPLE procedures together
makeglobal	Make a variable to be a global variable
makeparam	Change a variable to be a parameter
makeproc	Make a MAPLE procedure from formulae
makevoid	Don't return any values from a MAPLE procedure
maple2intrep	Convert a MAPLE procedure to an abstract syntax tree
MathML	Convert a MAPLE expression to MathML
optimize	Common subexpression optimization
packargs	Pack parameters of a MAPLE procedure into an array
packlocals	Pack locals of a MAPLE procedure into an array
packparams	Pack parameters of a MAPLE procedure into an array
prep2trans	Prepare a MAPLE procedure for translation
renamevar	Rename a variable in a MAPLE procedure
split	Prepare a MAPLE procedure for automatic differentiation
swapargs	Interchange two arguments in a MAPLE procedure

The MathML function is new to the package.

17.7.3 The *diffalg* package

The *diffalg* package is used for manipulating systems of (ordinary and partial) differential polynomial equations. It includes facilities for the reduction of the differential equations and the development of the solutions into formal power series. Functions include

belongs_to	Test if a differential polynomial belongs to a radical differential ideal
delta_leader	Return the difference of the derivation operators between the leaders of two differential polynomials
delta_polynomial	Return the delta-polynomial generated by two differential polynomials
denote	Convert a differential polynomial from an external form to another
derivatives	Return the derivatives occurring in a differential polynomial
differential_ring	Define a differential polynomial ring endowed with a ranking and a notation
differential_sprem	Sparse pseudo remainder of a differential polynomial
differentiate	Differentiate a differential polynomial
equations	Return the equations of a regular differential ideal
essential_components	Compute a minimal decomposition into regular differential ideals
field_extension	Define a field extension of $\mathbb{Q}$
greater	Compare the rank of two differential polynomials
inequations	Return the inequations of a regular differential

	ideal
initial	Return the initial of a differential polynomial
initial_conditions	Return the list of the initial conditions of a regular differential ideal
is_orthonomic	Test if a regular differential ideal is presented by an orthonomic system of equations
leader	Return the leader of a differential polynomial
power_series_solution	Expand the nonsingular zero of a regular
preparation_polynomial	Preparation polynomial differential ideal into integral power series
print_ranking	Print a message describing the ranking of a differential polynomial ring
rank	Return the rank of a differential polynomial
reduced	Test if a differential polynomial is reduced with respect to a set of differential polynomials
reduced_form	Compute a reduced form of a differential polynomial
rewrite_rules	Display the equations of a regular differential ideal
Rosenfeld_Groebner	Compute a representation of the radical of any finitely generated differential ideal as an intersection of regular differential ideals
separant	Return the separant of a differential polynomial

17.7.4 The *difforms* package

The differential forms package is *difforms.* Functions include

&^	Wedge product
d	Exterior differentiation
defform	Define a constant, scalar, or form
formpart	Find part of an expression that is a form
parity	Parity of an expression
scalarpart	Find part of an expression that is a scalar
simpform	Simplify an expression involving forms
wdegree	Degree of a form

17.7.5 The *Domains* package

New domains of computation can be defined using the *Domains* package. See `?Domains[examples]` for some examples.

17.7.6 The *finance* package

Finance calculations are done using the *finance* package. Functions include

amortization	Amortization table for a loan
annuity	Present value of an annuity
blackscholes	Present value of a call option

`cashflows`	Present value of a list of cash flows
`effectiverate`	Convert a stated rate to the effective rate
`futurevalue`	Future value of an amount
`growingannuity`	Present value of a growing annuity
`growingperpetuity`	Present value of a growing perpetuity
`levelcoupon`	Present value of a level coupon bond
`perpetuity`	Present value of a perpetuity
`presentvalue`	Present value of an amount
`yieldtomaturity`	Yield to maturity of a level coupon bond

17.7.7 The *genfunc* package

The *genfunc* package is used for manipulating rational generating functions (r.g.f.), which are related to sequences satisfying a linear recurrence. Functions include

`rgf_charseq`	Find characteristic sequence of a rational generating
`rgf_encode`	Encode rational generating functions
`rgf_expand`	Expand rational generating functions
`rgf_findrecur`	Find recurrence for terms in a sequence
`rgf_hybrid`	Find generating function of hybrid terms
`rgf_norm`	Normalize a rational generating function
`rgf_pfrac`	Compute complex partial fractions expansion of an r.g.f.
`rgf_relate`	Relate sequences with common factors in their generating
`rgf_sequence`	Extract information about a sequence from its r.g.f.
`rgf_simp`	Simplify an expression involving an r.g.f. sequence
`rgf_term`	Finds values of terms in a sequence
`termscale`	Determine the result of multiplying a generating function by a polynomial

17.7.8 The *geometry* package

The *geometry* package is used for doing two-dimensional Euclidean geometry. In Section 14.2 we used some functions in the *geometry* package to plot regular polygons. There are many other functions in this package:

`altitude`	Find the altitude of a given triangle
`Appolonius`	Find the Appolonius circles of three given circles
`area`	Compute the area of a triangle, square, circle, etc.
`AreCollinear`	Test if three points are collinear
`AreConcurrent`	Test if three lines are concurrent
`AreConcyclic`	Test if four points are concyclic
`AreConjugate`	Test if two triangles are conjugate for a circle
`AreHarmonic`	Test if a pair of points is the harmonic conjugate of another pair
`AreOrthogonal`	Test if two circles are orthogonal to each other
`AreParallel`	Test if two lines are parallel to each other
`ArePerpendicular`	Test if two lines are perpendicular to each other
`AreSimilar`	Test if two triangles are similar

`AreTangent`	Test if a line and a circle or two circles are tangent
`bisector`	Find the bisector of a given triangle
`center`	Find the center of a circle, an ellipse, or a hyperbola
`centroid`	Compute the centroid of a triangle or of a set of points
`circle`	Define a circle
`CircleOfSimilitude`	Find the circle of similitude of two circles
`circumcircle`	Find the circumcircle of a given triangle
`conic`	Define a conic
`convexhull`	Find the convex hull enclosing the given points
`coordinates`	Compute the coordinates of a given point
`CrossProduct`	Compute the cross product of two directed segments
`CrossRatio`	Compute the cross ratio of four points
`DefinedAs`	Return the endpoints or vertices of an object
`detail`	Give a detailed description of an object
`diagonal`	Return the length of the diagonal of a square
`diameter`	Compute the diameter of points on a plane
`dilatation`	Find the dilatation of a geometric object
`distance`	Find the distance between two points, or a point and a line
`draw`	Create a two-dimensional plot of an object
`dsegment`	Define a directed segment
`ellipse`	Define an ellipse
`Equation`	Equation of the geometric object
`EulerCircle`	Find the Euler circle of a given triangle
`EulerLine`	Find the Euler line of a given triangle
`excircle`	Find three excircles of a given triangle
`expansion`	Find the expansion of a geometric object
`ExternalBisector`	Find the external bisector of a given triangle
`FindAngle`	Find the angle between two lines or two circles
`foci`	Find the foci of an ellipse or a hyperbola
`form`	Return the form of the geometric object
`GergonnePoint`	Find the Gergonne point of a given triangle
`GlideReflection`	Find the glide-reflection of a geometric object
`homology`	Find the homology of a geometric object
`homothety`	Find the homothety of a geometric object
`HorizontalCoord`	Compute the horizontal coordinate of a given point
`HorizontalName`	Find the name of the horizontal axis
`hyperbola`	Define a hyperbola
`incircle`	Find the incircle of a given triangle
`intersection`	Find the intersections between two lines, a line and a circle, or two circles
`inversion`	Find the inversion of a point, line, or circle with respect to a given circle
`IsEquilateral`	Test if a given triangle is equilateral
`IsOnCircle`	Test if a point, a list, or set of points is on a circle

`IsOnLine`	Test if a point, a list, or a set of points is on a line
`IsRightTriangle`	Test if a given triangle is a right triangle
`line`	Define a line
`MajorAxis`	Find the length of the major axis of a given ellipse
`MakeSquare`	Construct squares
`medial`	Find the medial triangle of a given triangle
`median`	Find the median of a given triangle
`method`	Return the method of defining a triangle
`midpoint`	Find the midpoint of a segment joining two points
`MinorAxis`	Find the length of the minor axis of a given ellipse
`NagelPoint`	Find the Nagel point of a given triangle
`OnSegment`	Find the point that divides the segment joining two given points by a given ratio
`orthocenter`	Compute the orthocenter of a triangle
`parabola`	Define a parabola
`ParallelLine`	Find the line that goes through a given point and is parallel to a given line
`PedalTriangle`	Pedal triangle of a point with respect to a triangle
`PerpenBisector`	Find the perpendicular bisector of two given points
`PerpendicularLine`	Find the line that goes through a given point and is perpendicular to a given line
`point`	Define a point
`Polar`	Polar of a given point with respect to a given conic
`Pole`	Pole of a given line with respect to a given conic
`powerpc`	Power of a point with respect to a circle
`projection`	Find the projection of a given point on a given line
`RadicalAxis`	Find the radical axis of two given circles
`RadicalCenter`	Find the radical center of three given circles
`radius`	Compute the radius of a given circle
`randpoint`	Generate a random point on a line or a circle
`reciprocation`	Reciprocation of a point or line with respect to a circle
`reflection`	Reflection of an object with respect to a point or line
`RegularPolygon`	Define a regular polygon
`RegularStarPolygon`	Define a regular star polygon
`rotation`	Rotation of an object about a point
`segment`	Define a segment
`SensedMagnitude`	Find the sensed magnitude between two points
`sides`	Compute the sides of a given triangle or a given square
`similitude`	Find the insimilitude and outsimilitude of two circles
`SimsonLine`	Find the Simson line of a given triangle
`slope`	Compute the slope of a line
`SpiralRotation`	Find the spiral-rotation of a geometric object
`square`	Define a square
`stretch`	Find the stretch of a geometric object
`StretchReflection`	Find the stretch-reflection of a geometric object

`StretchRotation`	Find the stretch-rotation of a geometric object
`TangentLine`	Find the tangents of a point on a circle
`tangentpc`	Find the tangent of a point on a circle
`translation`	Find the translation of a geometric object
`triangle`	Define a triangle
`VerticalCoord`	Compute the vertical coordinate of a given point
`VerticalName`	Find the name of the vertical axis

17.7.9 The *geom3d* package

The *geom3d* package is used for doing three-dimensional Euclidean geometry. In Section 14.3 we described the functions in the *geom3d* package for plotting polyhedra.

The following transformations are available in the package

`rotation`	`translation`	`ScrewDisplacement`
`reflection`	`RotatoryReflection`	`GlideReflection`
`homothety`	`homology`	

Many of the functions in the package are analogous to those in the *geometry* package. Other functions in the package include

`Archimedean`	Define an Archimedean solid
`AreCoplanar`	Test if the given objects are on the same plane
`AreDistinct`	Test if given objects are distinct
`AreSkewLines`	Test if two lines are skew
`duality`	Define the dual of a given polyhedron
`faces`	Return the faces of a polyhedron
`facet`	Define a faceting of a given polyhedron
`gtetrahedron`	Define a general tetrahedron
`HarmonicConjugate`	Find the harmonic conjugate of a point with respect to two other points
`incident`	Vertices incident to a vertex of a polyhedron
`IsArchimedean`	Test if a polyhedron is Archimedean
`IsFacetted`	Test if the given polyhedron is of facetted form
`IsOnObject`	Test if a point (or points) is on an object
`IsQuasi`	Test if the given polyhedron is quasi-regular
`IsRegular`	Test if the given polyhedron is regular
`IsStellated`	Test if the given polyhedron is of stellated form
`IsTangent`	Test if a plane is tangent to a sphere
`plane`	Define a plane
`polar`	Polar of a point with respect to a sphere
`pole`	Pole of a plane with respect to a sphere
`powerps`	Power of a point with respect to a sphere
`QuasiRegularPolyhedron`	Define a quasi-regular polyhedron
`RadicalCenter`	Find the radical center of four given spheres
`RadicalLine`	Find the radical line of three given spheres
`RadicalPlane`	Find the radical plane of two given spheres

RegularPolyhedron	Define a regular polyhedron
schlafli	Return the Schlafli symbol of a given polyhedron
sphere	Define a sphere
stellate	Define a stellation of a given polyhedron
TangentPlane	Find the tangent plane of a point on a sphere
tname	Parameter name in parametric equations
volume	Volume of a sphere or regular polyhedron
xcoord	Compute the x-coordinate of a given point
xname	The name of the x-axis
ycoord	Compute the y-coordinate of a given point
yname	The name of the y-axis
zcoord	Compute the z-coordinate of a given point
zname	The name of the z-axis

17.7.10 The *Groebner* package

The Groebner basis package is *Groebner*. Groebner bases arise in certain skew polynomial rings and are quite useful for solving many problems in polynomial ideal theory. Functions include

fglm	Generalized FGLM algorithm
gbasis	Compute a reduced Groebner basis
gsolve	Preprocess an algebraic system for solving
hilbertdim	Compute the Hilbert dimension of an ideal
hilbertpoly	Compute the Hilbert polynomial of an ideal
hilbertseries	Compute the Hilbert series of an ideal
inter_reduce	Fully interreduce a list of polynomials
is_finite	Decide if an algebraic system has only finitely many solutions
is_solvable	Decide if algebraic system is consistent
leadcoeff	Compute the leading coefficient of a polynomial
leadmon	Compute the leading monomial of a polynomial
leadterm	Compute the leading term of a polynomial
MulMatrix	Multiplication matrix from a normal set
normalf	Normal form of a polynomial modulo an ideal
pretend_gbasis	Add a Groebner basis to the list of known ones
reduce	Full reduction of a polynomial
SetBasis	Normal set of a zero-dimensional Groebner basis
spoly	Compute the S-polynomial of two skew polynomials
termorder	Create a term order
testorder	Test whether two terms are in increasing order with respect to a given term order
univpoly	Compute the univariate polynomial of lowest degree in an ideal

The MulMatrix and SetBasis functions are new to the package.

17.7.11 The *liesymm* package

The *liesymm* package is used to determine the equations of the isogroup of a system of partial differential equations. It implements the Harrison-Estabrook procedure which uses Cartan's formulation in terms of differential forms. A detailed description of the package is given in a paper by Carminati, Devitt, and Fee.[5]

`&^`	Wedge product
`&mod`	Reduce a form modulo an exterior ideal
`annul`	Annul a set of differential forms
`autosimp`	Autosimp a set of differential forms
`close`	Compute the closure of a set of differential forms
`d`	The exterior derivative
`depvars`	Depvars a set of differential forms
`determine`	Find the determining equations for the isovectors of a pde
`dvalue`	Force evaluation of derivatives
`Eta`	Compute the coefficients of the generator of a finite point
`extvars`	Extvars a set of differential forms
`getcoeff`	Extract the coefficient part of a basis wedge product
`getform`	Extract the basis element of a single wedge product
`hasclosure`	Verify closure with respect to $d()$
`hook`	Inner product (hook)
`indepvars`	Indepvars a set of differential forms
`Lie`	The Lie derivative
`Lrank`	The Lie rank of a set of forms
`makeforms`	Construct a set of differential forms from a pde
`mixpar`	Order the mixed partials
`prolong`	Make substitutions for components of the extended isovector in terms of partials of the original isovector
`reduce`	Reduce a set of differential forms
`setup`	Define the coordinates
`TD`	An extended differential operator
`translate`	Partial derivative corresponding to a given name
`vfix`	Change variable dependencies in unevaluated derivatives
`wcollect`	Regroup the terms as a sum of products
`wdegree`	Compute the wedge degree of a form
`wedgeset`	Find the coordinate set
`wsubs`	Replace part of a wedge product

17.7.12 The *LREtools* package

The *LREtools* package is used for solving linear recurrence equations (LREs). Functions include

`autodispersion`	Compute self-dispersion of a polynomial
`constcoeffsol`	Find all solutions of LREs with constant coefficients
`delta`	Single or iterated differencing of an expression

`dispersion`	Compute dispersion of two polynomial polynomials
`divconq`	Find solutions of "divide and conquer" recurrence equations
`firstlin`	Find solutions of first-order linear recurrence equations
`hypergeomsols`	Find hypergeometric solution of an LRE
`polysols`	Find polynomial solutions of linear recurrence equations
`ratpolysols`	Find rational solutions of linear recurrence equations
`REcontent`	Content of a recurrence operator
`REcreate`	Create an RESol from a recurrence equation
`REplot`	2-dimensional plot of a sequence defined by a recurrence
`REreduceorder`	Apply the method of reduction of order to an LRE
`REtoDE`	Convert a recurrence into a differential equation
`REtodelta`	Return the difference operator associated to the LRE
`REtoproc`	Convert a recurrence into a procedure
`riccati`	Find solutions of Riccati recurrence equations
`shift`	Integer shift of an expression

Here `RESol` is a MAPLE data structure used to represent the solution of a recurrence equation.

17.7.13 The *Ore_algebra* package

The *Ore_algebra* package is for doing basic calculations in algebras of linear operators. An introduction to this package is available on Frédéric Chyzak's Mgfun Project page at
`http://pauillac.inria.fr/algo/chyzak/Mgfun.html`
Functions include

`annihilators`	Skew lcm of a pair of operators
`applyopr`	Apply an operator to a function
`diff_algebra`	Create an algebra of linear differential operators
`Ore_to_DESol`	Convert a differential operator to a DESol
`Ore_to_diff`	Convert a differential operator to a DE
`Ore_to_RESol`	Convert a shift operator to a DESol
`Ore_to_shift`	Convert a shift operator to a recurrence equation
`poly_algebra`	Create an algebra of commutative polynomials
`qshift_algebra`	Create an algebra of linear q-difference operators
`randpoly`	Random skew polynomial generator
`shift_algebra`	Create an algebra of linear difference operators
`skew_algebra`	Declare an Ore algebra
`skew_elim`	Skew elimination of an indeterminate
`skew_gcdex`	Extended skew gcd computation
`skew_pdiv`	Skew pseudo-division
`skew_power`	Power of an Ore algebra
`skew_prem`	Skew pseudo-remainder
`skew_product`	Inner product of an Ore algebra

Here `DESol` is a MAPLE data structure used to represent the solution of a differential equation.

17.7.14 The *PDEtools* package

The MAPLE function `pdsolve` is used for solving partial differential equations. Additional tools are in *PDEtools*, the partial differential equations package. There is a Web page for this package at
`http://lie.uwaterloo.ca/pdetools.htm`
Functions include

`build`	Build an explicit expression for the indeterminate function from the solution obtained using `pdsolve`
`casesplit`	Split into cases and decouple a system
`charstrip`	Find the characteristic strip corresponding to a given first-order PDE and divide it into uncoupled subsets
`dchange`	Change variables in mathematical expressions or procedures
`dcoeffs`	Obtain coefficients of a polynomial differential equation
`declare`	Declare a function for compact display
`difforder`	Evaluate the differential order of an algebraic expression
`dpolyform`	Polynomial form of a given system
`dsubs`	Perform differential substitutions into expressions
`mapde`	For mapping a PDE into a nicer PDE
`PDEplot`	Plot the solution to a first-order PDE
`separability`	Determine the conditions for sum or product separability
`splitsys`	Split sets of (algebraic or differential) equations into uncoupled subsets
`undeclare`	Undeclare a function for compact display

The `dpolyform` function is new to the package.

17.7.15 The *powseries* package

The *powseries* package is used for formal power series computations. Functions include

`compose`	Composition of formal power series
`evalpow`	General evaluator for expressions of formal power series
`inverse`	Multiplicative inverse of a formal power series
`multconst`	Multiplication of a power series by a constant
`multiply`	Multiplication of formal power series
`negative`	Negation of a formal power series
`powadd`	Addition of formal power series
`powcos`	Cosine of a formal power series
`powcreate`	Create formal power series
`powdiff`	Differentiation of a formal power series
`powexp`	Exponential of a formal power series
`powint`	Integration of a formal power series
`powlog`	Logarithm of a formal power series
`powpoly`	Create a formal power series from a polynomial
`powsin`	Sine of a formal power series
`powsolve`	Find power series solutions of linear ODEs

`powsqrt`	Square root of a formal power series
`quotient`	Quotient of two formal power series
`reversion`	Reversion of formal power series
`subtract`	Subtraction of two formal power series
`tpsform`	Truncated form of a formal series

17.7.16 The *process* package

The *process* package provides multiprocess MAPLE programming. It is only available on UNIX platforms. Functions include

`block`	`exec`	`fork`	`kill`	`pclose`
`pipe`	`popen`	`wait`		

17.7.17 The *simplex* package

Linear optimization uses the simplex algorithm. The corresponding MAPLE package is *simplex*. Functions include

`basis`	Variables that give a basis
`convexhull`	Convex hull that encloses the given points
`cterm`	Constants appearing on the rhs
`define_zero`	Define the zero tolerance for floats
`display`	Display a linear program in matrix form
`dual`	Compute the dual of a linear program
`feasible`	Determine if system is feasible or not
`maximize`	Maximize a linear program
`minimize`	Minimize a linear program
`pivot`	Construct a new set of equations given a pivot
`pivoteqn`	Return a sublist of equations given a pivot
`pivotvar`	Return a variable with positive coefficient
`ratio`	Return a list of ratios
`setup`	Construct a set of equations with variables on the lhs
`standardize`	Return a system of inequalities in standard form

17.7.18 The *Slode* package

The *Slode* package is used for finding formal power series solutions of ordinary linear differential equations. Functions include

`candidate_mpoints`	`candidate_points`	`DEdetermine`
`FPseries`	`FTseries`	`hypergeom_formal_sol`
`hypergeom_series_sol`	`mhypergeom_formal_sol`	`mhypergeom_series_sol`
`msparse_series_sol`	`polynomial_series_sol`	`rational_series_sol`
`series_by_leastsquare`		

17.7.19 The *Spread* package

The *Spread* package provides functions for accessing spreadsheet data on a programming level. Functions include

`CopySelection`	`CreateSpreadsheet`	`EvaluateCurrentSelection`
`EvaluateSpreadsheet`	`GetCellFormula`	`GetCellValue`
`GetFormulaeMatrix`	`GetMaxCols`	`GetMaxRows`
`GetSelection`	`GetValuesMatrix`	`InsertMatrixIntoSelection`
`IsStale`	`SetCellFormula`	`SetMatrix`
`SetSelection`		

17.7.20 The *sumtools* package

The *sumtools* package contains functions for computing indefinite and definite hypergeometric sums in closed form using algorithms due to Gosper and Zeilberger. A good reference is Petkovšek, Wilf, and Zeilberger's book.[6] Functions include

`extended_gosper`	Extended Gosper's algorithm for summation
`gosper`	Gosper's algorithm for summation
`hyperrecursion`	Koepf's extension of Zeilberger's algorithm
`hypersum`	Try to find a closed form for a hypergeometric sum
`hyperterm`	Input a hypergeometric term
`simpcomb`	Simplification of Gamma and related functions
`sumrecursion`	Find a recurrence for a hypergeometric sum using Zeilberger's algorithm
`sumtohyper`	Express an indefinite sum as a Hypergeometric function

Zeilberger has his own implementation *EKHAD* available as a package on the Web at
`http://www.math.temple.edu/~zeilberg/programs.html`
On this page you will see many other related packages.

17.7.21 The *tensor* package

The *tensor* package deals with tensors, their operations, and their use in general relativity both in the natural basis and in a moving frame. Functions include

`act`	Act on either a tensor, spin, or curvature table
`antisymmetrize`	Antisymmetrize the components of a tensor over any of of its indices
`change_basis`	Transform a tensor from the natural basis to a noncoordinate basis
`Christoffel1`	Compute the Christoffel symbols of the first kind
`Christoffel2`	Compute the Christoffel symbols of the second kind
`commutator`	Commutator of two contravariant vector fields

`compare`	Compare two objects of the same type
`conj`	Complex conjugation
`connexF`	Compute the covariant components of the connection coefficients in a rigid frame
`contract`	Contract a tensor over one or more pairs of indices
`convertNP`	Convert the connection coefficients or the Riemann tensor into Newman-Penrose formalism
`cov_diff`	Covariant derivative of a tensor_type
`create`	Create a new tensor_type object
`d1metric`	Compute the first partials of the covariant tensor
`d2metric`	Compute the second partials of the covariant tensor
`directional_diff`	Compute the directional derivative
`display_allGR`	Display nonzero components of all GR tensors
`displayGR`	Display nonzero components of a GR tensor
`dual`	Perform the dual operation on the indices of a tensor
`Einstein`	Compute covariant components of the Einstein tensor
`entermetric`	Enter metric tensor components
`exterior_diff`	Exterior derivative of a completely antisymmetric covariant tensor
`exterior_prod`	Exterior product of two covariant antisymmetric tensors
`frame`	Frame that brings the metric to diagonal signature metric
`geodesic_eqns`	Euler-Lagrange equations for the geodesic curves
`get_char`	Return the index character field of a tensor_type
`get_compts`	Return the components field of a tensor_type
`get_rank`	Return the rank of a tensor_type
`invars`	Scalar invariants of Riemann tensor of a space-time
`invert`	Form the inverse of any second rank tensor_type
`Jacobian`	Jacobian of a coordinate transformation
`Killing_eqns`	Compute component expressions for Killings equations
`Levi_Civita`	Covariant and contravariant Levi-Civita pseudo-tensors
`Lie_diff`	Lie derivative of a tensor
`lin_com`	Linear combination of any number of tensor_types
`lower`	Lower a contravariant index
`npcurve`	Newman-Penrose curvature in Debever's formalism
`npspin`	Newman-Penrose spin coefficients in Debever's formalism
`partial_diff`	Compute the partial derivatives of a tensor_type
`permute_indices`	Permutation of the indices of a tensor_type
`petrov`	Find the Petrov classification of the Weyl tensor
`prod`	Inner and outer tensor product
`raise`	Raise a covariant index
`Ricci`	Compute the covariant Ricci tensor
`Ricciscalar`	Compute the Ricci scalar
`Riemann`	Compute the covariant Riemann curvature tensor
`RiemannF`	Covariant Riemann curvature tensor in a rigid frame
`symmetrize`	Symmetrize the components of a tensor over any of its

	indices
`tensorsGR`	Compute general relativity curvature tensors in a coordinate basis
`transform`	Transform a tensor under a new coordinate system
`Weyl`	Compute the covariant Weyl tensor

Here GR is an abbreviation for general relativity. Peter Musgrave, Denis Pollney, and Kayll Lake have developed another MAPLE tensor package, called *GRTensorII*. It is available from
`http://grtensor.phy.queensu.ca/`

17.8 New packages

We give an overview of the packages that are new for MAPLE 7.

17.8.1 The *CurveFitting* package

The package *CurveFitting* is used for fitting curves to data points. Functions include

`BSpline`	B-spline basis function
`BSplineCurve`	B-spline curve
`LeastSquares`	Least squares approximation
`PolynomialInterpolation`	Interpolating polynomial
`RationalInterpolation`	Interpolating rational function
`Spline`	Natural spline
`ThieleInterpolation`	Thiele's interpolating continued fraction function

17.8.2 The *ExternalCalling* package

The *ExternalCalling* package has facilities for linking to programs outside MAPLE. There are two functions:

`DefineExternal`	Create a link to an external function
`ExternalLibraryName`	Name of the relevant external shared object

17.8.3 The *LinearFunctionalSystems* package

The *LinearFunctionalSystems* package is used for finding polynomial, rational function, and formal power series solutions of linear functional systems of equations with polynomial coefficients. Functions include

`AreSameSolution`	Test if solutions of a system are equivalent
`CanonicalSystem`	Canonical system equivalent to given system
`ExtendSeries`	Extend the number of terms in a series solution
`HomogeneousSystem`	Homogeneous system equivalent to given system
`IsSolution`	Test a solution
`MatrixTriangularization`	Equivalent matrix recurrence system
`PolynomialSolution`	Polynomial solutions if they exist
`Properties`	Properties of the system
`RationalSolution`	Rational function solutions if they exist
`SeriesSolution`	Formal power series solutions if they exist
`UniversalDenominator`	Common denominator of rational solutions

17.8.4 The *LinearOperators* package

The *LinearOPerators* package is used to solve equations involving differential or difference operators. Functions include

`Apply`	Apply an Ore polynomial to a function
`dAlembertianSolver`	d'Alembertian solution of a nonhomogeneous equation
`DEToOrePoly`	Convert lhs of DE to Ore polynomial
`FactoredAnnihilator`	Factor the annihilator of an expression
`FactoredGCRD`	Greatest common right divisor in completely factored form
`FactoredMinimalAnnihilator`	Completely factored minimal annihilator
`FactoredOrePolyToDE`	Convert a factored Ore polynomial to a DE
`FactoredOrePolyToRE`	Convert a factored Ore polynomial to a recurrence equation
`FactoredOrePolyToOrePoly`	Expand a factored Ore polynomial
`IntegrateSols`	Check for primitive element and perform accurate integration
`MinimalAnnihilator`	Minimal annihilator
`OrePolyToDE`	Convert an Ore polynomial to a DE
`OrePolyToRE`	Convert Ore polynomial to a recurrence equation

17.8.5 The *ListTools* package

The *ListTools* package contains many functions for manipulating lists:

`BinaryPlace`	Find largest index n so that $L[n]$ precedes x
`BinarySearch`	Perform binary search of list
`Categorize`	Categorize elements of a list with respect to a proc
`DotProduct`	Dot product of two lists
`FindRepititions`	Find repeated elements in a list
`Flatten`	Convert lists of lists to a single list
`FlattenOnce`	Do flatten once
`Group`	Group a list into sublists relative to a proc
`Interleave`	Interleave a number of lists
`Join`	Insert an object between each element of a list
`JoinSequence`	Insert a sequence between each element of a list
`MakeUnique`	Remove repeated elements from a list
`Pad`	Pad the elements of a list
`PartialSums`	Return a list of partial sums
`Reverse`	Reverse the order of the list
`Rotate`	Cyclically shift elements of a list
`Sorted`	Test whether a list is sorted relative to a proc or order
`Split`	Split a list into a sequence of lists relative to a proc
`Transpose`	Transpose a list of lists as if it were a matrix

17.8.6 The *MathML* package

MathML is a new markup language for representing mathematical expressions in Web documents. It is still under development and has not yet been implemented by Web browsers. The *MathML* package contains functions for converting MAPLE expressions to MathML and vice versa. Functions include

`Export`	Convert a MAPLE expression into MathML
`ExportContent`	Convert MAPLE expression into content-only MathML
`ExportPresentation`	Convert MAPLE into presentation-only MathML
`Import`	Convert MathML into MAPLE
`ImportContent`	Convert content-only MathML into MAPLE

17.8.7 The *OrthogonalSeries* package

The *OrthogonalSeries* packages contains functions for manipulating infinite series of classical orthogonal polynomials. Functions include

`Add`	Add a linear combination of two series
`ApplyOperator`	Apply a differential or difference operator
`ChangeBasis`	Expand in terms of a new basis
`Coefficients`	Extract a coefficient from a series
`ConvertToSum`	Convert series object to a sum
`Copy`	Make copy of a series
`Create`	Create a series of orthogonal polynomials
`Degree`	Degree of a finite or infinite series
`Derivate`	Take the derivative of a series
`DerivativeRepresentation`	Series of differentiated orthogonal polynomials
`Evaluate`	Evaluate a finite orthogonal series
`Multiply`	Multiply two series
`PolynomialMultiply`	Multiply a series by a polynomial
`ScalarMultiply`	Multiply a series by a scalar
`SimplifyCoefficients`	Simplify the coefficients of an orthogonal series
`Truncate`	Truncate a series

17.8.8 The *RandomTools* package

The *RandomTools* package contains functions for generating random objects of certain types called flavors. The available functions are

`AddFlavor`	Add a flavor template to generate random objects
`Generate`	Generate a random object
`GetFlavor`	Return definition of flavor
`GetFlavors`	Return names of all known flavors
`HasFlavor`	Check if a flavor is known
`RemoveFlavor`	Remove a flavor template

Possible flavors include

`choose`	`complex`	`exprseq`	`float`	`identical`
`intger`	`list`	`listlist`	`negative`	`negint`
`nonnegative`	`nonnegint`	`nonposint`	`nonpositive`	`nonzero`
`nonzeroint`	`polynom`	`posint`	`positive`	`rational`
`set`	`structured`	`truefalse`		

17.8.9 The *RationalNormalForms* package

The *RationalNormalForms* package is useful when dealing with summation problems involving hypergeometric terms. Functions include

`AreSimilar`	Test if the ratio of two hypergeometric terms is a rational function
`IsHypergeometricTerm`	Test whether a term is hypergeometric
`MinimalRepresentation`	First and second minimal representations of a hypergeometric term
`PolynomialNormalForm`	Polynomial normal form of a rational function
`RationalCanonicalForm`	Rational canonical forms of a rational function

17.8.10 The *RealDomain* package

When working in the real domain we use the *RealDomain* package. When the package is loaded, the following functions are redefined so that their domain is the set of real numbers:

`Im`	`Re`	`^`	`arccos`	`arccosh`	`arccot`	`arccoth`
`arccsc`	`arccsch`	`arcsec`	`arcsech`	`arcsin`	`arcsinh`	`arctan`
`arctanh`	`cos`	`cosh`	`cot`	`coth`	`csc`	`csch`
`eval`	`exp`	`expand`	`limit`	`ln`	`log`	`sec`
`sech`	`signum`	`simplify`	`sin`	`sinh`	`solve`	`sqrt`
`surd`	`tan`	`tanh`				

See Section 3.1.6 for some examples.

17.8.11 The *Sockets* package

The *Sockets* package is for programmers who want to do network communication in MAPLE. Functions include:

`Address`	Find IP address of hostname or vice versa
`Close`	Close a TCP/IP connection
`Configure`	Set configuration options for socket connection
`GetHostName`	Return name of local host
`GetLocalHost`	Return hostname of local endpoint of socket connection
`GetLocalPort`	Return port number of local endpoint of socket connection
`GetPeerHost`	Return hostname of remote endpoint of socket connection
`GetPeerPort`	Return port number of remote endpoint of socket connection

`GetProcessID`	Return process ID of the calling process
`LookupService`	Return port number of specified Internet service
`Open`	Open a client TCP/IP connection
`ParseURL`	Parse a URL into its components
`Peek`	Check for data on a socket
`Read`	Read text data from a socket connection
`ReadBinary`	Read binary data from a network connection
`ReadLine`	Read a line of text from a socket
`Serve`	Establish a MAPLE server
`Status`	Return status of all open socket connections
`Write`	Write text data to a socket connection
`WriteBinary`	Write binary data to a socket connection

17.8.12 The *SolveTools* package

SolveTools is a package for programmers interested in routines useful in solving systems of algebraic equations. Functions include

`Basis`	Simplest common basis for a list of expressions
`Complexity`	Complexity of an expression
`GreaterComplexity`	Compare the complexity of two expressions
`RationalCoefficients`	Rational coefficients in a linear combination
`SortByComplexity`	Sort expressions by their complexity

17.8.13 The *StringTools* package

The *StringTools* is a package for programmers wanting fancy tools for manipulating strings. Functions include

`AndMap`	Determine if a proc applies to all elements of a string
`Capitalize`	Capitalize the first letter of each word
`Char`	ASCII character corresponding to given code number
`CharacterMap`	Change all instances of a character in a string
`Chomp`	Remove end-of-line character from string
`CommonPrefix`	Length of longest common prefix of two strings
`CommonSuffix`	Length of longest common suffix of two strings
`Compare`	Compare two strings lexicographically
`CompareCI`	Compare two case-insensitive strings lexicographically
`Drop`	Remove a prefix from a string
`Explode`	Convert a string to a list of characters
`FirstFromLeft`	Locate first occurence of a character from the left
`FirstFromRight`	Locate first occurence of a character from the right
`FormatMessage`	Format a string
`Group`	Divide a string into groups relative to a property
`Implode`	Convert a list of characters to a string
`IsAlhpa`	Determine if character is alphabetic
`IsAlhpaNumeric`	Determine if character is alphabetic or a digit

`IsASCII`	Determine if character is in the ASCII character set
`IsBinaryDigit`	Determine if character is a binary digit
`IsControlCharacter`	Determine if a control character
`IsDigit`	Determine if character is a decimal digit
`IsGraphic`	Determine if alphanumeric or a punctuation character
`IsHexDigit`	Determine if character is a hexadecimal digit
`IsIdentifier`	Determine if character is valid MAPLE identifier
`IsLower`	Determine if character is lower case
`IsOctalDigit`	Determine if character is an octal digit
`IsPrefix`	Test for initial substring
`IsUpper`	Determine if character is upper case
`IsSuffix`	Test for terminal substring
`Join`	Join a list of strings
`LeftFold`	Apply a proc iteratively to characters of a string
`Levenshtein`	Levenshtein distance bewtween two strings
`Lowercase`	Change each character to lower case
`Map`	Map a proc onto a string
`OrMap`	Determine if a proc applies to any character of a string
`Ord`	ASCII code number of a character
`Random`	Return a random string
`RightFold`	Apply a proc iteratively to characters of a string from the right
`RegMatch`	Determine if a string matches a regular expression
`RegSub`	Perform character substitutions
`Remove`	Remove characters from a string
`Select`	Select characters from a string
`SelectRemove`	Split a string using select and remove
`Soundex`	Soundex function
`Split`	Split a string relative to a separating character
`Squeeze`	Remove extra spaces
`Substitute`	Substitute first occurrence of a string by another
`SubstituteAll`	Substitute all occurrences of a string by another
`SubString`	Extract a substring
`Take`	Extract a prefix from a string
`TrimLeft`	Remove leading white space
`TrimRight`	Remove trailing white space
`Trim`	Remove leading and trailing white space
`Uppercase`	Change each character to upper case

and the functions:

`LongestCommonSubstring` `LongestCommonSubSequence`

17.8.14 The *Units* package

The unit conversion package is *Units*. Programming level functions include

`AddBaseUnit`	Add a base unit and associated dimension function
`AddDimenions`	Add or rename a dimension
`AddSystem`	Add or modify a system of units

`AddUnit`	Add or modify a unit
`GetDimension`	Dimension as a product of powers of base dimensions
`GetDimensions`	List all known dimensions
`GetSystem`	List units in a system of units
`GetSystems`	List all known systems of units
`GetUnit`	Return information for specified unit
`GetUnits`	List all unit names
`HasDimension`	Test whether a dimension exists
`HasSystem`	Test whether a system of units exists
`HasUnit`	Test whether a unit exists
`RemoveDimension`	Remove a dimension
`RemoveSystem`	Remove a system of units
`UseContexts`	Set a default context
`UseSystem`	Set a default system of units
`UsingContexts`	List the default system of units
`UsingSystem`	Return the default system of units

See Section 3.4 for some practical examples at the base level.

17.8.15 The *XMLTools* package

XMLTools is a package providing programmers tools for manipulating MAPLE's XML documents. XML is an abbreviation for extensible markup language. XML is a language for creating data and documents for the Web. MathML (see Section 17.8.6) is an XML application. Functions include

```
AddAtrributes                   AddChild
AttrCont                        AtrributeCount
AtrributeNames                  Atrributes
AtrributeValue                  AtrributeValueWithDefault
CData                           CDataData
CleanXML                        Comment
CommentText                     ContentModel
ContentModelCount               Element
ElementName                     ElementStatistics
Equal                           FirstChild
FromString                      GetAttribute
GetChild                        HasAttribute
HasChild                        IsCData
IsComment                       IsProcessingInstruction
IsTree                          JoinEntities
LastChild                       MakeElement
Print                           PrintToFile
PrintToString                   ProcessAttributes
ProcessingInstruction           ProcessingInstructionData
ProcessingInstructionname       ReadFile
RemoveAttribute                 RemoveAttributes
RemoveChild                     RemoveContent
```

SecondChild	SeparateEntities
Serialize	StripAtrributes
StripComments	SubsAtrribute
SubsAttributeName	ThirdChild
ToString	WriteFile

APPENDIX A MAPLE RESOURCES

At the time of writing this book, the main MAPLE Web site was `http://www.maplesoft.com`

On this page there are four links:

• Waterloo Maple Corporate Site (`http://www.maplesoft.com/main.html`). This page contains the latest product, support, and contact information.

• Maple Application Center (`http://www.mapleapps.com`). The Maple Application Center contains links to resources contributed by MAPLE users from all over the world.

• Maple Student Center (`http://www.maple4students.com`). This page contains links to free on-line MAPLE resources for students, including tutorials for different courses.

• Registration Web Site (`http://register.maplesoft.com`). This is the place to go to register your MAPLE product.

The MAPLE Application Center

The URL for the Maple Application Center is `http://www.mapleapps.com`. It contains links to MAPLE resources contributed by users all over the world. It supersedes the MAPLE Share library.

Go to `http://www.mapleapps.com` using your favorite Web browser. Click on the button [I need an online tutorial for maple]. This will bring you to the Maple Tutorials page `http://www.mapleapps.com/tutorial.html`. At present there are two tutorials:

• Maple Essentials

• Introduction to Maple for Physics Students

Click on Maple Essentials. This will bring you to MAPLE's online tutorial. The tutorial is in HTML format and contains information on numerical calculations, algebraic calculations, graphing, and solving equations. It is also possible to download the complete tutorial in *mws* form.

Clicking on Introduction to Maple for Physics Students will bring you to a more advanced tutorial designed by Ross L. Spencer from Brigham Young University. It covers plotting, calculus, complex numbers and functions, linear algebra, solving equations, ordinary differential equations and programming. There is a link for downloading the MAPLE worksheets for the complete tutorial.

Go back to the Maple Application Center page and click on Maple Power Tools. This will bring you to a page with three links:

- Education
- Research
- Application

Clicking on Education will bring you to a page listing eleven packages:

• Calculus I — A complete set of MAPLE worksheets covering first semester calculus, developed by the Department of Mathematics, University of Wisconsin-Milwaukee.

• Calculus II — A continuation of the previous package.

• Calculus III — A collection of 25 demos developed at St. Louis University for their Calculus III classes.

• Vector Calculus — A MAPLE package covering differential operations, curve analysis, coordinate system conversions, multiple integrals and line and surface integrals.

• 100 Calculus Projects — A collection of 100 student projects developed at IUPUI for Calculus I and II.

• Introduction to Maple for Physics Students (see above)

• Advanced Engineering Mathematics — A package of 273 MAPLE worksheets to accompany Robert Lopez's book [7] of the same title.

• MathClass — MAPLE tools for constructing textbook-quality mathematical diagrams.

• Maple Essentials (see above)

• Post-Secondary Mathematics Education Pack — A collection of 49 MAPLE modules by Gregory A. Moore of Cerritos College for enlivening the teaching of mathematics at all levels.

• Matrix Algebra Education Pack — A package of 30 modules by Wlodzislaw Kostecki of The Papua New Guinea University of Technology, which covers MAPLE's linalg package.

Go back to the Maple Powertools page (`http://www.mapleapps.com/powertools/powertools.html`) and click on Research. This will bring you to a page listing four packages:

• Finite Elements — A set of MAPLE packages by Artur Portela of the New University of Lisbon for analyzing physical structures using symbolic finite-element models.

• Nonlinear Programming — Contains a package by Jason Schattman for finding local extrema of nonlinear functions subject to constraints.

• Statistics Supplement — This package is for use with Zavan Karian's [8] book *Probability and Statistics: Explorations with Maple.*

• Vector Calculus (see above)

Go back to the Maple Powertools page and click on Application. This will bring you to a page listing three packages:

• Finite Elements (see above)

• Multibody Dynamics Dynaflex — A MAPLE package developed by John McPhee and Pengfei Shi of The University of Waterloo that automatically generates the kinematic and dynamic equations in symbolic form for 3-D flexible multibody systems, given only a description of the system as input.

Return to the Maple Application Center page (`http://www.mapleapps.com`). On the second part of this page you will see a host of links under ten categories:

• Mathematics — Abstract algebra, calculus, chaos theory, combinatorics, complex analysis, cryptography, differential equations, differential geometry, engineering mathematics, game theory, geometry, graph theory, group theory, knot theory, linear algebra, logic, number theory, numerical analysis, operations research, PDEs, real analysis, tensors, topology, and vector calculus.

• Education — Elementary school, precalculus calculus, vector calculus, DEs, real analysis, physics, engineering, quantum mechanics, operations research, economics, statistics and case studies.

• Science — Astrophysics, biochemistry, biology, chemistry, dynamical systems, physics, and quantum mechanics.

• Engineering — Chemical, civil & structural, control, electrical, finite element modeling, fluid dynamics, heat transfer, manufacturing, engineering mathematics, mechanical, modeling & simulation, and nuclear.

• Graphics — Animations, animations gallery, graphics gallery, and applied graphics.

• Maple Tools — Animations, applied graphics, games, Maple functionality, Maple programming, and Maple 7 demos.

• Finance — Economics and financial engineering.

• Communications — Capacity modeling, cryptography and signal processing.

• Computer Science — C code generation, cryptography, error correction, FORTRAN, graph theory, logic, Maple programming, numerical analysis, and theory of computation.

• Statistics & Data Analysis — Maple maps, statistics and stochastic modeling.

The MAPLE Student Center

The Maple Student Center is at `http://www.maple4students.com`. Go to this page using your favorite Web browser. You will find links to the two MAPLE tutorials Maple Essentials and Introduction to Maple for Physics Students, which were mentioned in the previous section.

Under the heading I Need Help With My Classes! there is a menu:

Choose a tutorial
Calculus I
Calculus II
Calculus III
Vector Calculus
Differential Equations
Linear Algebra
Complex Variables
Real Analysis
Engineering
Physics
Other

These basically correspond to the educational Powertools at the Maple Application Center, which were mentioned in the previous section. Selecting Other will bring you to the Maple Application page.

The MAPLE Share Library

Before MAPLE 6, the Share library was the place to find MAPLE packages written by other users. The packages in the Share library are now scattered about the Maple Application Center. To find out what happened to the Share library go to

`http://www.mapleapps.com/packages/whathappenedtoshare.html`

There is a link on this page for downloading the Share library. It is still available by anonymous ftp at `ftp.maplesoft.com`. Just look in the subdirectory *pub/maple/share*.

Interesting URLs

In this section we list some interesting MAPLE Web sites.

`http://www.math.ufl.edu/~frank/maple-book/mbook.html`

This is the Web page for The Maple Book. It contains links to MAPLE *mws* and *txt* files that are mentioned in the book. There are MAPLE text files containing all the MAPLE commands used in the book. See Section 12.3.

`http://math.la.asu.edu/~kawski/maple.html`

Matthias Kawski's (Arizona State University) MAPLE page. Contains numerous MAPLE worksheets on many mathematical topics.

`http://daisy.uwaterloo.ca/SCG/index.html`

The home page of the Symbolic Computation Group, the brains behind the MAPLE software.

`http://daisy.uwaterloo.ca/SCG/MUG.html`

The home page of the MUG (MAPLE Users Group).

`ftp://daisy.uwaterloo.ca/pub/maple/MUG`

An ftp listing of the digests of the MAPLE Users Group. This is a collection of E-mails about MAPLE dating back to 1989.

`http://www.math.ncsu.edu/MapleInfo/`

The NCSU MAPLE Information page.

`http://web.mit.edu/afs/athena.mit.edu/software/maple/www/home.html`

MAPLE at MIT.

`http://www.indiana.edu/ statmath/math/maple/`

MAPLE at Indiana Univeristy.

`http://www.cecm.sfu.ca/CAG/`

Computer Algebra Group at Simon Fraser University.

`http://www.math.utsa.edu/mirrors/maple/maplev.html`

A German/English MAPLE resource page.

`http://www-math.math.rwth-aachen.de/MapleAnswers/`

U. Klein's compilation of hundreds of answers posed to MUG.

`http://www.ms.uky.edu/ carl/hand98.html`

Carl Eberhart's (University of Kentucky) on-line MAPLE handbook.

Appendix B Glossary of Commands

`@` Function composition operator

SYNTAX: `f@g`
DESCRIPTION: Gives the composition of the functions f and g.
EXAMPLE:
```
> (sin@cos)(x);
```

`%` The ditto operator

SYNTAX: `%`
DESCRIPTION: Refers to value of the previous expression computed.
EXAMPLE:
```
> int(1/(1+x^3),x); diff(%,x);
```

`animate` Animation of a two-dimensional plot
[*plots*]

SYNTAX: `animate(F(x,t),x=a..b,t=c..d)`
DESCRIPTION: Animation of $F(x,t)$ on the interval $[a,b]$ with frames $c \leq t \leq d$.
EXAMPLE:
```
> with(plots): animate(sin(x*t),x=-10..10,t=1..2);
```

`animate3d` Animation of a three-dimensional plot
[*plots*]

SYNTAX: `animate3d(F(x,y,t),x=a..b,y=c..d,t=p..q)`
DESCRIPTION: Animation of $F(x,y,t)$ for $a \leq x \leq b$, $c \leq y \leq d$ with frames $p \leq t \leq q$.
EXAMPLE:
```
> with(plots): animate3d(cos(x+t*y),x=0..Pi,y=-Pi..Pi,t=1..2);
```

`assign` Assignment of solution sets

SYNTAX: `assign(S)`
DESCRIPTION: Assigns the variables given in the set S.
EXAMPLE:
```
> S:={y=-1,x=2}: assign(%); x,y;
```

`asympt` Asymptotic expansion

SYNTAX: `asympt(f(x),x,n)`
DESCRIPTION: Gives the asymptotic expansion to order n of $f(x)$ as $x \to \infty$.
EXAMPLE:
```
> asympt(GAMMA(x)^2/GAMMA(2*x)*4^x/sqrt(Pi),x,3);
```

`C` Convert to C code
[codegen]

SYNTAX: `C(expr)`
DESCRIPTION: Converts the expression into C code
EXAMPLE:
```
> with(codegen): F:=exp((1+x+x^2)^3); C(F);
```

`changevar` Perform a substitution in an integral
[student]

SYNTAX: `changevar(u=g(x),int(f(x),x),u)`
DESCRIPTION: Performs the substitution $u = g(x)$ on the given integral.
EXAMPLE:
```
> with(student): Int(x^2/sqrt(1-x^6),x):
> changevar(u=x^3,%,u);
```

`coeff` Coefficient in a polynomial

SYNTAX: `coeff(p(x),x,k)`
DESCRIPTION: Returns the coefficient of x^k in the polynomial $p(x)$.
EXAMPLE:
```
> expand((1+x+x^2)^10): coeff(%,x,10);
```

`collect` Collect coefficients of like powers

SYNTAX: `collect(expr,x)`
DESCRIPTION: Writes the expression as a polynomial in x.
EXAMPLE:
```
> (x+1)^3*y-(y+1)^3*x: collect(%,x);
```

`combine` Combine terms

SYNTAX: `combine(expr)`
DESCRIPTION: Combines terms in the expression.
EXAMPLE:
```
> combine(exp(2*x)^3*exp(y));
```

`contourplot` Two-dimensional contour plot

SYNTAX: `contourplot(f(x,y),x=a..b,y=c..d)`
DESCRIPTION: Produces level curves of the function $f(x, y)$ with x, y in the specified ranges.
EXAMPLE:
```
> with(plots): contourplot(sin(x*y),x=0..Pi, y=0..Pi);
```

`convert` Convert data type

SYNTAX: `convert(expr,type)`
DESCRIPTION: Converts the expression to the new *type*.
EXAMPLE:
```
> series(sqrt(1-x),x,4): convert(%,polynom);
```

`degree` Degree of a polynomial

SYNTAX: `degree(p(x),x)`

DESCRIPTION: Returns the degree of the polynomial in x.

EXAMPLE:

```
> degree((x+y)^6*(y-x^2)^10,x);
```

`denom` Denominator of an expression

SYNTAX: `denom(expr)`

DESCRIPTION: Returns the denominator of the expression.

EXAMPLE:

```
> denom((x*sin(x)-cos(x))/x^2);
```

`det` Determinant of a matrix
[linalg]

SYNTAX: `det(A)`

DESCRIPTION: Determinant of the matrix A.

EXAMPLE:

```
> with(linalg): A:=matrix(4,4,(i,j)->x^(i*j));
> det(A); factor(%);
```

`diff` Differentiation

SYNTAX: `diff(z,x)`

DESCRIPTION: Returns the (partial) derivative $\left(\frac{\partial z}{\partial x}\right)$ $\frac{dz}{dx}$.

EXAMPLE:

```
> diff(sin(x^2*y),x);
```

`display` Display a list of plots
[plots]

SYNTAX: `display(L)`

DESCRIPTION: Displays the plot structures in the list L.

EXAMPLE:

```
> with(plots): P1:=plot(sin(x),x=0..Pi,style=POINT):
> P2:=plot(x,x=0..Pi): display([P1,P2]);
```

`dsolve` Solve ordinary differential equations

SYNTAX: `dsolve(deqn,function)`

DESCRIPTION: Solves the given differential equation for the unknown function.

EXAMPLE:

```
> dsolve(diff(y(x),x$2)-y(x)=sin(x), y(x));
```

`evalf` Evaluate using floating-point arithmetic

SYNTAX: `evalf(expr,n)`

DESCRIPTION: Evaluates the `expression` to n digits.

EXAMPLE:

```
> evalf(exp(-Pi),20);
```

`expand` Expand an expression

SYNTAX: `expand(expr)`
DESCRIPTION: Expands the expression.
EXAMPLE:

```
> expand((2*x+1)*(3*x-5));
```

`factor` Factor a polynomial

SYNTAX: `factor(p)`
DESCRIPTION: Factors the polynomial p.
EXAMPLE:

```
> factor(x^3+x^2*y-x*y^2-y^3);
```

`floor` Greatest integer function

SYNTAX: `floor(r)`
DESCRIPTION: Returns the greatest integer less than or equal to r.
EXAMPLE:

```
> floor(-11/3);
```

`fortran` Convert to Fortran code
[*codegen*]

SYNTAX: `fortran(expr)`
DESCRIPTION: Converts the expression into Fortran code.
EXAMPLE:

```
> with(codegen):  F:=exp((1+x+x^2)^3); fortran(F);
```

`fsolve` Solve using floating-point arithmetic

SYNTAX: `fsolve(eqns,vars)`
DESCRIPTION: Finds an approximate solution to the given set of equations.
EXAMPLE:

```
> fsolve(cos(x)=x/2,x);
```

`gausselim` Gaussian elimination
[*linalg*]

SYNTAX: `gausselim(A)`
DESCRIPTION: Reduces the matrix A to row-echelon form.
EXAMPLE:

```
> with(linalg):  A:=matrix([[1,2,3,4],[2,3,4,5],[5,6,7,8]]);
> gausselim(A);
```

`ifactor` Prime factorization of an integer

SYNTAX: `ifactor(n)`
DESCRIPTION: Computes the prime factorization of the integer n.
EXAMPLE:

```
> ifactor(999);
```

`implicitplot` 2-D plot of a function defined implicitly
[*plots*]

SYNTAX: `implicitplot(f(x,y)=c,x=a..b,y=c..d)`
DESCRIPTION: Plots the set of points (x, y) satisfying $f(x, y) = c$ in the indicated ranges.
EXAMPLE:

```
> with(plots):
> implicitplot((x^2)^(1/3)+(y^2)^(1/3)=1, x=-1..1, y=-1..1);
```

`implicitplot3d` 3-D plot of a function defined implicitly
[*plots*]

SYNTAX: `implicitplot3d(f(x,y,z)=c,x=a..b,y=c..d,z=e..f)`
DESCRIPTION: Plots the set of points (x, y, z) satisfying $f(x, y, z) = c$ in the indicated ranges.
EXAMPLE:

```
> with(plots):
> implicitplot3d(x^2+y^2+z^2=1,x=-1..1,y=-1..1,z=-1..1);
```

`int` Compute an integral

SYNTAX: `int(f(x),x)`
DESCRIPTION: Computes $\int f(x)\,dx$.
SYNTAX: `int(f(x),x=a..b)`
DESCRIPTION: Computes the definite integral $\int_a^b f(x)\,dx$.
EXAMPLE:

```
> int(x^2/sqrt(1+x^2),x=1..sqrt(3));
```

`inverse` Inverse of a matrix
[*linalg*]

SYNTAX: `inverse(A)`
DESCRIPTION: Returns the inverse of the square matrix A.
EXAMPLE:

```
> with(linalg): A:=matrix(3,3,(i,j)->1/2^(i*j)); inverse(A);
```

`isolve` Integer solutions to equations

SYNTAX: `isolve(eqns,var)`
DESCRIPTION: Finds integer solutions to the given set of equations (if they exist).
EXAMPLE:

```
> isolve({x^3+x*y=2,x^2+y^2=2},{x,y});
```

`kernel` Basis for the nullspace
[*linalg*]

SYNTAX: `kernel(A)`
DESCRIPTION: Returns a basis for the nullspace of the matrix A.

EXAMPLE:

```
> with(linalg): A:=matrix(5,5,(i,j)->7^(i+j)); kernel(A);
```

`latex` Convert to LaTeX

SYNTAX: `latex(expr)`

DESCRIPTION: Converts the expression into LaTeX.

EXAMPLE:

```
> latex(Int(1/x,x));
```

`lhs` Left-hand side of an equation

SYNTAX: `lhs(eqn)`

DESCRIPTION: Gives the left-hand side of the given equation.

EXAMPLE:

```
> e:=x^2+y^2=r^2: lhs(e);
```

`limit` Compute a limit

SYNTAX: `limit(f(x),x=a)`

DESCRIPTION: Computes the limit $\lim_{x \to a} f(x)$.

EXAMPLE:

```
> limit((cos(x)-1)/x^2,x=0);
```

`map` Map a function onto a list

SYNTAX: `map(f,L)`

DESCRIPTION: For the list $L = [a_1, a_2, \ldots, a_n]$, it gives $[f(a_1), f(a_2), \ldots, f(a_n)]$.

EXAMPLE:

```
> L := [seq(10^i-1,i=1..6)]; map(ifactor,L);
```

`matrix` Define a matrix

SYNTAX: `matrix(m,n,f)`

DESCRIPTION: Defines an $m \times n$ matrix whose ijth entry is $f(i,j)$.

EXAMPLE:

```
> A:=matrix(4,4,(i,j)->x^(i+j));
```

`modp` Reduce modulo p

SYNTAX: `modp(m,n)`

DESCRIPTION: Reduces the integer m modulo n.

EXAMPLE:

```
> modp(13*19^5,34);
```

`normal` Normalize a rational function

SYNTAX: `normal(expr)`

DESCRIPTION: Simplifies the expression by clearing common factors.

EXAMPLE:

```
> normal((1-q^7)*(1-q^6)/(1-q^2)/(1-q));
```

`numer` Numerator of an expression

SYNTAX: `numer(expr)`
DESCRIPTION: Returns the numerator of the expression.
EXAMPLE:

```
> numer((x*sin(x)-cos(x))/x^2);
```

`op` Extract operands of an expression

SYNTAX: `op(expr)`
DESCRIPTION: Converts the `expr`ession into a list of operands.
SYNTAX: `op(n,expr)`
DESCRIPTION: Extracts the nth operand in the `expr`ession.
EXAMPLE:

```
> w:=x^3+x*y+y:  op(w);  op(2,w);
```

`plot` Two-dimensional plot of a function

SYNTAX: `plot(f(x),x=a..b)`
DESCRIPTION: Plots the function $y = f(x)$, $a \leq x \leq b$.
EXAMPLE:

```
> plot(x*sin(x),x=0..Pi);
```

`plot3d` Three-dimensional plot of a function

SYNTAX: `plot3d(f(x,y),x=a..b,y=c..d)`
DESCRIPTION: Plots the function $z = f(x, y)$, $a \leq x \leq b$, $c \leq y \leq d$.
EXAMPLE:

```
> plot3d(sin(x*y),x=0..Pi,y=0..Pi);
```

`polarplot` Plot a polar curve
[*plots*]

SYNTAX: `polarplot(f(t),t=a..b)`
DESCRIPTION: Plots the polar curve $r = f(\theta)$, $a \leq \theta \leq b$.
EXAMPLE:

```
> with(plots):  polarplot(sin(t),t=0..2*Pi);
```

`product` Find the product

SYNTAX: `product(f(i),i=a..b)`
DESCRIPTION: Computes the product $\prod_{i=a}^{b} f(i)$.
EXAMPLE:

```
> product((a+i-1),i=1..6);
```

`radsimp` Simplify radicals

SYNTAX: `radsimp(expr)`
DESCRIPTION: Simplifies the `expr`ession containing radicals.

EXAMPLE:

```
> radsimp(sqrt(3)*sqrt(15));
```

`rand` Generate random numbers

SYNTAX: `rand(a..b)`

DESCRIPTION: Produces a function that returns a random integer between a and b.

EXAMPLE:

```
> R9 := rand(0..9); R9(); R9(); R9();
```

`rationalize` Rationalize the denominator

SYNTAX: `rationalize(expr)`

DESCRIPTION: Rationalizes the denominator in the `expr`ession.

EXAMPLE:

```
> (1+sqrt(2))/(sqrt(2)-sqrt(3)):  rationalize(%);
```

`rhs` Right-hand side of an equation

SYNTAX: `rhs(eqn)`

DESCRIPTION: Gives the right-hand side of the given equation.

EXAMPLE:

```
> e:=x^2+y^2=r^2:  rhs(e);
```

`seq` Create a sequence

SYNTAX: `seq(f(i),i=a..b)`

DESCRIPTION: This creates the sequence $f(a), f(a+1), \ldots, f(b)$.

EXAMPLE:

```
> seq(x+(y-x)*i/4,i=0..4);
```

`simplify` Simplify an expression

SYNTAX: `simplify(expr)`

DESCRIPTION: Simplifies the `expr`ession.

EXAMPLE:

```
> simplify((sin(x)+cos(x))^2);
```

`solve` Solve equations

SYNTAX: `solve(eqns,var)`

DESCRIPTION: Finds solutions to the given set of equations (if they exist).

EXAMPLE:

```
> solve({x^2+x*y-y=17,y^2-x-y=9},{x,y});
```

`spacecurve` Plot space curve
[*plots*]

SYNTAX: `spacecurve([f(t),g(t),h(t)],t=a..b);`

DESCRIPTION: Plots the space curve parametrized by $x = f(t)$, $y = g(t)$, $z = h(t)$, $a \leq t \leq b$.

EXAMPLE:

```
> with(plots): spacecurve([sin(t),cos(t),t,t=0..2*Pi]);
```

`subs` Substitute into an expression

SYNTAX: `subs(x=a,expr)`

DESCRIPTION: Replaces x by a in the `expression`.

EXAMPLE:

```
> t^2+t+1: subs(t=1+sqrt(5),%);
```

`sum` Summation

SYNTAX: `sum(f(i),i=a..b)`

DESCRIPTION: Computes the sum $\sum_{i=a}^{b} f(i)$.

EXAMPLE:

```
> sum(i^2,i=1..100);
```

`taylor` Taylor series

SYNTAX: `taylor(f(x),x=a,n)`

DESCRIPTION: Computes the Taylor series expansion to order n of the function $f(x)$ near $x = a$.

EXAMPLE:

```
> taylor(tan(x),x=0,10);
```

`type` Test the type of an expression

SYNTAX: `type(expr,t)`

DESCRIPTION: Tests whether the `expression` is of type t.

EXAMPLE:

```
> R := (1-q^6)*(1-q^5)*(1-q^4)/(1-q)/(1-q^2)/(1-q^3);
> P := normal(R); type(P, polynom);
```

`value` Value of an inert expression

SYNTAX: `value(expr)`

DESCRIPTION: Computes the value of the inert `expression`.

EXAMPLE:

```
> Int(1/x,x): value(%);
```

`unapply` Convert to a function

SYNTAX: `unapply(expr,x)`

DESCRIPTION: Converts the `expression` into a function of x.

EXAMPLE:

```
> F:=expand((1+x+x^2)^10): f:=unapply(F,x); f(x);
```

`whattype` Basic type of expression

SYNTAX: `whattype(expr)`

DESCRIPTION: Returns the basic type of the given `expression`.

EXAMPLE:

```
> L := [seq(i,i=1..10)]; whattype(L);
```

APPENDIX C FURTHER READING

A fairly complete list of MAPLE books can be found on the Web at http://www.maplesoft.com/publications/books/index.html.

Below is a list of some recent books on MAPLE.

Introductory books

Maple 7 Learning Guide, Waterloo Maple, 2001, 288 pages.

Cornil, J.M. and Testud, P., *An Introduction to Maple* (trans. from French), Springer-Verlag, 2000, 496 pages.

Heck, A., *Introduction to Maple*, Springer-Verlag, 1996, 699 pages.

Kamerich, E., *A Guide to Maple*, Springer-Verlag, 1999, 325 pages.

Schwartz, D., *Introduction to Maple*, Prentice-Hall, 1999, 225 pages.

Reference books

Abell, M. and Braselton, J., *Maple V by Example*, 2nd ed., 1998, 656 pages.

Monagan, M.B., Geddes, K.O., et al., *Maple 7 Programming Guide*, Waterloo Maple, 2001, 628 pages.

Von zur Gathen, J., and Gerhard, J., *Modern Computer Algebra*, Cambridge University Press, 1999, 750 pages.

Wright, F., *Computing with Maple*, Chapman & Hall/CRC, 2001, 512 pages.

Maple and Calculus

Gresser, J.T., *A Maple Approach to Calculus*, Prentice Hall, 1998, 284 pages.

Smith, R. and Minton R.B., *Insights into Calculus Using Maple*, McGraw-Hill Higher Education, 2001, 70 pages.

Maple and Differential Equations

Betounes, D., *Differential Equations: Theory and Applications With Maple* (with CD-ROM), Springer-Verlag, 680 pages.

Davis, J.H., *Differential Equations with Maple: An Interactive Approach* (with CD-ROM), Birkhauser, 2001, 392 pages.

Lynch, S.J., *Dynamical Systems With Applications Using Maple*, Birkhauser, 2001, 398 pages.

Stavroulakis, I.P. and Tersian, S.A., *Partial Differential Equations: An Introduction With Mathematica and Maple*, World Scientific, 308 pages.

Maple and Linear Algebra

Auer, J.W., *Essentials of Linear Algebra Using Maple V*, Marnie Heus, 1999, 427 pages.

Herman, E.A., King, J.R., Pepe, M.D., and Moore, R.T., *Linear Algebra: Modules for Interactive Learning Using Maple 6* (with CD-ROM), Addison-Wesley, 2001, 496 pages.

Maple, Science, and Engineering

Enns, R.H., and McGuire, G.C., *Nonlinear Physics With Maple for Scientists and Engineers*, Birkhauser, 2000, 656 pages.

Kreyszig, E. and Normington, E.J., *Maple Computer Guide*, Supplement for Erwin Kreyszig's *Advanced Engineering Mathematics*, John Wiley, 2001, 245 pages.

Parlar, M., *Interactive Operations Research with Maple*, Birkhauser, 2000, 484 pages.

Scott, Bill, *Maple for Environmental Sciences: A Helping Hand*, Springer, 2001.

Stroeker, R.J., Hoogerheide, L.F., and Kasshoek, J.F., *Discovering Mathematics With Maple: An Interactive Exploration for Mathematicians, Engineers and Econometricians* (with CD-ROM), Birkhauser, 1999, 248 pages.

Richards, D., *Advanced Mathematical Methods with Maple*, Cambridge University Press, 2001, 896 pages.

Other

Kilma, R.E., Sigmon, N., and Stitzinger, E., *Applications of Abstract Algebra with Maple*, CRC Press, 2000, 272 pages.

Karian, Z.A. and Tanis, E.A., *Probability and Statistics: Explorations With Maple*, Prentice Hall, 1999.

Oprea, J., *The Mathematics of Soap Films: Explorations with Maple*, AMS, 2000, 266 pages.

Prisman, E.Z., *Pricing Derivative Securities* (with CD-ROM), Academic Press, 2000, 760 pages.

Rovenski, V., *Geometry of Curves and Surfaces with Maple*, Birkhauser, 2000, 310 pages.

Vivaldi, F., *Experimental Mathematics with Maple*, Chapman & Hall/CRC, 2001, 240 pages.

REFERENCES

1. Cheb-Terrab, E. S., Duarte L. G. S., and da Mota, L. A. C. P., Computer algebra solving of second order ODEs using symmetry methods, *Computer Phys. Comm.* **108** (1998), 90.
2. Cheb-Terrab, E. S. and Roche, A. D., Integrating factors for second order ODEs, *J. Symbolic Comput.* **27** (1999), 501–519.
3. Borwein, J. M. and Borwein, P. B., *Pi and the AGM*, John Wiley & Sons, New York, 1998.
4. Whittaker, E. T. and Watson, G. N., *A Course of Modern Analysis*, Cambridge University Press, Cambridge, 1996.
5. Carminati, J., Devitt, J. S., and Fee, G. J., Isogroups of differential equations using algebraic computing, *J. Symbolic Comput.* **14** (1992), 103–120.
6. Petkovšek, M., Wilf, H. S., and Zeilberger, D., $A = B$, A. K. Peters, Wellesley, MA, 1996.
7. Lopez, R. J., *Advanced Engineering Mathematics*, Addison-Wesley, Boston, 2001.
8. Karian, Z. A., and Tanis, E. A., *Probability and Statistics: Explorations With Maple*, Prentice Hall, New York, 1999.

REFERENCES

INDEX